The Role of Water in ATP Hydrolysis Energy Transduction by Protein Machinery

Makoto Suzuki
Editor

The Role of Water in ATP Hydrolysis Energy Transduction by Protein Machinery

Editor
Makoto Suzuki
Institute of Engineering Education
Tohoku University
Sendai, Miyagi
Japan

ISBN 978-981-10-8458-4 ISBN 978-981-10-8459-1 (eBook)
https://doi.org/10.1007/978-981-10-8459-1

Library of Congress Control Number: 2018933015

© Springer Nature Singapore Pte Ltd. 2018
This work is subject to copyright. All rights are reserved by the Publisher, whether the whole or part of the material is concerned, specifically the rights of translation, reprinting, reuse of illustrations, recitation, broadcasting, reproduction on microfilms or in any other physical way, and transmission or information storage and retrieval, electronic adaptation, computer software, or by similar or dissimilar methodology now known or hereafter developed.
The use of general descriptive names, registered names, trademarks, service marks, etc. in this publication does not imply, even in the absence of a specific statement, that such names are exempt from the relevant protective laws and regulations and therefore free for general use.
The publisher, the authors and the editors are safe to assume that the advice and information in this book are believed to be true and accurate at the date of publication. Neither the publisher nor the authors or the editors give a warranty, express or implied, with respect to the material contained herein or for any errors or omissions that may have been made. The publisher remains neutral with regard to jurisdictional claims in published maps and institutional affiliations.

Printed on acid-free paper

This Springer imprint is published by the registered company Springer Nature Singapore Pte Ltd.
part of Springer Nature
The registered company address is: 152 Beach Road, #21-01/04 Gateway East, Singapore 189721, Singapore

Preface

The adenosine triphosphate (ATP) hydrolysis reaction is an energetic source of life. How is the reaction utilized in a cell? Recent studies on this issue require some essential corrections to several basic statements in life sciences textbooks.

Although an expression like "The energy from the ATP hydrolysis reaction is utilized in various cellular functions" often appears in a lot of the literature, it can be misleading. In a muscle cell, for example, myosin plays an essential role in the force generation through the following steps; it binds and hydrolyzes ATP and then generates force upon release of the reaction products in the presence of actin filaments. However, it is an experimental fact that the release of enough energy to generate force was not observed during the ATP hydrolysis step.

From this energetic viewpoint, the power-stroke model, which is widely accepted as the molecular mechanism of actomyosin motility, is required for reconsideration. In the mechanisms of protein motors working in water, the solvent water has been commonly thought of as just a background. Nevertheless, in 1956 Szent-Györgyi advocated the energetic importance of water in muscle contraction. However, the concrete role of water has thus far still not been revealed. Through recent studies on ATP hydrolysis energy and ATP-driven protein machinery, it is being revealed that the change in hydration state occurs in cooperation with the structural change of protein molecules. Furthermore, new findings unveiling the protein motor mechanism are being reported.

This book covers the above-mentioned novel scientific findings as reviewed by experts. I believe that this book will help the advancement of the life sciences through its detailing of the role of water and the energetics of protein machinery.

Sendai, Japan
November 2017

Makoto Suzuki
Professor Emeritus of Tohoku University

Contents

Part I Basis of ATP Hydrolysis Reaction

1 **Free Energy Analyses for the ATP Hydrolysis in Aqueous Solution by Large-Scale QM/MM Simulations Combined with a Theory of Solutions** 3
Hideaki Takahashi

2 **Role of Metal Ion Binding and Protonation in ATP Hydrolysis Energetics** 25
Shun-ichi Kidokoro

3 **Spatial Distribution of Ionic Hydration Energy and Hyper-Mobile Water** 33
George Mogami, Makoto Suzuki and Nobuyuki Matubayasi

4 **Theoretical Studies of Strong Attractive Interaction Between Macro-anions Mediated by Multivalent Metal Cations and Related Association Behavior: Effective Interaction Between ATP-Binding Proteins Can Be Regulated by Hydrolysis** 53
Ryo Akiyama

5 **Statistical Mechanical Integral Equation Approach to Reveal the Solvation Effect on Hydrolysis Free Energy of ATP and Its Analogue** 69
Norio Yoshida and Fumio Hirata

6 **A Solvent Model of Nucleotide–Protein Interaction—Partition Coefficients of Phosphates Between Water and Organic Solvent** 87
Hideyuki Komatsu

Part II Basis of Protein-Ligand and Protein-Protein Interactions

7 Energetics of Myosin ATP Hydrolysis by Calorimetry 103
Takao Kodama

8 Orchestrated Electrostatic Interactions Among Myosin, Actin, ATP, and Water 113
Mitsunori Takano

9 Protonation/Deprotonation of Proteins by Neutron Diffraction Structure Analysis 123
Ichiro Tanaka, Katsuhiro Kusaka and Nobuo Niimura

10 All-Atom Analysis of Free Energy of Protein Solvation Through Molecular Simulation and Solution Theory 141
Nobuyuki Matubayasi

11 Uni-directional Propagation of Structural Changes in Actin Filaments 157
Taro Q. P. Uyeda, Kien Xuan Ngo, Noriyuki Kodera and Kiyotaka Tokuraku

12 Functional Mechanisms of ABC Transporters as Revealed by Molecular Simulations 179
Tadaomi Furuta and Minoru Sakurai

13 Statistical Thermodynamics on the Binding of Biomolecules 203
Tomohiko Hayashi

Part III Functioning Mechanisms of Protein Machinery

14 Ratchet Model of Motor Proteins and Its Energetics 231
Yohei Nakayama and Eiro Muneyuki

15 Single-Molecule Analysis of Actomyosin in the Presence of Osmolyte 245
Mitsuhiro Iwaki, Kohji Ito and Keisuke Fujita

16 Novel Intermolecular Surface Force Unveils the Driving Force of the Actomyosin System 257
Makoto Suzuki, George Mogami, Takahiro Watanabe and Nobuyuki Matubayasi

17 Extremophilic Enzymes Related to Energy Conversion 275
Satoshi Wakai and Yoshihiro Sambongi

18 Functioning Mechanism of ATP-Driven Proteins Inferred on the Basis of Water-Entropy Effect 303
Masahiro Kinoshita

19 Controlling the Motility of ATP-Driven Molecular Motors Using High Hydrostatic Pressure 325
Masayoshi Nishiyama

20 Modulation of the Sliding Movement of Myosin-Driven Actin Filaments Associated with Their Distortion: The Effect of ATP, ADP, and Inorganic Phosphate 339
Kuniyuki Hatori and Satoru Kikuchi

Part I
Basis of ATP Hydrolysis Reaction

Chapter 1
Free Energy Analyses for the ATP Hydrolysis in Aqueous Solution by Large-Scale QM/MM Simulations Combined with a Theory of Solutions

Hideaki Takahashi

Abstract We conducted a set of molecular simulations referred to as QM/MM-ER, which combines a hybrid QM/MM with a theory of solutions, to elucidate the microscopic mechanism for the free energy release ΔG_{hyd} associated with hydrolyses of ATP (adenosine triphosphate) or PPi (pyrophosphoric acid) in aqueous solutions. A particular interest is placed on an experimental fact that ΔG_{hyd} stays almost constant irrespective of the number of excess charges on these solute molecules. In the QM/MM-ER simulations the free energy ΔG_{hyd} was decomposed into the contributions ΔG_{ele} and ΔG_{sol} which are, respectively, the free energies due to the electronic states and the solvations of the solutes. It was revealed that ΔG_{ele} is largely negative on the hydrolyses; that is, the products (ADP and Pi) are much stable in the electronic free energies than the reactants. This is attributed mostly to the reduction of the Coulomb repulsion among the excess electrons on ATP or PPi associated with the fragmentation. On the contrary, ΔG_{sol} was found to be highly positive indicating the reactant states are much favorable for hydrations than the products, which can be qualitatively understood in terms of the Born's solvation model. Thus, a drastic compensation takes place between the two free energy contributions ΔG_{ele} and ΔG_{sol} resulting in a modest free energy release ΔG_{hyd} on hydrolyses. A set of classical molecular dynamics simulations for hydrolyses in ethanol was also performed to examine the effect of the dielectric constant of the solvent on the energetics. It was shown that the superb balance between ΔG_{ele} and ΔG_{sol} established in water is seriously degraded in the ethanol solution.

Keywords Free energy · Hydrolysis of ATP · QM/MM · Theory of solutions

---

H. Takahashi (✉)
Department of Chemistry, Graduate School of Science, Tohoku University, Sendai, Miyagi 980-8578, Japan
e-mail: hideaki@m.tohoku.ac.jp

© Springer Nature Singapore Pte Ltd. 2018

M. Suzuki (ed.), *The Role of Water in ATP Hydrolysis Energy Transduction by Protein Machinery*, https://doi.org/10.1007/978-981-10-8459-1_1

1.1 Introduction

ATP (adenosine triphosphate) plays a decisive role in living systems as a main energy source for biological processes (Voet et al. 2013; Berg et al. 2015; Nelson and Cox 2013; Meyerhof and Lohmann 1932). Actually, the free energy associated with ATP hydrolysis is consumed by proteins to exhibit their specific functions. Thus, the conversion of ATP to ADP (adenosine diphosphate) is a reaction of principal importance in biology. Actually, a lot of experimental and theoretical works have been devoted to clarify the energetics of hydrolysis (George et al. 1970; Kodama 1985; Alberty and Goldberg 1992; Pepi et al. 2004; Colvin et al. 1995; Klähn et al. 2006; Grigorenko et al. 2006; Ross 2006; Ruben et al. 2008; Arabi and Matta 2009; Kamerlin and Warshel 2009; Hong et al. 2012; Wang et al. 2015; Sun et al. 2017). However, the microscopic mechanism underlying the free energy release in aqueous solution is not understood on the molecular basis.

In standard textbooks of biochemistry (Voet et al. 2013; Berg et al. 2015; Nelson and Cox 2013), one finds notions that the origins of the free energy would be: (1) Product state of the hydrolysis is more favorable than the reactant, (2) the electronic repulsive energy is released by the fragmentation of the highly charged triphosphate ion, and (3) Pi (phosphoric acid) has a larger resonance stabilization than the triphosphate ion. To the extent of our knowledge none of these issues have not been examined on the basis of molecular theories. In 1969 a preliminary theoretical calculation was done by Boyd and Lipscomb (1969) who applied the extended Hückel theory to ATP and related molecules and provided the population analyses on these molecules. Their work, however, did not incorporate the solvation effects of the surrounding water molecules which will significantly affect the energetics of hydrolysis. The calculations were, of course, limited only to the solute molecules due to the lack of computational resources. It was in 2009 that the energetics of the hydrolysis of a model ATP was investigated in the presence of explicit water molecules by Kamerlin and Warshel (2009) who conducted the quantum mechanical calculations combined with molecular mechanical (QM/MM) simulations (Warshel 1991; Gao and Xia 1992; Ruiz-Lopez 2003; Canuto 2008; Rivail et al. 2015). They deduced a conclusion that the fragmentation of the charged solute makes a significant contribution to the overall energetics. In 2015 Wang et al. performed a large-scale QM/MM simulations utilizing NWChem (Valiev et al. 2010) software to compute the potential of mean force (PMF) along the reaction path of ATP with Mg^{2+} in aqueous solution represented by a sufficient number of MM water molecules. Their interest was, however, placed rather on the reaction pathway and the role of Mg^{2+} and not on the energetics of the free energy release. In 2017 Sun et al. also performed large QM/MM simulations conducting the CP2K code (Hutter et al. 2013) combined with a metadynamics (Laio and Parrinello 2002) method to simulate the protein-mediated hydrolysis of ATP. The major objectives were, however, to elucidate the role of the protein in hydrolysis and to assess the performance of their approach. Thus, there exists no preceding work to clarify the energetics of ATP hydrolysis, which constitutes the major motivation to the present work.

The mechanism of the free energy release has a close relevance with an experimental fact that the free energies ΔG_{hyd} associated with the hydrolyses of ATP or PPi (pyrophosphoric acid) are almost constant irrespective of the charge states of these molecules (George et al. 1970). ATP or PPi has five ionic states depending on the number N_{ex} of excess electrons on these molecules ($N_{\mathrm{ex}} = 0, 1, 2, 3, \text{and } 4$). The experiment by George (1970) revealed that ATP molecules with $N_{\mathrm{ex}} = 3$ and 4, respectively, show $\Delta G_{\mathrm{hyd}} = -9.9$ and -12.8 kcal/mol, and ΔG_{hyd} for PPi with excess electrons ($N_{\mathrm{ex}} = 0, 1, 2, 3, \text{and } 4$) stay within the range from -7.5 to -10.4 kcal/mol. The constancy of ΔG_{hyd} is surprising because the electronic structures as well as the hydration effects are quite different among these ionic states. Such a property of ATP would be necessitated by proteins to ensure their stable operations against the change of pH in the environment.

The major purpose of the present chapter is to assess the mechanism of the free energy release of ATP as well as PPi regarded as a prototypical model of ATP by performing a set of QM/MM simulations combined with a statistical theory of solutions. We include the whole of the ATP molecule in a QM region to exclude the possible artifact arising from the models employed. Such a large-scale QM/MM simulation for the full ATP is made possible by conducting a sophisticated software "Vmol" (Takahashi et al. 2000, 2001a, b) on a massively parallel computer. Our primary attention will be focused on the interplay between the electrostatic repulsion within the solutes and the hydration effects. To this end, we analyze the free energy ΔG_{hyd} by decomposing it into contributions due to electronic free energy ΔG_{ele} and the solvation free energy ΔG_{sol}; thus, $\Delta G_{\mathrm{hyd}} = \Delta G_{\mathrm{ele}} + \Delta G_{\mathrm{sol}}$. In the next section we provide a brief review of our theoretical approach and the computational techniques.

1.2 Theoretical Method

In this section we first provide a concise review of the real-space grid approach (Chelikowsky et al. 1994a, b; Hirose et al. 2005; Takahashi et al. 2000, 2001a, b) which is a key to the massively parallel implementation of the Kohn–Sham (KS) density functional theory (DFT) (Kohn and Sham 1965; Parr and Yang 1989). In second subsection we formulate the equations related to the QM/MM approach for later references. The third subsection is devoted to describe the outline of the method referred to as QM/MM-ER (Takahashi et al. 2004, 2012) which combines the QM/MM approach with the theory of energy representation (Matubayasi and Nakahara 2000, 2002) to calculate the solvation free energy $\Delta\mu$ of a solute. In the development of QM/MM-ER $\Delta\mu$ is decomposed into the contributions due to the two-body and the many-body interactions in the QM/MM potential energy. For each free energy contribution we separately provide a concise formulation in terms of the distribution functions of the solute–solvent interaction on the basis of the density functional theory of solutions. In the last subsection we will address how to apply the QM/MM-ER method to the present problem.

1.2.1 Real-Space Grid Approach

The method of the real-space grid (RSG) approach (Chelikowsky et al. 1994a, b; Hirose et al. 2005) has been established in the field of theoretical condensed matter physics. A one-electron wave function for KS-DFT is represented by a set of the probability amplitudes defined on the RSGs. In the following we use the atomic units (a.u.) to describe the equations related to the electronic structure calculations. In the real-space representation the Kohn–Sham equation is explicitly given by

$$\left(-\frac{1}{2}\nabla^2 + \upsilon_{\mathrm{H}}[n](\boldsymbol{r}) + \upsilon_{\mathrm{xc}}[n](\boldsymbol{r}) + \upsilon_{\mathrm{nuc}}(\boldsymbol{r})\right)\varphi_i(\boldsymbol{r}) = \varepsilon_i\varphi_i(\boldsymbol{r}) \tag{1.1}$$

where φ_i and ε_i are, respectively, the eigenfunction and eigenvalue of the ith orbital and $\boldsymbol{r}$ is the spatial coordinate of the electron. n in Eq. (1.1) denotes the electron density and is given by

$$n(\boldsymbol{r}) = 2\sum_{i}^{occ} |\varphi_i(\boldsymbol{r})|^2 . \tag{1.2}$$

The factor 2 in Eq. (1.2) counts the contribution from α and β spins assuming a closed shell. The terms in the parenthesis in the left-hand side of Eq. (1.1) represent the operator for kinetic, Hartree, exchange correlation, and electron-nuclei potentials, respectively. The Hartree potential υ_{H} is the electrostatic potential by the electron density $n(\boldsymbol{r})$ and explicitly written as

$$\upsilon_{\mathrm{H}}[n](\boldsymbol{r}) = \int \frac{n(\boldsymbol{r}')}{|\boldsymbol{r}-\boldsymbol{r}'|}d\boldsymbol{r}' \tag{1.3}$$

The exchange-correlation potential υ_{xc} is defined as the functional derivative of the exchange-correlation energy with respect to electron density (Parr and Yang 1989); thus,

$$\upsilon_{xc}[n](\boldsymbol{r}) = \frac{\delta E_{xc}[n]}{\delta n(\boldsymbol{r})} \tag{1.4}$$

Throughout this chapter we omit the spin index for the sake of simplicity. In the RSG approach proposed by Chelikowsky et al. (1994a, b) the electron-nuclei potential υ_{nuc} is represented by the pseudopotential (Bachelet et al. 1982; Kleinman and Bylander 1982) to alleviate the steep behavior of the wave function around the atomic core. Hence, the operator is actually *non-local* and the operation on a wave function φ_i can be written as

$$\upsilon_{\mathrm{nuc}}(\boldsymbol{r})\varphi_i(\boldsymbol{r}) = \int d\boldsymbol{r}' \sum_{A} \upsilon_A^{\mathrm{non-loc}}(\boldsymbol{r},\boldsymbol{r}')\varphi_i(\boldsymbol{r}') \tag{1.5}$$

where A is the index attached to a nucleus. In the parallel implementation of the KS-DFT with RSG, a rectangular cell containing a set of grids is introduced and it is divided into subcells to which the CPUs in a parallel computer will be assigned. Thus, we usually divide the whole system in the real space instead of the orbital space. When an operator in the Kohn–Sham Hamiltonian is fully local in the real-space representation, there is no need to communicate the data of the wave functions among the processors. Thus, the non-locality of the pseudopotential will possibly spoil the parallel efficiency. However, the potential $\upsilon_A^{\text{non-loc}}(\boldsymbol{r}, \boldsymbol{r}')$ in Eq. (1.5) is being constructed so that it is fully attenuated outside the sphere with radius of a small cutoff distance from a nucleus A. Hence, the use of the pseudopotential does not seriously affect the parallel efficiency of the RSG approach. Actually, in our previous development we implemented the parallel QM/MM method employing RSG approach (Takahashi et al. 2001b), which showed a rather high efficiency on a machine with a distributed memory architecture. In a recent work we further sophisticated the parallel algorithm, by which the massively parallel computations were made possible.

We also make a remark on the construction of the Hartree potential $\upsilon_{\text{H}}(\boldsymbol{r})$ in Eq. (1.3). With the real-space grid approach, the potential can be efficiently obtained by the fast Fourier transform (FFT) of the electron density $n(\boldsymbol{r})$. However, FFT algorithm is not usually suitable for parallel computation and hence cannot manage a large QM cell. A possible solution to this is to consult the Poisson equation (Hirose et al. 2005) explicitly written as

$$\nabla^2 \upsilon_{\text{H}}(\boldsymbol{r}) = -4\pi n(\boldsymbol{r}). \tag{1.6}$$

This equation can be solved by representing the Laplacian in Eq. (1.6) with the finite difference of the potential υ_{H} on the grid points. Then, the solution of Eq. (1.6) can be cast into the minimization of the quantity $1/2 \langle \upsilon_{\text{H}} | L | \upsilon_{\text{H}} \rangle + 4\pi \langle \upsilon_{\text{H}} | n \rangle$ with respect to υ_{H}, where L represents the Laplacian operator. Thus, the solution of Eq. (1.6) can be easily parallelized as in the kinetic energy operator in Eq. (1.1) though one has to determine the boundary condition of the Hartree potential.

We, so far, described only the outline of the RSG method. We refer the readers to the previous papers (Chelikowsky et al. 1994a, b; Takahashi et al. 2000, 2001a, b) for the accuracy and efficiency of the approach. We close this subsection by presenting a perspective of the RSG approach in the field of theoretical chemistry. In recent sophistications of the exchange-correlation functional the exact Hartree–Fock exchange plays a decisive role (Becke 1993a, b). Actually, it is well known that the exact exchange added as an ingredient to the pure DFT functional remarkably improves the computational accuracy of various properties of molecules. It is, thus, desirable to implement the hybrid exchange functional in the RSG approach. However, the exact exchange potential $\upsilon_x^{\text{exact}}$ has a fully non-local form in the real space; thus,

$$\upsilon_x^{\text{exact}}(\boldsymbol{r}, \boldsymbol{r}') = \frac{n_1(\boldsymbol{r}, \boldsymbol{r}')}{|\boldsymbol{r} - \boldsymbol{r}'|} \tag{1.7}$$

where $n_1(\boldsymbol{r}, \boldsymbol{r}')$ represents the first-order density matrix for a spin. The non-locality of v_x^{exact} is serious, which will degrade the efficiency in the parallel computation as well as in a serial calculation. To overcome the difficulty is the key to the implementation of the hybrid functional in the RSG method. We will undertake the work to solve this problem in the near future.

1.2.2 QM/MM Approach

The QM/MM approach (Warshel 1991; Gao and Xia 1992; Ruiz-Lopez 2003; Canuto 2008; Rivail et al. 2015; Takahashi et al. 2001b) is well established and extensively utilized in the studies of condensed phase such as solutions and biological systems. Though QM/MM method is very simple, it offers a versatile framework for various applications. In this subsection we also provide a concise review of the QM/MM approach for later references. The basic equation for the QM/MM approach is given by

$$E_{\text{tot}} = E_{\text{QM}} + E_{\text{QM/MM}} + E_{\text{MM}}. \tag{1.8}$$

where E_{tot} denotes the total energy of a QM/MM system and E_{QM} and E_{MM} are the energies of the QM and MM subsystems, respectively. The energy $E_{\text{QM/MM}}$ in Eq. (1.8) describes the interaction between QM and MM subsystems, and it plays a role to couple the two subsystems. $E_{\text{QM/MM}}$ is given by the sum of the electrostatic contribution E_{ele} and the van der Waals interaction E_{vdW}; thus,

$$E_{\text{QM/MM}} = E_{\text{ele}} + E_{\text{vdW}}. \tag{1.9}$$

In the present application E_{vdW} is evaluated using the Lennard–Jones potential (Allen and Tidesley 1987). The energy E_{QM} in Eq. (1.8) can be obtained by solving the Schrödinger equation under the influence of the electrostatic potential v_{pc} due to the point charges $\{q_i\}$ on the interaction sites in the MM subsystem. Explicitly, v_{pc} is expressed as

$$v_{\text{pc}}(\boldsymbol{r}) = \sum_i \frac{q_i}{|\boldsymbol{r} - \boldsymbol{s}_i|} \tag{1.10}$$

where $\boldsymbol{s}_i$ denotes the position vector of ith interaction site. Then, we solve the equation

$$\left(\hat{H}_0 + \hat{v}_{\text{pc}}\right)|\Psi_{\text{sol}}\rangle = \left(E_{\text{QM}} + E_{\text{ele}}\right)|\Psi_{\text{sol}}\rangle \tag{1.11}$$

where H_0 is the Hamiltonian for the isolated QM subsystem and Ψ_{sol} denotes the ground state eigenfunction of the QM system in solution. In Eq. (1.11) E_{QM} and E_{sol} are, respectively, given by

$$E_{\mathrm{QM}} = \langle \Psi_{\mathrm{sol}} | \hat{H}_0 | \Psi_{\mathrm{sol}} \rangle \tag{1.12}$$

and

$$E_{\mathrm{ele}} = \langle \Psi_{\mathrm{sol}} | \hat{\upsilon}_{\mathrm{pc}} | \Psi_{\mathrm{sol}} \rangle . \tag{1.13}$$

We note E_{QM} in Eq. (1.12) is always larger than the ground state energy E_0 of the isolated solute since Ψ_{sol} is deformed by υ_{pc}. In the following we refer to the energy difference $E_{\mathrm{QM}} - E_0$ as the distortion energy E_{dist}.

1.2.3 QM/MM-ER Method

Free energy is of primary importance among a various statistical properties since it governs the major path of chemical event. In 2004 we developed a method (Takahashi et al. 2004, 2012) to compute solvation free energy of a solute in a solution by combining the QM/MM method (Takahashi et al. 2001b) with a theory of solutions (Hansen and McDonald 2006). Explicitly, our QM/MM approach is combined with a theory of energy representation (ER) (Matubayasi and Nakahara 2000, 2002) where the distribution function of the solute–solvent interaction potential serves as a fundamental variable in density functional theory of solutions to construct the solvation free energy of the solute. The approach, referred to as QM/MM-ER (Takahashi et al. 2004, 2012), has been extensively utilized for various applications, and the efficiency and the accuracy of the method have been well established (Takahashi et al. 2004, 2005a, b, 2008, 2009, 2011a, b, 2012; Hori et al. 2006; Hori and Takahashi 2007). In the following we make a concise review of QM/MM-ER for later references.

First, we provide a formulation for the solvation free energy $\Delta\mu$ in the framework of the QM/MM approach (Takahashi et al. 2004). In the following we assume that a solute molecule is described by a quantum mechanical (QM) method and the solvent is represented by a classical force field. Then, the free energy $\Delta\mu$ can be expressed as

$$\exp\left(-\beta\Delta\mu\right) = \frac{\int dX \exp\left[-\beta\left(E_{\mathrm{QM}}[X] - E_0 + E_{\mathrm{QM/MM}}[X] + E_{\mathrm{MM}}[X]\right)\right]}{\int dX \exp\left(-\beta E_{\mathrm{MM}}[X]\right)} \tag{1.14}$$

where β is the reciprocal of the Boltzmann constant k_{B} multiplied by temperature T and X collectively represents the coordinates $\{x_i\}$ of the solvent molecules. E_0 in Eq. (1.14) is the energy of the QM solute at isolation. We note the dependence of each term on X is explicitly shown in Eq. (1.14). To apply the standard ER method being constructed on the assumption that solute–solvent interaction is pairwise, we decompose the free energy into two terms $\Delta\overline{\mu}$ and $\delta\mu$; thus,

$$\begin{aligned}\exp(-\beta\Delta\mu) = & \exp\left(-\beta\Delta\overline{\mu}\right) \times \exp(-\beta\delta\mu) \\ = & \frac{\int dX \exp\left[-\beta\left(E_{\mathrm{QM/MM}}\left[\widetilde{n},X\right] + E_{\mathrm{MM}}[X]\right)\right]}{\int dX \exp\left(-\beta E_{\mathrm{MM}}[X]\right)} \\ & \times \frac{\int dX \exp\left[-\beta\left(E_{\mathrm{dist}}[X] + E_{\mathrm{QM/MM}}[X] + E_{\mathrm{MM}}[X]\right)\right]}{\int dX \exp\left[-\beta\left(E_{\mathrm{QM/MM}}\left[\widetilde{n},X\right] + E_{\mathrm{MM}}[X]\right)\right]}. \end{aligned} \tag{1.15}$$

In Eq. (1.15) we introduce a frozen electron density $\widetilde{n}$ independent of the instantaneous solvent configuration $\boldsymbol{X}$. E_{dist} is the distortion energy of the QM solute as described at the end of the previous subsection and is defined by $E_{\mathrm{QM}}[X] - E_0$ in Eq. (1.14). Then, the free energy $\Delta\overline{\mu}$ is responsible for the free energy due to the two-body interaction between solute and solvent. The residual of $\Delta\mu$ is the free energy $\delta\mu$ due to the electron density fluctuation of the solute around $\widetilde{n}$. $\delta\mu$ arises from the many-body interaction in the QM/MM system. For the evaluation of $\Delta\overline{\mu}$ the standard version of the theory of energy representation can be applied straightforwardly. It is worthy of note that $\Delta\overline{\mu}$ also includes the cavitation free energy needed to make an exclusion volume in the MM solvent to accommodate the QM solute. The frozen density $\widetilde{n}$ can be arbitrarily chosen in principle. In the present application we adopt the average electron density of the solute of interest in solution. Explicitly, $\widetilde{n}$ is the ensemble average of the electron density $n(\boldsymbol{r})$ and is obtained by

$$\widetilde{n}(\boldsymbol{r}) = \frac{\int dXn[X](\boldsymbol{r}) \exp\left[-\beta\left(E_{\mathrm{dist}}[X] + E_{\mathrm{QM/MM}}[X] + E_{\mathrm{MM}}[X]\right)\right]}{\exp\left[-\beta\left(E_{\mathrm{dist}}[X] + E_{\mathrm{QM/MM}}[X] + E_{\mathrm{MM}}[X]\right)\right]}. \tag{1.16}$$

The functional for $\Delta\overline{\mu}$ can be formulated exactly in terms of the energy distribution functions (Matubayasi and Nakahara 2000, 2002); thus,

$$\begin{aligned}\Delta\overline{\mu} = -k_B T \int d\varepsilon \Big[& \rho(\varepsilon) - \rho_0(\varepsilon) + \beta\rho(\varepsilon)\omega(\varepsilon) \\ & -\beta\int_0^1 d\lambda\omega(\varepsilon;\lambda)\left(\rho(\varepsilon) - \rho_0(\varepsilon)\right)\Big]. \end{aligned} \tag{1.17}$$

where ε describes the pairwise interaction potential υ between the solute with density $\widetilde{n}$ and a solvent molecule. The energy distribution functions $\rho(\varepsilon)$ and $\rho_0(\varepsilon)$ are, respectively, defined by

$$\rho(\varepsilon) = \left\langle \sum_i \delta\left(\varepsilon - \upsilon\left(\boldsymbol{x}_i\right)\right) \right\rangle \tag{1.18}$$

and

$$\rho_0(\varepsilon) = \left\langle \sum_i \delta\left(\varepsilon - \upsilon\left(\boldsymbol{x}_i\right)\right) \right\rangle_0 \tag{1.19}$$

where the notations $\langle\cdots\rangle$ and $\langle\cdots\rangle_0$ stand for the ensemble averages taken in the solution and the pure solvent systems, respectively. λ in Eq. (1.17) describes the coupling strength of the solute with solvent. Hence, $\lambda = 1$ designates the solution system and $\lambda = 0$ corresponds to the pure solvent. $\omega(\varepsilon)$ in Eq. (1.17) plays an essential role in the density functional theory of solutions, and it describes the mean potential for the solvent arising from the correlation among the solvent molecules. $\omega(\varepsilon)$ is explicitly written as

$$\omega(\varepsilon) = -k_B T \ln\left(\frac{\rho(\varepsilon)}{\rho_0(\varepsilon)}\right) - \varepsilon. \tag{1.20}$$

In the actual calculation of Eq. (1.17) the integration with respect to the coupling parameter λ is being evaluated by a functional combining PY and HNC approximations (Hansen and McDonald 2006).

The free energy contribution $\delta\mu$ in Eq. (1.15) due to many-body interaction can be computed separately utilizing the functional (Takahashi et al. 2012),

$$\delta\mu = \int d\eta \left(k_B T \ln\left(\frac{Q(\eta)}{Q_0(\eta)}\right) + \eta\right) W(\eta) \tag{1.21}$$

where η is the energy coordinate given by

$$\eta = E_{\mathrm{QM}}[\boldsymbol{X}] + E_{\mathrm{QM/MM}}(n[\boldsymbol{X}], \boldsymbol{X}) - E_{\mathrm{QM/MM}}(\widetilde{n}, \boldsymbol{X}). \tag{1.22}$$

We note the energy coordinate η is not subtracted by the energy E_0 of a QM solute at isolation. Since highly charged anionic species are unstable in gaseous environment due to the repulsion among the negative charges, we do not adopt E_0 as the standard of energy in the present applications. The functions $Q(\eta)$ and $Q_0(\eta)$ are the distributions of η in the solution and the reference systems and defined as

$$Q(\eta) = \frac{\int d\boldsymbol{X} \delta(\eta - H) \exp\left[-\beta\left(E_{\mathrm{dist}}[\boldsymbol{X}] + E_{\mathrm{QM/MM}}[\boldsymbol{X}] + E_{\mathrm{MM}}[\boldsymbol{X}]\right)\right]}{\int d\boldsymbol{X} \exp\left[-\beta\left(E_{\mathrm{dist}}[\boldsymbol{X}] + E_{\mathrm{QM/MM}}[\boldsymbol{X}] + E_{\mathrm{MM}}[\boldsymbol{X}]\right)\right]} \tag{1.23}$$

and

$$Q_0(\eta) = \frac{\int d\boldsymbol{X} \delta(\eta - H) \exp\left[-\beta\left(E_{\mathrm{QM/MM}}(\widetilde{n}, \boldsymbol{X}) + E_{\mathrm{MM}}[\boldsymbol{X}]\right)\right]}{\int d\boldsymbol{X} \exp\left[-\beta\left(E_{\mathrm{QM/MM}}(\widetilde{n}, \boldsymbol{X}) + E_{\mathrm{MM}}[\boldsymbol{X}]\right)\right]}. \tag{1.24}$$

The quantity H in Eqs. (1.23) and (1.24) denotes the value of right-hand side of Eq. (1.22). $W(\eta)$ in Eq. (1.21) is just a weight function that can be chosen somewhat arbitrarily as long as it is being normalized. As was derived in reference Takahashi et al. (2012) Eq. (1.21) involves no approximation although it is quite simple. The details of the construction of the free energy functional for $\Delta\overline{\mu}$ were provided in references Matubayasi and Nakahara (2002), Takahashi et al. (2004), and those for $\delta\mu$ were presented in reference Takahashi et al. (2012).

1.2.4 Free Energy of Hydrolysis

Now, we are ready to calculate the free energy change ΔG_{hyd} associated with a hydrolysis of P-O bond in an aqueous solution. To illustrate the procedure to evaluate ΔG_{hyd} we take the reaction of $\mathrm{ATP}^{4-} + \mathrm{H_2O} \rightarrow \mathrm{ADP}^{3-} + \mathrm{H_2PO_4^-}$ as an example. ΔG_{hyd} for this reaction can be expressed as

$$\Delta G_{\mathrm{hyd}}\left(\mathrm{ATP}^{4-}\right) = G\left(\mathrm{ADP}^{3-}\right) + G\left(\mathrm{H_2PO_4^-}\right) - G\left(\mathrm{ATP}^{4-}\right) - G\left(\mathrm{H_2O}\right) \quad (1.25)$$

where $G(\mathrm{S})$ represents the free energy of a solute S in solution. Within the framework of QM/MM-ER method the free energy $G(\mathrm{S})$ can be given by sum of the contributions G_{sol} due to solvation and G_{ele} from electronic state; thus,

$$G(\mathrm{S}) = G_{\mathrm{sol}}(\mathrm{S}) + G_{\mathrm{ele}}(\mathrm{S}). \quad (1.26)$$

G_{sol} in Eq. (1.26) corresponds to the free energy $\Delta\overline{\mu}$ in Eq. (1.17), and G_{ele} originates from the many-body interaction and can be obtained by Eqs. (1.21)–(1.24). In constructing the distributions of Eqs. (1.23) and (1.24) the ground state energy E_0 of the solute at isolation is usually taken as a standard of energy. In the present application, however, the average energy $\langle E_{\mathrm{QM}}\rangle$ in the solution system is adopted as a standard of energy since the solutes with highly negative charge are unstable in the gaseous environment. Accordingly, the free energy ΔG_{hyd} can also be expressed in terms of the free energy differences G_{sol} and G_{ele},

$$\Delta G_{\mathrm{hyd}} = \Delta G_{\mathrm{sol}} + \Delta G_{\mathrm{ele}} \quad (1.27)$$

where ΔG_{sol}, for instance, is given by

$$\Delta G_{\mathrm{sol}} = G_{\mathrm{sol}}\left(\mathrm{ADP}^{3-}\right) + G_{\mathrm{sol}}\left(\mathrm{H_2PO_4^-}\right) - G_{\mathrm{sol}}\left(\mathrm{ATP}^{4-}\right) - G_{\mathrm{sol}}\left(\mathrm{H_2O}\right). \quad (1.28)$$

Equation (1.27) itself provides a way to decompose the hydration free energy ΔG_{hyd}.

1.3 Computational Details

In this section we provide the computational details for the QM/MM-ER method applied to the hydrolyses of PPi (pyrophosphoric acids) and ATP. First, we list below the chemical reactions treated in the present applications for later references. The reactions from R1 to R5 are the hydrolyses of PPi with excess electrons from 0 to 4. Similarly, the reactions from R6 to R8 are hydrolyses of ATP with excess charges of 0, 3, and 4, respectively.

Hydrolyses of PPi

R1	$H_4P_2O_7 + H_2O \rightarrow H_3PO_4 + H_3PO_4$
R2	$H_3P_2O_7^- + H_2O \rightarrow H_2PO_4^- + H_3PO_4$
R3	$H_2P_2O_7^{2-} + H_2O \rightarrow H_2PO_4^- + H_2PO_4^-$
R4	$HP_2O_7^{3-} + H_2O \rightarrow HPO_4^{2-} + H_2PO_4^-$
R5	$P_2O_7^{4-} + H_2O \rightarrow HPO_4^{2-} + HPO_4^{2-}$

Hydrolyses of ATP

R6	$\text{ATP H}_4 + H_2O \rightarrow \text{ADP H}_3 + H_3PO_4$
R7	$\text{ATP H}^{3-} + H_2O \rightarrow \text{ADP H}^{2-} + H_2PO_4^-$
R8	$\text{ATP}^{4-} + H_2O \rightarrow \text{ADP}^{3-} + H_2PO_4^-$

In the following we provide the computational setup for the QM/MM-ER simulations.

The geometries of the solute molecules are optimized by Gaussian 09 package (Frisch et al. 2010) with the theoretical level of B3LYP (Becke 1993b; Lee et al. 1988) /aug-cc-pVDZ (Dunning 1989). To mimic the effect of the solvent molecules the polarizable continuum model (PCM) method (Tomasi et al. 2005) is also utilized. The molecular structure of the solutes is being fixed during the QM/MM simulations. The wave functions of an ATP molecule are enclosed in a rectangular real-space cell of the sizes $x = 26.7$ Å, $y = 18.2$ Å, and $z = 12.1$ Å. The x, y, and z axes are uniformly discretized by 176, 120, and 80 grids, which leads to a grid width of 0.152 Å for each axis. We also introduce the double grids (Ono and Hirose 1999) around the atomic cores to realize the steep behaviors of the wave functions. The probability amplitude on each grid is evaluated by solving the Kohn–Sham equation of Eq. (1.1). The exchange-correlation energy of the QM solute was evaluated by the BLYP functional (Becke 1988; Lee et al. 1988). To carry out the parallel computation, the rectangular QM cell was divided into 128 domains by slicing the cell into 8, 4, and 4 layers along the x, y, and z directions, respectively. The data communications among the processors are commanded by invoking the routines in the MPI library.

The QM cell is embedded in the center of a spherical water droplet Ω which consists of 2160 molecules represented with the SPC/E model (Berendsen et al. 1987), where the radius a of the sphere Ω is set at $a = 23.0$ Å. To evaluate the free energy $\Delta\bar{\mu}$ due to two-body interaction we construct the energy distribution functions in the solution (Eq. 1.18) and in the pure solvent systems (Eq. 1.19) for 200 ps and 400 ps, respectively. For the free energy $\delta\mu$ due to many-body interaction we carried out 100-ps QM/MM simulations to construct the distribution functions Eq. (1.23) in the solution and (1.24) in the reference systems. The average energy $\langle E_{\text{QM}} \rangle$ taken as the standard of energy in the construction of these distributions is corrected with zero-point vibrational energy and with the thermal vibrational and rotational free energies. The average electron distribution $\tilde{n}$ defined in Eq. (1.16) is obtained through a 50-ps QM/MM simulation. The internal structure of the MM water molecules is kept fixed during the simulations. The free energy contribution ΔG_{Born} due to the electrostatic interaction between an ionic solute and the water molecules outside the sphere Ω is evaluated by the Born's equation (Born 1920)

$$\Delta G_{\mathrm{Born}} = -165.9 \times \frac{N_{\mathrm{ex}}^2 \left(1 - \frac{1}{\varepsilon}\right)}{a} \tag{1.29}$$

where N_{ex} is the number of excess electrons on an anion of interest and ε is the dielectric constant of the water solvent. The adequacy of Eq. (1.29) was fully examined in the Supporting Info. in reference Takahashi et al. (2017). All the QM/MM simulations combined with the theory of solutions are performed utilizing the code "Vmol" (Takahashi et al. 2000, 2001a, b) developed originally in our group.

We also performed QM/MM-ER simulations for hydrolyses of PPi with $N_{\mathrm{ex}} = 0, 1, 2, 3,$ and 4. For these molecules we employ the cubic simulation box with periodic boundary conditions. Specifically, for the molecules with $N_{\mathrm{ex}} = 0$ and 1 the cell size is set at $L = 24.6$ Å which contains 494 SPC/E molecules. For the rest of the PPi molecules, we use the larger simulation box with size $L = 49.3$ Å containing 3994 solvent molecules. The QM molecules related to the PPi hydrolyses are enclosed in a cubic cell with size $L = 10.6$ Å, where we adopt the same grid width as that used in the simulations for ATP. The energy distribution functions are constructed for the water molecules inside the spheres with radius $a = 11.0$ Å for the smaller simulation cells and $a = 22.0$ Å for the larger cells. The free energy contributions from the water molecules outside the spheres are also evaluated by the Born's equation.

The OPLS parameter set (Jorgensen et al. 1996) is utilized to describe the Lennard–Jones potential E_{vdW} in Eq. (1.8) between QM and MM regions. The size parameters for the oxygen sites of ATP and PPi are modified so that the QM/MM pair potential reproduces the potential given by a full QM calculation. The details of the method of optimization were presented in Supporting Info. in reference Takahashi et al. (2017). The Newtonian equations of motion for the molecular dynamics simulations are solved numerically using the velocity Verlet algorithm (Allen and Tidesley 1987) with a time step 1 fs. The internal coordinates of the solvent molecules as well as the QM solutes are kept fixed during the simulations. The thermodynamic condition is set at $T = 300$ K and $\rho = 1.0\ \mathrm{g/cm^3}$ throughout the simulations.

1.4 Results and Discussion

In Fig. 1.1a we provide the energy distribution functions $\rho(\varepsilon)$ and $\rho_0(\varepsilon)$ for PPi^{4-} defined, respectively, in Eqs. (1.18) and (1.19). It is shown in the figure that the pair interaction between the solute with average electron distribution $\widetilde{n}(\boldsymbol{r})$ defined by Eq. (1.16) and the SPC/E solvent has distribution up to −40 kcal/mol suggesting a large ion–dipole interaction. The distribution functions for the pair potential for ATP^{4-} are also presented in Fig. 1.1b. PPi and ATP treated in these figures have the same excess electrons $N_{\mathrm{ex}} = 4$. However, the distribution on the energy coordinate

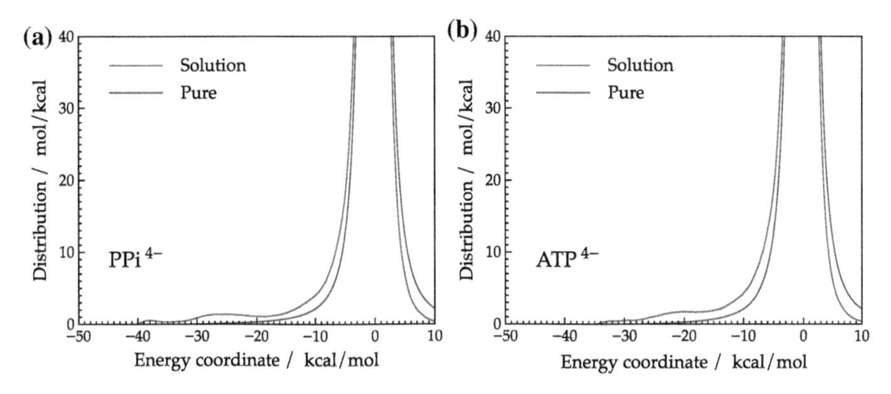

Fig. 1.1 The energy distribution functions $\rho(\varepsilon)$ in solution and $\rho_0(\varepsilon)$ in pure solvent of the pair potential ε between solute and solvent for PPi^{4-} (**a**) and ATP^{4-} (**b**) in aqueous solutions

ranging from −40 to −35 kcal/mol apparently disappears in ATP^{4-}. This implies the fact that the four excess electrons are more delocalized in ATP than PPi since ATP has more phosphate groups than PPi to accommodate the electrons. With these distribution functions for each solute we evaluate the free energy G_{sol} due to two-body interaction utilizing the functional of Eq. (1.17). Then, $G_{sol}(PPi^{4-})$ and $G_{sol}(ATP^{4-})$ are obtained as −718.6(−119.1) and −658.2(−113.9) kcal/mol, respectively, where the values in the parentheses are the Born's correction term given by Eq. (1.29). We, thus, find that the solvation free energy of PPi with $N_{ex} = 4$ is much lower than that of ATP with the same number of excess electrons.

Next, we consider the free energy G_{ele} of the electronic state which includes the effect of the electron density fluctuation of the solute around the average density $\widetilde{n}$ in response to the motion of the solvent molecules. Actually, it is expected that the density fluctuation would be significant for the solute with large N_{ex}. Figure 1.2 shows the density difference $n[\boldsymbol{X}](\boldsymbol{r}) - \widetilde{n}(\boldsymbol{r})$ of ATP^{4-} at an instantaneous solvent configuration $\boldsymbol{X}$. It is apparent in the figure that the density polarization from $\widetilde{n}$ occurs over the whole of the solute molecule. The free energy contribution due to the density fluctuation is incorporated in the functional of Eq. (1.21). This constitutes a distinct advantage of the QM/MM-ER approach since self-consistent reaction field (SCRF) method cannot consider the contribution in principle.

The distribution functions $Q(\eta)$ in the solution and $Q_0(\eta)$ in the reference system of the energy η defined in Eq. (1.22) are shown in Figs. 1.3a and b. The average $\langle E_{QM}\rangle$ of the energy E_{QM} given by Eq. (1.12) is taken as the standard for these distributions. The degree of the electron density fluctuation of the QM solute in solution is reflected in the width of the distribution $Q(\eta)$ or $Q_0(\eta)$. The width of the distribution $Q(\eta)$ as well as $Q_0(\eta)$ for ATP^{4-} is almost comparable to that for the neutral solute ATP H_4, suggesting the significant fluctuation of the electron density even in the neutral ATP molecule. Another notable feature in Figs. 1.3a and b is that the distributions of ATP^{4-} are shifted toward the lower energy coordinate than those of ATP H_4 by ~6 kcal/mol implying the larger polarization in ATP^{4-} than ATP H_4.

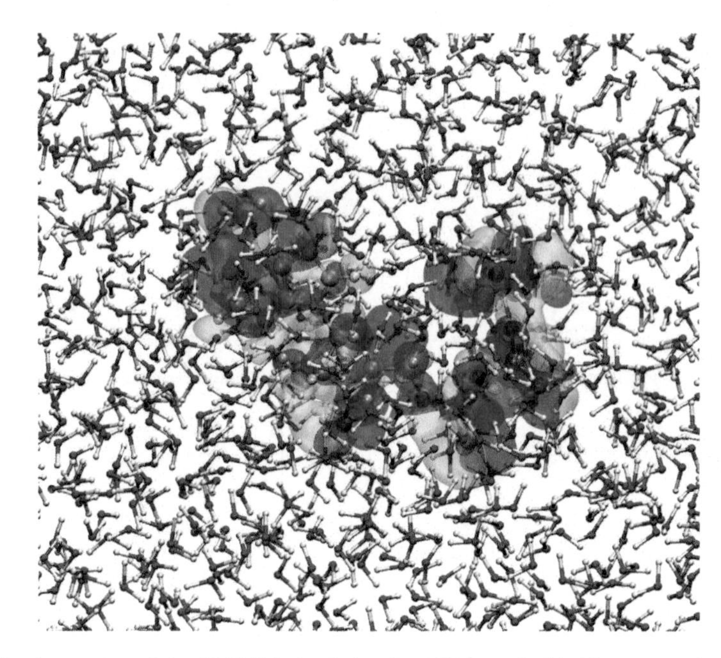

Fig. 1.2 A snapshot of the QM/MM simulation for ATP^{4-} embedded in a water droplet. The contour surfaces of the electron density difference $n[\boldsymbol{X}](\boldsymbol{r}) - \widetilde{n}(\boldsymbol{r})$ of the solute are depicted. The yellow and blue transparent surfaces indicate increase and decrease in the density by 2.0×10^{-4} a.u.$^{-3}$, respectively. The phosphate groups are located in the right-hand side of the figure

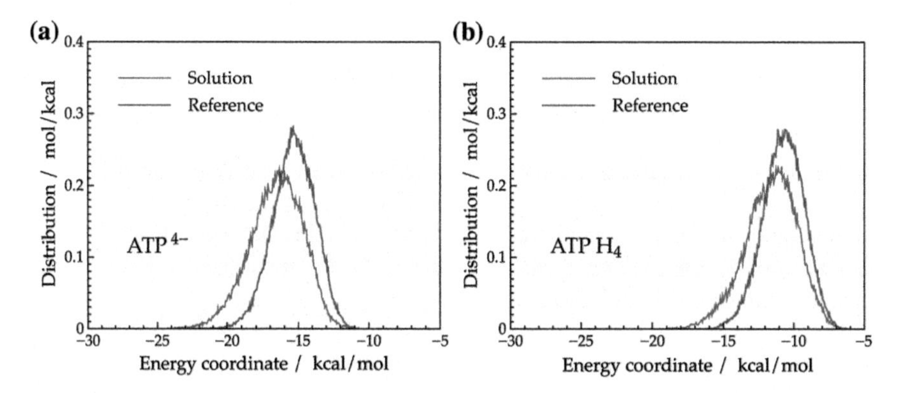

Fig. 1.3 The energy distribution functions $Q(\eta)$ in solution and $Q_0(\eta)$ in reference system of the energy η defined in Eq. (1.22) for the solute ATP^{4-} (**a**) and ATP H_4 (**b**) in aqueous solutions

By substituting the distributions in Fig. 1.3a to the free energy functional of Eq. (1.21) the free energy G_{ele} for ATP^{4-} is evaluated as -15.8 kcal/mol where $\langle E_{QM} \rangle = -339.94113$ a.u. is used as the standard of energy. Similarly, G_{ele} for ATP H_4 is obtained as -11.1 kcal/mol using the standard $\langle E_{QM} \rangle = -342.72864$ a.u. The free energy components G_{ele} and G_{sol} are summarized in Tables 1.1 and 1.2 for the solutes relevant to the hydrolyses of PPi and ATP, respectively.

Table 1.1 The free energy components G_{ele} and G_{sol} for the species relevant to the hydrolyses of PPi (the reactions from R1 to R5 listed in Computational Details). The value in the parenthesis at the entry of G_{ele} is the average $\langle E_{QM} \rangle$ used as the standard of the energy. E_{QM} is corrected with the zero-point vibrational energy and includes the rotational and vibrational free energies

Species	G_{ele} kcal/mol (a.u.)	G_{sol} kcal/mol
$P_2O_7^{4-}$	−9.3(−124.46274)	−718.6
$HP_2O_7^{3-}$	−8.6(−125.42087)	−426.5
$H_2P_2O_7^{2-}$	−6.6(−126.28095)	−193.8
$H_3P_2O_7^-$	−6.4(−126.93057)	−74.6
$H_4P_2O_7$	−4.7(−127.40830)	−47.1
H_2O	−1.7(−17.12174)	−9.7
HPO_4^{2-}	−4.9(−71.02249)	−229.3
$H_2PO_4^-$	−4.0(−71.74383)	−84.2
H_3PO_4	−1.9(−72.26784)	−35.1

Table 1.2 The free energy components G_{ele} and G_{sol} for the species relevant to the hydrolyses of ATP (the reactions from R6 to R8 listed in Computational Details). The value in the parenthesis at the entry of G_{ele} is the average $\langle E_{QM} \rangle$ used as the standard of the energy. E_{QM} is corrected with the zero-point vibrational energy and includes the rotational and vibrational free energies

Species	G_{ele} kcal/mol (a.u.)	G_{sol} kcal/mol
ATP^{4-}	−15.8(−339.94113)	−658.2
ATP H^{3-}	−15.0(−340.88468)	−360.1
ATP H_4	−11.1(−342.72864)	−41.1
ADP^{3-}	−13.5(−285.69306)	−430.0
ADP H^{2-}	−11.9(−286.54555)	−198.8
ADP H_3	−10.1(−287.68348)	−27.4
H_2O	−1.5(−17.08730[a])	−8.6
$H_2PO_4^-$	−4.5(−71.60914[a])	−80.6
H_3PO_4	−3.2(−72.13322[a])	−27.7

[a]We note that the version of our program "Vmol" (Takahashi et al. 2000, 2001a, b) used in the ATP hydrolyses was different from that employed in the PPi reactions. Explicitly, the Lagrange interpolation method was used in the double-grid technique for the QM solutes in the PPi hydrolyses, while cubic interpolation was utilized in the simulations for ATP. This led to the difference between Tables 1.1 and 1.2 in the energies of $\langle E_{QM} \rangle$ for the species H_2O, $H_2PO_4^-$, and H_3PO_4

By consulting these tables it is possible to evaluate the free energy ΔG_{hyd} of hydrolysis for a reaction listed in Computational Details. For instance, ΔG_{hyd} for R8 is obtained as −16.3 kcal/mol adopting the values in Table 1.2. We, thus, construct the free energies ΔG_{hyd} for the reactions R1–R8 and summarize them in Table 1.3 where we also provide experimental results to make comparisons. It is shown in the Table that ΔG_{hyd} given by present work agrees with those obtained by experiments although ΔG_{hyd} for some reactions are overestimated as compared to the experiments. The discrepancies can be attributed mainly to the delocalization error inherent in the exchange-correlation functional based on the local density approximation (LDA) (Parr and Yang 1989). As described excellently in a review (Cohen et al. 2008) for DFT the excess electron tends to delocalize over the system due to the nature of the exchange-hole model in LDA, which leads to a substantial underestimation of the strength of the solute–solvent interaction. This is also true for a functional corrected with the generalized gradient approximation (GGA) (Parr and Yang 1989) such as BLYP (Becke 1988; Lee et al. 1988). The degree of the delocalization will be more serious in the reactant state of the hydrolyses with larger excess electrons. Thus, it is likely that the product states will be relatively more stabilized than the reactants. The addition of the exact exchange, i.e., Hartree–Fock exchange, to the exchange functional as an ingredient may alleviate the overestimation in ΔG_{hyd}. The effect of the inclusion of the exact exchange cannot be studied due to the computational cost associated with the operation of the non-local Hartree–Fock exchange potential. Anyway, the constancy of ΔG_{hyd} is realized in our simulations with respect to the number N_{ex} of excess electrons on PPi or ATP. The free energy changes ΔG_{ele} and ΔG_{sol} associated with hydrolyses are plotted in Fig. 1.4 to elucidate the roles of the electronic and solvation effects in the reaction. It is clearly exhibited in the figure that the contribution of ΔG_{ele} is largely negative and in contrast ΔG_{sol} gives completely opposite contribution to the free energy release. As a result a drastic compensation takes place between these free energies ΔG_{ele} and ΔG_{sol}, which yields

Table 1.3 The free energies ΔG_{hyd} for the hydrolyses of PPi and ATP in aqueous solution

Reaction	ΔG_{hyd} kca/mol	$\Delta G_{\mathrm{hyd}}(\mathrm{exp})^{a}$ kcal/mol
PPi hydrolyses		
R1	−14.3	−9.5
R2	−7.3	−7.5
R3	−17.8	−7.7
R4	−16.2	−7.1
R5	−17.8	−10.4
ATP hydrolyses		
R6	−6.6	–
R7	−25.2	−9.9
R8	−16.3	−10.7

[a]Experimental values in reference George et al. (1970)

modest value of ΔG_{hyd}. It is quite surprising that the constancy of ΔG_{hyd} with respect to the number N_{ex} of the excess electrons is realized irrespective of the significant differences in electronic states and hydrations among the species with different N_{ex}.

In Fig. 1.4 we also provide ΔG_{sol} evaluated by the Born's equation in Eq. (1.29). The radius a of an ionic solute is determined from a sphere containing the same volume V as the solute. The value V is evaluated with the "Volume" option in the Gaussian 09 package (Frisch et al. 2010). It is worthy of note that the Born's equation can reproduce at least the qualitative behavior of ΔG_{sol} given by QM/MM-ER. With this fact in mind it is possible to elucidate the underlying mechanism for the destabilization in ΔG_{sol}. As described in Eq. (1.29) the free energy due to the electrostatic interaction is proportional to the square of N_{ex}. Hence, the fragmentation of a solute into species with smaller numbers of excess electrons will lead to substantial destabilization in the electrostatic free energy. Of course, the degree of the destabilization is more significant for species with larger number of N_{ex}. The variation of ΔG_{ele} can also be understood with a qualitative discussion. The stabilization in ΔG_{ele} can be merely ascribed to the release of the Coulombic repulsion among the excess charges. For instance, we consider the reaction R5 where PPi with four excess electrons dissociates into two fragments with two excess electrons. Since the distance between two phosphate ions is approximately 3 Å, the release of the electronic repulsion due to the dissociation can be evaluated as −452 kcal/mol assuming that two point charges with charge $2e$ reside on the two phosphorus atoms separately. Such a crude estimation gives a much lower value than the actual value $\Delta G_{ele} = -288$ kcal/mol obtained by the QM/MM-ER method. The origin of the difference can be attributed mainly to the delocalization of the excess charges over the reactant and also to the larger

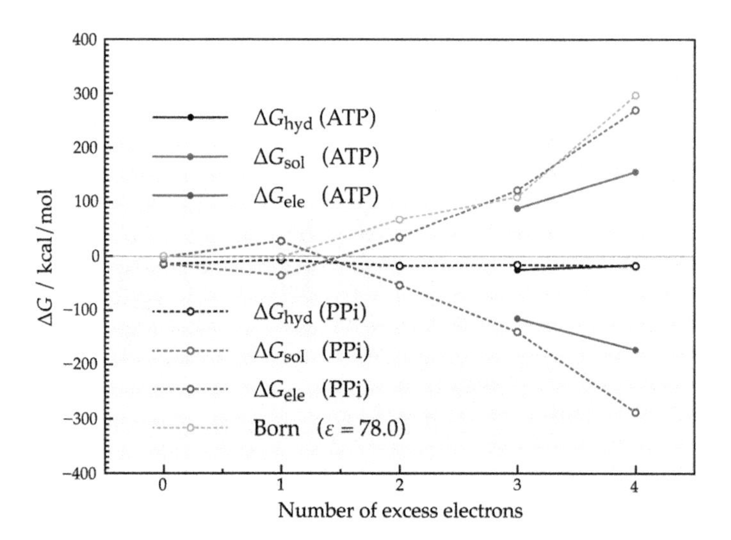

Fig. 1.4 The free energy changes ΔG_{hyd} and its components ΔG_{ele} and ΔG_{sol} ($\Delta G_{hyd} = \Delta G_{ele} + \Delta G_{sol}$) associated with hydrolyses of PPi and ATP in aqueous solutions are shown as functions of the excess electrons. ΔG_{sol} estimated by Born's equation is also presented for comparisons

relaxation energy in the reactant state than the product. Anyway, it would be reasonable to state that the behavior of ΔG_{ele} is mainly dominated by the Coulomb repulsion among the excess electrons.

We next consider the energetics of the ATP hydrolyses in another solvent with a smaller dielectric constant. Our interest is placed on how the constancy of the ΔG_{hyd} will be affected by the property of the solvent. To this end we perform purely classical simulations for the systems of ATP with $N_{ex} = 0, 3$ and 4 embedded in ethanol solvents. The point charges assigned to the atoms in ATP molecules with various values of N_{ex} are those determined by the ESP (electrostatic potential) procedure for the corresponding average electron densities obtained through QM/MM simulations. The explicit values of the charges are provided in the Supporting Info. in reference Takahashi et al. (2017). The results are summarized in Fig. 1.5. We recognize in the figure that purely classical MD simulation overestimates the absolutes of the free energies ΔG_{sol} for ATP in water solvent as compared with those given by the QM/MM simulations. These discrepancies can be attributed to the fact that the electron densities are reduced to sets of point charges placed on the atomic sites since the shrink of the charge width will lead to the enhancement of the electrostatic interaction. It is also shown in the figure that the absolutes of ΔG_{sol} for ATP in ethanol are decreased substantially as compared with those in water. This can be readily understood by means of the Born's equation of Eq. (1.29). The decrease in the dielectric constant ($\varepsilon = 24.3$) directly causes the reduction of the absolute of the solvation free energy of an anion, and hence, the difference ΔG_{sol} also decreases accordingly. Assuming that ΔG_{ele} is not affected by the change of the dielectric constant of the

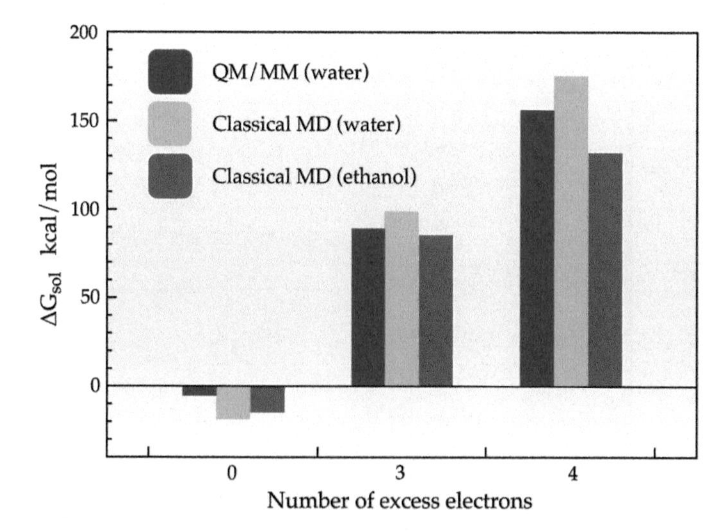

Fig. 1.5 The free energy component ΔG_{sol} with respect to the change of the number N_{ex} of excess electrons. The bars with color of deep blue show the results given by QM/MM simulations. The rest of them are yielded by purely classical MD simulations. The terms in the parentheses in the legend denote the solvent used in the simulations

solvent, the free energy $\Delta G_{\rm hyd}$ for ATP in ethanol will become highly negative in contrast to that in water. Actually, $\Delta G_{\rm hyd}$ in ethanol are estimated as −15.8, −28.9, and −40.5 kcal/mol for ATP molecules with $N_{\rm ex} = 0, 3$ and 4, respectively. As mentioned above the point charge representation of the solute will tend to overestimate the solvation free energy for anions in particular. Therefore, it is reasonably expected that the actual values $\Delta G_{\rm hyd}$ for ATP in ethanol will certainly be more negative than those given by classical MD simulations. It is, thus, shown that the constancy of $\Delta G_{\rm hyd}$ with respect to the variation of $N_{\rm ex}$ is established as a consequence of the interplay between the electronic state and the hydration contributions. Such a superb balance is being realized under the influence of water solvent.

1.5 Conclusion

By conducting sets of QM/MM-ER simulations we computed the free energy $\Delta G_{\rm hyd}$ of hydrolyses for ATP and PPi molecules in aqueous solutions to clarify the microscopic mechanism of the free energy release of these solutes. A particular interest was placed on the constancy of $\Delta G_{\rm hyd}$ with respect to the number of excess electrons. We found drastic cancelation takes place between the free energy contributions $\Delta G_{\rm ele}$ due to electronic state and $\Delta G_{\rm sol}$ due to solvation. Explicitly, it was revealed that $\Delta G_{\rm ele}$ is largely negative upon the dissociation due to the reduction of the electrostatic repulsion among the excess charges. On the contrary, it was found $\Delta G_{\rm sol}$ makes an opposite contribution to $\Delta G_{\rm hyd}$. Such a destabilization of $G_{\rm sol}$ on the hydrolysis can be easily understood by consulting the Born's equation which clearly shows that the fragmentation of an ion into pieces with smaller charges is energetically unfavorable. In this sense, the notion relevant to ATP hydrolysis in a standard biochemistry textbook is rather problematic since it states that the product (ADP + Pi) is more favorable for hydration than the reactant (ATP). Anyway, it was revealed that a large destabilization in $\Delta G_{\rm sol}$ is perfectly compensated with a stabilization in $\Delta G_{\rm ele}$ and it leads to a modest free energy release on hydrolysis. We also examined the effect of the solvent on the energetics of hydrolysis by performing a set of purely classical MD simulations for ATP ($N_{\rm ex} = 0, 3$ and 4) immersed in ethanol. We, then, found that $\Delta G_{\rm hyd}$ becomes highly negative in ethanol as a consequence of the decrease in the free energy $\Delta G_{\rm sol}$ in the solvent with a small dielectric constant. Thus, we conclude that the constancy of $\Delta G_{\rm hyd}$ is established on the exquisite balance between two opposite effects, which are being mediated by water solvent.

References

Alberty RA, Goldberg RN (1992) Biochemistry 31(43):10610. https://doi.org/10.1021/bi00158a025

Allen MP, Tidesley DJ (1987) Computer simulation of liquids. Oxford University Press, Oxford

Arabi AA, Matta CF (2009) J Phys Chem A 113(14):3360. https://doi.org/10.1021/jp811085c
Bachelet GB, Hamann DR, Schlüter M (1982) Phys Rev B 26(8):4199. https://doi.org/10.1103/physrevb.26.4199
Becke AD (1988) Phys Rev A 38(6):3098
Becke AD (1993a) J Chem Phys 98(2):1372
Becke AD (1993b) J Chem Phys 98(7):5648
Berendsen HJC, Grigera JR, Straatsma TP (1987) J Phys Chem 91(24):6269. https://doi.org/10.1021/j100308a038
Berg JM, Tymoczko JL, Gatto JGL, Stryer L (2015) Biochemistry, 8th edn. W. H Freeman and Company, New York
Born M (1920) Z Phys 1(1):45. https://doi.org/10.1007/bf01881023
Boyd DB, Lipscomb WN (1969) J Theor Biol 25:403
Canuto S (2008) Challenges and advances in computational chemistry and physics, vol 6. Springer, Heidelberg
Chelikowsky JR, Troullier N, Wu K, Saad Y (1994a) Phys Rev B 50(16):11355
Chelikowsky JR, Troullier N, Saad Y (1994b) Phys Rev Lett 72(8):1240
Cohen AJ, Mori-Sanchez P, Yang W (2008) Science 321(5890):792. https://doi.org/10.1126/science.1158722
Colvin ME, Evleth E, Akacem Y (1995) J Am Chem Soc 117(15):4357. https://doi.org/10.1021/ja00120a017
Dunning TH (1989) J Chem Phys 90(2):1007. https://doi.org/10.1063/1.456153
Frisch MJ, Trucks GW, Schlegel HB, Scuseria GE, Robb MA, Cheeseman JR, Scalmani G, Barone V, Mennucci B, Petersson GA (2010) Gaussian 09, revision C. 01. Gaussian, Inc., Wallingford, CT
Gao J, Xia X (1992) Science 258(5082):631. https://doi.org/10.1126/science.1411573
George P, Witonsky RJ, Trachtman M, Wu C, Dorwart W, Richman L, Richman W, Shurayh F, Lentz B (1970) Biochimica et Biophysica Acta (BBA). Bioenergetics 223(1):1. https://doi.org/10.1016/0005-2728(70)90126-x
Grigorenko BL, Rogov AV, Nemukhin AV (2006) J Phys Chem B 110(9):4407. https://doi.org/10.1021/jp056395w
Hansen P, McDonald IR (2006) Theory of simple liquids, 3rd edn. Academic Press, London
Hirose K, Ono T, Fujimotmo Y, Tsukamoto S (2005) First-principles calculations in real-space formalism. Imperial college press, London
Hong J, Yoshida N, Chong SH, Lee C, Ham S, Hirata F (2012) J Chem Theory Comput 8(7):2239. https://doi.org/10.1021/ct300099e
Hori T, Takahashi H, Nakano M, Nitta T, Yang W (2006) Chem Phys Lett 419(1–3):240. https://doi.org/10.1016/j.cplett.2005.11.096
Hori T, Takahashi H, Ichi Furukawa S, Nakano M, Yang W (2007) J Phys Chem B 111(3):581. https://doi.org/10.1021/jp066334d
Hutter J, Iannuzzi M, Schiffmann F, VandeVondele J (2013) Wiley Interdiscip Rev Comput Mol Sci 4(1):15. https://doi.org/10.1002/wcms.1159
Jorgensen WL, Maxwell DS, Tirado-Rives J (1996) J Am Chem Soc 118(45):11225. https://doi.org/10.1021/ja9621760
Kamerlin SCL, Warshel A (2009) J Phys Chem B 113(47):15692. https://doi.org/10.1021/jp907223t
Klähn M, Rosta E, Warshel A (2006) J Am Chem Soc 128(47):15310. https://doi.org/10.1021/ja065470t
Kleinman L, Bylander DM (1982) Phys Rev Lett 48(20):1425
Kodama T (1985) Physiol Rev 65:467
Kohn W, Sham LJ (1965) Phys Rev 140(4A):A1133
Laio A, Parrinello M (2002) Proc Natl Acad Sci USA 99(20):12562. https://doi.org/10.1073/pnas.202427399
Lee C, Yang W, Parr RG (1988) Phys Rev B 37(2):785

Matubayasi N, Nakahara M (2000) J Chem Phys 113(15):6070. https://doi.org/10.1063/1.1309013
Matubayasi N, Nakahara M (2002) J Chem Phys 117(8):3605
Meyerhof O, Lohmann K (1932) Biochem Z 253:431
Nelson DL, Cox MM (2013) Lehninger principles of biochemistry, 6th edn. W. H Freeman and Company, New York
Ono T, Hirose K (1999) Phys Rev Lett 82(25):5016
Parr RG, Yang W (1989) Density-functional theory of atoms and molecules. Oxford University Press, New York
Pepi F, Ricci A, Rosi M, Stefano MD (2004) Chem—A Eur J 10(22):5706. https://doi.org/10.1002/chem.200400293
Rivail JL, Ruiz-Lopez M, Assfeld X (2015) Challenges and advances in computational chemistry and physics, vol 21. Springer, Heidelberg
Ross J (2006) J Phys Chem B 110(13):6987. https://doi.org/10.1021/jp0556862
Ruben EA, Plumley JA, Chapman MS, Evanseck JD (2008) J Am Chem Soc 130(11):3349. https://doi.org/10.1021/ja073652x
Ruiz-Lopez MF (2003) J Mol Struct THEOCHEM 632:1
Sun R, Sode O, Dama JF, Voth GA (2017) J Chem Theory Comput 13(5):2332. https://doi.org/10.1021/acs.jctc.7b00077
Takahashi H, Hori T, Wakabayashi T, Nitta T (2000) Chem Lett 3:222
Takahashi H, Hori T, Wakabayashi T, Nitta T (2001a) J Phys Chem A 105(17):4351. https://doi.org/10.1021/jp004348s
Takahashi H, Hori T, Hashimoto H, Nitta T (2001b) J Comp Chem 22(12):1252
Takahashi H, Matubayasi N, Nakahara M, Nitta T (2004) J Chem Phys 121(9):3989
Takahashi H, Omi A, Morita A, Matubayasi N (2012) J Chem Phys 136:214503
Takahashi H, Maruyama K, Karino Y, Morita A, Nakano M, Jungwirth P, Matubayasi N (2011a) J Phys Chem B 115(16):4745. https://doi.org/10.1021/jp2015676
Takahashi H, Iwata Y, Kishi R, Nakano M (2011b) Int J Quantum Chem 111(7–8):1748. https://doi.org/10.1002/qua.22814
Takahashi H, Satou W, Hori T, Nitta T (2005a) J Chem Phys 122(4):044504. https://doi.org/10.1063/1.1839858
Takahashi H, Kawashima Y, Nitta T, Matubayasi N (2005b) J Chem Phys 123(12):124504. https://doi.org/10.1063/1.2008234
Takahashi H, Miki F, Ohno H, Kishi R, Ohta S, Ichi Furukawa S, Nakano M (2009) J Math Chem 46(3):781. https://doi.org/10.1007/s10910-009-9544-2
Takahashi H, Ohno H, Kishi R, Nakano M, Matubayasi N (2008) Chem Phys Lett 456(4–6):176. https://doi.org/10.1016/j.cplett.2008.03.038
Takahashi H, Umino S, Miki Y, Ishizuka R, Maeda S, Morita A, Suzuki M, Matubayasi N (2017) J Phys Chem B 121(10):2279. https://doi.org/10.1021/acs.jpcb.7b00637
Tomasi J, Mennucci B, Cammi R (2005) Chem Rev 105(8):2999. https://doi.org/10.1021/cr9904009
Valiev M, Bylaska E, Govind N, Kowalski K, Straatsma T, Dam HV, Wang D, Nieplocha J, Apra E, Windus T, de Jong W (2010) Comp Phys Comm 181(9):1477. https://doi.org/10.1016/j.cpc.2010.04.018
Voet D, Voet JG, Pratt CW (2013) Fundamentals of biochemistry, 4th edn. Wiley, Hoboken
Wang C, Huang W, Liao JL (2015) J Phys Chem B 119(9):3720
Warshel A (1991) Computer modeling of chemical reactions in enzymes and solutions. Wiley, New York

Chapter 2
Role of Metal Ion Binding and Protonation in ATP Hydrolysis Energetics

Shun-ichi Kidokoro

Abstract Based on the thermodynamic parameters by Alberty (Biochemistry 31:10610–10615, 1992), the thermodynamic properties such as the standard transformed Gibbs energy change of ATP hydrolysis were systematically evaluated. From the calculation, it was found that the release of Mg^{2+} ion induced by the hydrolysis at pH 7.0, 298.15 K, and 1 atm affects both Gibbs energy and enthalpy significantly. The contribution of the proton binding and ionic strength was also examined.

Keywords Gibbs energy · Enthalpy · Ion binding number · pH
Ionic strength

2.1 Introduction

The chemical energy of ATP, "currency of energy" for all the livings in the earth, is utilized by its hydrolysis in solution. The chemical energy, the Gibbs energy change of the reaction, in one solution condition is determined by the chemical potential, partial mole Gibbs energy, of all the components of the reaction there. Table 2.1, from a literature (Alberty 1992), shows the standard Gibbs energy of formation and the standard enthalpy of formation at 298.15 K and zero ionic strength ($I = 0$), of several substances relating the ATP hydrolysis to ADP. Based on the procedure described in the literature, we can calculate the standard transformed Gibbs energy of formation and the standard transformed enthalpy of formation as a function of pH, pMg, and ionic strength, I (Alberty 1992, 1998).

The substances in the table may suggest the importance of two kinds of ions, Mg^{2+} and proton (H^{+}), which can bind to the basic three substances, ATP^{4-}, ADP^{3-}, and HPO_4^{2-}. However, it has not been shown in detail the dependence of the

S. Kidokoro (✉)
Department of Bioengineering, Nagaoka University of Technology,
Kamitomioka 1603-1, Nagaoka, Niigata 940-2188, Japan
e-mail: kidokoro@nagaokaut.ac.jp

© Springer Nature Singapore Pte Ltd. 2018

M. Suzuki (ed.), *The Role of Water in ATP Hydrolysis Energy Transduction by Protein Machinery*, https://doi.org/10.1007/978-981-10-8459-1_2

Table 2.1 Standard formation properties of substances at 298.15 K, 1 atm, and $I = 0$ (Alberty 1992)

Substance	$\Delta_f H^0$/ kJ mol^{-1}	$\Delta_f G^0$/ kJ mol^{-1}	z	N_H	N_{Mg}
ATP^{4-}	−2997.91	−2573.49	−4	12	0
$HATP^{3-}$	−2991.61	−2616.87	−3	13	0
H_2ATP^{2-}	−3006.61	−2643.58	−2	14	0
$MgATP^{2-}$	−3442.01	−3064.07	−2	12	1
$MgHATP^-$	−3441.71	−3092.89	−1	13	1
Mg_2ATP	−3898.21	−3534.72	0	12	2
ADP^{3-}	−2005.24	−1711.55	−3	12	0
$HADP^{2-}$	−1999.64	−1752.53	−2	13	0
H_2ADP^-	−2017.24	−1777.42	−1	14	0
$MgADP^-$	−2453.24	−2193.39	−1	12	1
MgHADP	−2454.14	−2222.10	0	13	1
HPO_4^{2-}	−1299.00	−1096.10	−2	1	0
$H_2PO_4^-$	−1302.60	−1137.30	−1	2	0
$MgHPO_4$	−1753.80	−1566.87	0	1	1
H_2O	−285.83	−237.19	0	2	0
H^+	0.00	0.00	+1	–	–
Mg^{2+}	−467.00	−455.30	+2	–	–

Gibbs energy and enthalpy of ATP hydrolysis, and the binding number change of the ions on the concentration of Mg^{2+} and proton, and ionic strength. This chapter will illustrate these dependence systematically and will discuss the significance of the ions on ATP energetics.

2.2 Theory

The standard transformed Gibbs energy of formation and the standard transformed enthalpy of formation at 298.15 K and 1 atm were calculated using the standard Gibbs energy of formation and the standard enthalpy of formation in Table 2.1 as described in the literature (Alberty 1992, 1998). The mole fractions of each chemical species, f_i, were then calculated from the standard transformed Gibbs energy of formation as

$$f_i(\mathrm{pH, pMg, I}) = \exp\left(-\frac{\Delta_f G_i'^0(\mathrm{pH, pMg, I})}{RT}\right) / \sum_j \exp\left(-\frac{\Delta_f G_j'^0}{RT}\right) \quad (1)$$

for ATP, ADP, and Pi, respectively, and the standard transformed Gibbs energy of formation, $\Delta_f G'^0$, and the standard transformed enthalpy of formation, $\Delta_f H'^0$, of ATP, ADP, and phosphate, Pi, were evaluated as

$$\begin{aligned} \Delta_f G'^0(\text{pH}, \text{pMg}, \text{I}) &= -RT \ln\left[\sum_j \exp\left(-\frac{\Delta_f G_j'^0}{RT}\right)\right] \\ \Delta_f H'^0(\text{pH}, \text{pMg}, \text{I}) &= \sum_j f_j \Delta_f H_j'^0 \end{aligned} \tag{2}$$

Using these functions, the standard transformed Gibbs energy change of ATP hydrolysis, $\Delta_r G''^0$, and the standard transformed enthalpy change, $\Delta_r H''^0$, were calculated as

$$\begin{aligned} \Delta_r G'^0(\text{pH}, \text{pMg}, \text{I}) &= \Delta_f G'^0_{\text{ADP}} + \Delta_f G'^0_{\text{Pi}} - \Delta_f G'^0_{\text{ATP}} - \Delta_f G'^0_{\text{H}_2\text{O}} \\ \Delta_r H'^0(\text{pH}, \text{pMg}, \text{I}) &= \Delta_f H'^0_{\text{ADP}} + \Delta_f H'^0_{\text{Pi}} - \Delta_f H'^0_{\text{ATP}} - \Delta_f H'^0_{\text{H}_2\text{O}} \end{aligned} \tag{3}$$

Using the same procedure, the binding number of Mg^{2+}, N_{Mg}, and proton, N_H, were calculated as a function of pH, pMg, and I for ATP, ADP, and Pi, respectively. The apparent pK values were evaluated in the following. The pK_{Mg}1 of ATP or pK_{Mg} of ADP and Pi were determined as the pMg where $N_{Mg} = 1.5$ for ATP and $N_{Mg} = 0.5$ for ADP and Pi, and pK_{Mg}2 was the pMg where $N_{Mg} = 0.5$ for ATP. The pK_a2 was determined as the pH where $N_H = 13.5$ for ADP and $N_H = 1.5$ for Pi. The pK_a3 was the pH where $N_H = 13.5$ for ATP and $N_H = 12.5$ for ADP. The pK_a4 was the pH where $N_H = 12.5$ for ATP.

The three-dimensional (3D) maps were illustrated with a graphic software, Origin 2017 (Origin Lab, Northampton, MA). The concentration of proton, (H^+), and Mg^{2+}, (Mg^{2+}) were designated by pH and pMg, respectively:

$$\begin{aligned} \text{pH} &= -\log_{10}[(\text{H}^+)/\text{M}] \\ \text{pMg} &= -\log_{10}\left[(\text{Mg}^{2+})/\text{M}\right] \end{aligned} \tag{4}$$

2.3 Results and Discussions

2.3.1 ATP Hydrolysis Energetics at pH 7.0

Figure 2.1a shows the three-dimensional (3D) map of the transformed standard Gibbs energy change of ATP hydrolysis at pH 7.0 as the function of ionic strength, I, and the concentration of $Mg^{2+,}$ pMg. The large pMg dependence of the Gibbs energy change is thermodynamically produced by the binding number change of Mg^{2+} ion based on the following equation:

$$\left(\frac{\partial \Delta_r G'0}{\partial \text{pMg}}\right)_{T,\text{pH}} = (\ln 10) RT \Delta_r N_{\text{Mg}} \tag{5}$$

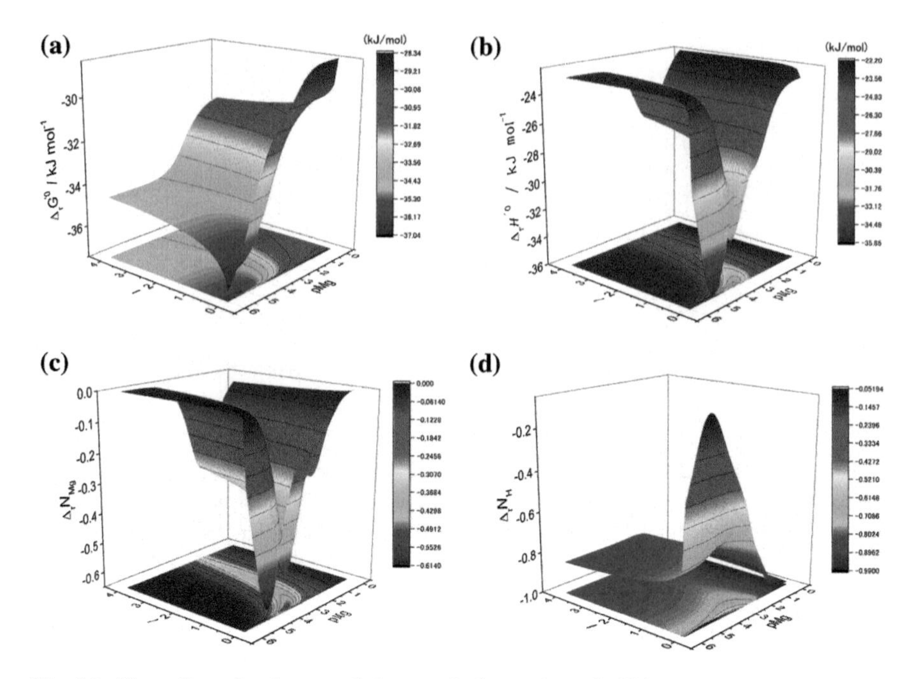

Fig. 2.1 Three-dimensional map of the standard transformed Gibbs energy change (**a**), the standard transformed enthalpy change (**b**), the Mg^{2+} binding number change (**c**), and the H^+ binding number change (**d**) due to ATP hydrolysis at pH 7.0, 298.15 K, and 1 atm as the function of the ionic strength (I) and the concentration of Mg^{2+} ion (pMg). These values were calculated with the standard formation parameters of Table 2.1

Here it is worthwhile to note that $(\ln 10)RT$ is 5.7 kJ/mol at 298.15. As seen in Fig. 2.1c, the Mg^{2+} binding number change due to the ATP hydrolysis, $\Delta_r N_{Mg}$, is negative in the range of pMg 0–6 indicating Mg^{2+} ion is released into the solution by ATP hydrolysis. This release thermodynamically reduces the Gibbs energy change in the condition of lower Mg^{2+} concentration (= higher pMg) by around several kJ/mol.

This effect becomes stronger by reducing the ionic strength because the binding number of Mg ion to each species is decreased by increasing the ionic strength. Even in the high ionic strength, $I > 1$, however, this effect remains clearly indicating the significance of the Mg^{2+} binding on the ATP energetics in almost all the biological systems.

Figure 2.1b shows the 3D map of the standard transformed enthalpy change of ATP hydrolysis. Its dependence on pMg and I resembled that of the Mg^{2+} binding number change, indicating the enthalpy profile was almost determined by the metal ion binding number. As the dissociation enthalpy of Mg^{2+} ion from ATP, ADP, and Pi is negative, for example, −18 kJ/mol (Wilson 1991), −10 kJ/mol for the first dissociation and −15 kJ/mol for the second dissociation (Nakamura et al. 2013) for ATP, it seems reasonable that one Mg^{2+} ion release accompanying ATP hydrolysis reduces the total hydrolysis enthalpy by several kJ/mol. The effect on the hydrolysis enthalpy becomes more than 10 kJ/mol in the low ionic strength and is several kJ/mol even in the higher ionic strength ($I > 1$). The Mg^{2+} binding is significant also for the hydrolysis enthalpy.

The metal ion binding affects the proton binding number change as shown in Fig. 2.1d. The change produces the pH dependence of Gibbs energy change as

$$\left(\frac{\partial \Delta_r G'^0}{\partial \mathrm{pH}}\right)_{T,\mathrm{pMg}} = (\ln 10) RT \Delta_r N_H \quad (6)$$

The proton binding number change is negative at pH 7.0, namely the proton is released by ATP hydrolysis into the solution. Then the chemical energy of ATP hydrolysis becomes larger at higher pH as in almost the same way of Mg^{2+} ion above mentioned. Roughly speaking, however, the H^+ ion has the tendency to be less released in the condition that the Mg^{2+} ion is highly released. This tendency indicates the negative cooperativity between H^+ and Mg^{2+} ions.

2.3.2 Apparent pK_a's for ATP, ADP, and Pi

In the previous section, we found that the binding number change of Mg^{2+} is significant to determine the ATP energetics, and the change is determined by each binding constants, namely pK_a's, for ATP, ADP, and Pi. Based on the thermodynamic parameters in Table 2.1, they were calculated as the functions of pMg and I as shown in Fig. 2.2 The pK_a's decrease by decreasing pMg, indicating the Mg^{2+} ion binding to ATP, ADP, and Pi induces the dissociation of H^+ ion from the molecules. We should be aware that the apparent pK_a's of the molecule decrease drastically by increasing the concentration of Mg^{2+} ion. The ionic strength dependence of pK_a's was not small especially in the condition of lower concentration of Mg^{2+} ion. There, pK_a's decrease drastically by increasing ionic strength.

As all the apparent pK_a's in Fig. 2.2 become less than 7 at higher concentration of Mg^{2+}, it may be necessary to consider the apparent pK_a3 of Pi at that condition, which we have no thermodynamic parameters of formation until now. In order to calculate the contribution, the standard thermodynamic functions of formation for $MgPO_4^-$ should be obtained. If the contribution of the species is not small, the proton binding number of Pi may be decreased, and the change of the proton binding number may also be reduced.

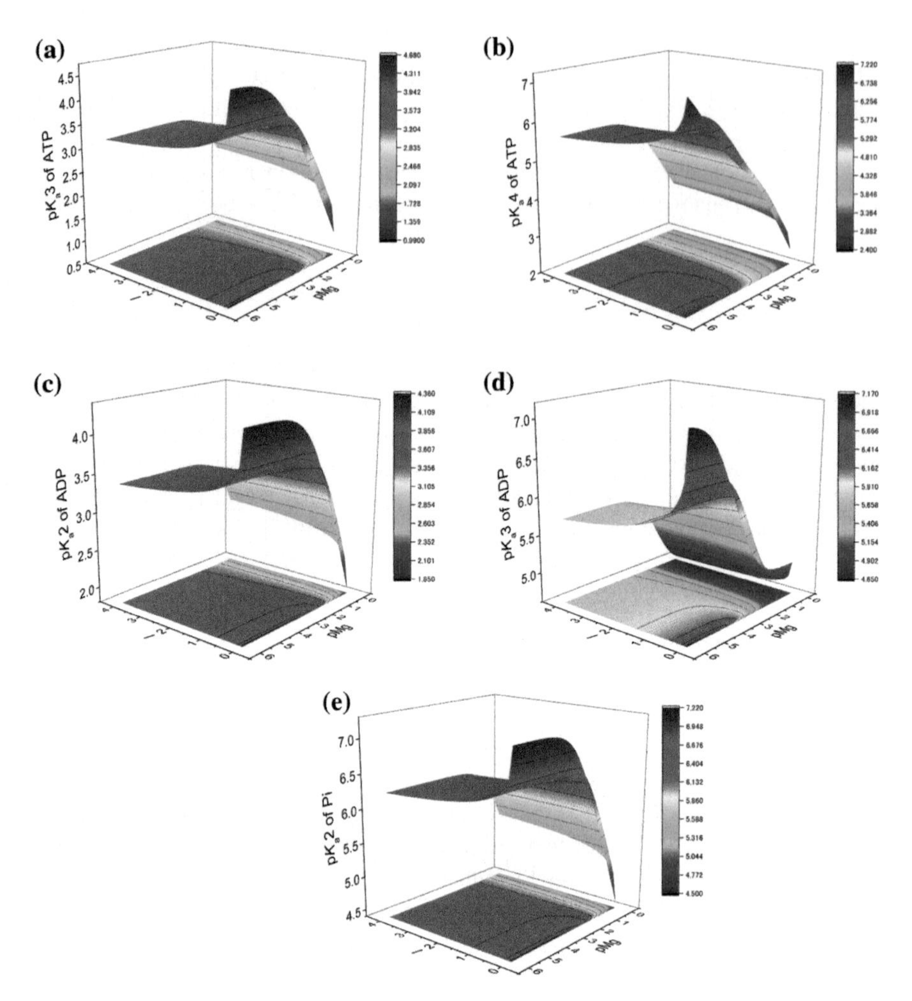

Fig. 2.2 3D map of the apparent pK_a's evaluated as the function of pMg and I at 298.15 K and 1 atm. For ATP, pK_a3 (**a**) and pK_a4 (**b**) were illustrated; for ADP, pK_a2 (**c**) and pK_a3 (**d**); for Pi, pK_a2 (**e**)

2.3.3 Apparent pK_{Mg}'s for ATP, ADP, and Pi

While we explored the Mg^{2+} binding effect on proton binding in the previous section, the proton binding effect on the Mg^{2+} binding was described in this section. As many proteins bind or recognize ATP as the Mg^{2+} bound form, the pK_{Mg}'s of ATP may be biologically important.

Figure 2.3 illustrated the pK_{Mg}'s as the function of pH and I. As expected from the results of the previous section, the Mg^{2+} binding is inhibited by the proton binding, and all the pK_{Mg}'s decrease by decreasing pH. The dependence of ionic

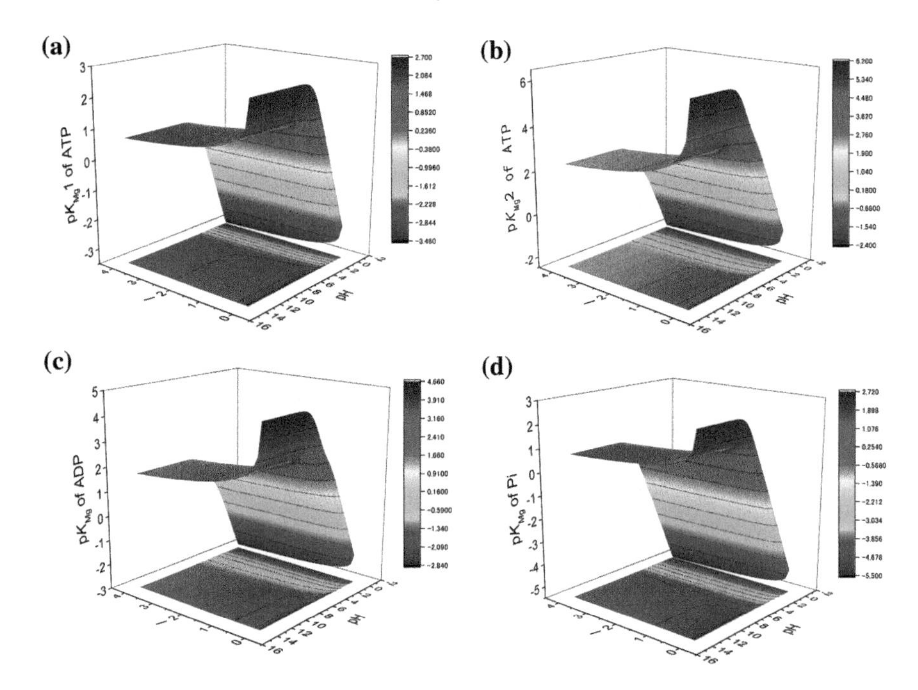

Fig. 2.3 3D map of the apparent pK_{Mg}'s evaluated as the function of pH and I at 298.15 K and 1 atm. The pK_{Mg}1 (**a**) and pK_{Mg}2 (**b**) for ATP, and pK_{Mg}'s for ADP (**c**) and Pi (**d**) were illustrated

strength on pK_{Mg}'s seemed to be very resemble, especially in the lower proton concentration to that on pK_a's previously seen in Fig. 2.2.

2.3.4 *Effect of Other Metal Ions on ATP Hydrolysis Energetics*

In the previous sections, the Mg^{2+} ion binding, determined by pK_{Mg}'s, affects the energetics of ATP hydrolysis. Several other kinds of metal ions were found to bind ATP molecules (Wilson 1991). From the results, the Ca^{2+} ion binds strongly to ATP. Because the concentration of the ion in the cell is high, the real ATP energetics in the cell may be affected strongly by Ca^{2+} ion binding. In addition, we recently observed that two Ca^{2+} ions were bound to one ATP molecule as Mg^{2+} ions by isothermal titration calorimetry (ITC) (data not shown).

As the binding of monovalent metal ions such as Na^+ and K^+ was weak (Wilson 1991), the binding effect was usually neglected. Based on the results of Sect. 2.3.3, it may not be neglected for these metal ions to affect the binding of divalent metal ions such as Mg^{2+} and Ca^{2+}.

Recently, we have characterized the new ATP-Mg^{2+} complex, $Mg(ATP)_2$, and determined the thermodynamic parameters of the complex (Nakamura et al. 2013). When the concentration of ATP molecules is high, we should consider such kind of new chemical species in order to discuss the ATP hydrolysis energetics.

2.4 Conclusions

The systematic evaluation of the transformed thermodynamic functions based on the standard thermodynamic functions of formation by Alberty (1992) indicated the Mg^{2+} ion could release accompanying the ATP hydrolysis at pH 7.0, 298.15 K, producing the significant pMg dependence of ATP hydrolysis Gibbs energy and enthalpy. The apparent pK_a's of ATP, ADP, and Pi were also affected by the Mg^{2+} binding. It strongly suggested that the contribution of the chemical species, $MgPO_4^-$, not considered in this calculation, should be considered in under the condition of high Mg^{2+} concentration. These results also indicated the significance of binding of divalent metal ions in the real system, especially Ca^{2+} ion to predict the ATP hydrolysis energetics. The binding effect of monovalent metal ions such as Na^+ and K^+ may contribute the binding of the divalent cations by direct interaction to ATP, ADP, and Pi as proton. Further study on the metal ion binding and protonation may be necessary to calculate the real ATP energetics in the livings because the proton and metal ions affect the thermodynamics significantly.

References

Alberty RA (1992) Biochemistry 31:10610–10615
Alberty RA (1998) Arch Biochem Biophys 353:116–130
Nakamura S et al (2013) Thermochim Acta 563:82–89
Wilson JE, Chin A (1991) Anal Biochem 193:16–19

Chapter 3
Spatial Distribution of Ionic Hydration Energy and Hyper-Mobile Water

George Mogami, Makoto Suzuki and Nobuyuki Matubayasi

Abstract In this chapter, we provide the following two topics.

1: We carry out DRS measurements for divalent metal chloride and trivalent metal chloride solutions and clarify the hydration states. All the tested solutions have hyper-mobile water (HMW) with higher dielectric relaxation frequency f_1 (~20 GHz) than that of bulk water (12.6 GHz at 10 °C), and dispersion amplitude of HMW is aligned to Hofmeister series. According to the correlation between an intensity of HMW signal and water structure entropy, HMW can be a scale for the water structure.

2: We carry out the spatial-decomposition analysis of energetics of hydration for a series of ionic solutes in combination with molecular dynamics (MD) simulation. The hydration analysis is conducted on the basis of a spatial-decomposition formula for the excess partial molar energy of the ion that expresses the ther-

---

G. Mogami (✉) · M. Suzuki
Department of Materials Processing, Graduate School of Engineering, Tohoku University, 6-6-02, Aramaki Aza Aoba, Aoba-ku, Sendai 980-8579, Japan
e-mail: mogami-g@material.tohoku.ac.jp

M. Suzuki
e-mail: makoto.suzuki.c5@tohoku.ac.jp

M. Suzuki
Biological and Molecular Dynamics, Institute of Multidisciplinary Research for Advanced Materials (IMRAM), Tohoku University, Katahira 2-1-1, Aoba-Ku, Sendai 980-8577, Japan

M. Suzuki
Department of Biomolecular Engineering, Graduate School of Engineering, Tohoku University, 6-6-07 Aoba, Aramaki, Aoba-Ku, Sendai 980-8579, Japan

N. Matubayasi
Division of Chemical Engineering, Graduate School of Engineering Science, Osaka University, Toyonaka, Osaka 560-8531, Japan
e-mail: nobuyuki@cheng.es.osaka-u.ac.jp

N. Matubayasi
Elements Strategy Initiative for Catalysts and Batteries, Kyoto University, Katsura, Kyoto 615-8520, Japan

© Springer Nature Singapore Pte Ltd. 2018

M. Suzuki (ed.), *The Role of Water in ATP Hydrolysis Energy Transduction by Protein Machinery*, https://doi.org/10.1007/978-981-10-8459-1_3

modynamic quantity as an integral over the whole space of the ion–water and water–water interactions conditioned by the ion–water distance. In addition, we examine the correlation between the electric field formed by ion and the number of HMW around ion.

Keywords Dielectric relaxation spectroscopy · Hyper-mobile water
Ionic hydration energy · Spatial distribution · Hofmeister effect

3.1 Introduction

A variety of chemical reactions, including ATP hydrolysis in aqueous solution, are accompanied by association or dissociation of ions resulting in charge separation. When the valence of an ion differs, the change in the free energy of ionic hydration is comparable to that of a covalent bond (Philip et al. 1970). The effects of ions on the structure and dynamics of the solvent water have been studied extensively (Robinson and Stokes 1959; Ohtaki and Radnai 1993; Jungwirth and Tobias 2006; Bakker 2008; Marcus 2009). The ion-specific effect depends on the ionic radius; the Hofmeister series describes these relative effects by organizing ions in the order of their ability to affect the solubility of proteins (Lo Nostro and Ninham 2012). The Hofmeister effect is also evident in the viscosity *B*-coefficient: ions with positive and negative *B*-coefficients are classified as water structure makers and breakers, respectively (Gurney 1953; Marcus 2009). The Hofmeister effect is caused by the structure of the water molecules around the ions; thus, one of the biggest challenges is establishing a clear and universal index for water structure.

Dielectric relaxation spectroscopy (DRS) (Kaatze and Feldman 2006; Buchner 2008; Kaatze 2013) is a promising technique for measuring ionic hydration. This method can detect the first layer of water molecules as well as the outer layers; furthermore, it can assess the dynamics of a water–ions collective mode (Miyazaki et al. 2008; Mogami et al. 2013). However, the accuracy of the energetics of the water molecules observed by DRS is under debate. Therefore, full elucidation of the physical meaning of DRS data at a molecular level is necessary to apply the technique to gather practical information regarding hydration. A statistical mechanical study using molecular dynamics (MD) simulation of water molecules around a monatomic ion ($Z = 0, \pm 1, \pm 2$) revealed that the cross-correlation term of molecular polarization is essential to enable relaxation to occur faster than that of bulk water; this is related to hyper-mobile water (HMW), which is described in more detail in the following paragraph.

In previous study (Mogami et al. 2013), we reported high-resolution DRS of aqueous solutions of sodium and potassium halides, i.e., NaX and KX, where X = F, Cl, Br, and I, at concentration of 0.05 and 0.1 M in the frequency range of 0.2–26 GHz. In this study, the spectrum of each solution was simulated using a combination of two Debye components: one with a relaxation frequency, f_1,

(~20 GHz) higher than that of bulk water (12.6 GHz at 10 °C) and one with a sub-gigahertz frequency, f_c. The latter Debye component was assigned to the relaxation of the counterion cloud around an ion, and the former Debye component was assigned to HMW. The dispersion amplitude of HMW increased with increasing ionic radii in the order of the Hofmeister series. HMW was also found around an actin filament (Kabir et al. 2003; Suzuki et al. 2004), which is a skeletal muscle protein that facilitates muscle contraction with the myosin protein. Recently, it was proposed that increasing or decreasing the amount of HMW by changing the structure of actin could induce a novel driving force for the actomyosin system (Chap. 16; Suzuki et al. 2017). Since HMW could be directly related to protein function, it is important to examine the thermodynamic properties of HMW.

In Sect. 3.2, we describe our DRS analyses of divalent metal chloride and trivalent metal chloride solutions to clarify their hydration states. The dielectric relaxation (DR) spectra of monovalent alkali halide solutions were previously measured (Mogami et al. 2013); thus, we analyzed these data for comparison using the same conditions. In addition, since HMW was expected to be detected in all tested solutions in amounts that correlate with the Hofmeister series, we compared the DRS results with thermodynamic data for ionic hydration (Marcus 1997); this enabled us to determine the relationship between the DRS and thermodynamic data for the hydration structure.

A thermodynamic quantity including hydration data based on spectroscopic measurements, such as DRS, provides information integrated over the whole system, for example. Thus, a rigorous framework is necessary to fill the gap between the intuitive picture of the local perturbation of water around ions and an experimentally accessible quantity. In a previous study, a scheme to describe spatial decomposition was developed to rigorously bridge local correlations of molecules and macroscopic observables (Matubayasi et al. 1994, 1998; Matubayasi and Levy 1996; Kubota et al. 2012; Tu et al. 2014a, b); this was achieved by formulating an integral expression for the macroscopic observable in terms of the molecular correlation functions throughout the space.

In Sect. 3.3, we describe a spatial-decomposition analysis of the energetics of hydration for a series of ionic solutes in combination with MD simulations.

3.2 Hydration Measurement by Dielectric Relaxation Spectroscopy

3.2.1 Preparation of Aqueous Solutions of Salts

All reagents used in the present study were purchased from Wako Pure Chemical Industries (Osaka, Japan). As for NaX and KX (X = F, Cl, Br, I), the DR spectra reported in a previous study (Mogami et al. 2013) were reanalyzed to compare the results under the same analysis conditions. $MgCl_2$ (97.0%), $FeCl_2 \cdot 4H_2O$ (99.9%),

$MnCl_2 \cdot 4H_2O$ (99.9%), $CaCl_2$ (95.0%), and $FeCl_3 \cdot 6H_2O$ (99.9%) were dissolved in water purified by Milli-Q (Millipore, Billerica, MA, USA) at concentrations of 0.02–0.1 M. Partial-specific volumes (v_B) were calculated from the solution densities measured using an Anton-Paar density meter DMA-58 and DMA5000 M (Graz, Austria). Notably, the density and v_B values of all aqueous solutions agreed with the literature data (Laliberté 2009).

3.2.2 Dielectric Spectroscopy: Experimental Method

All measurements were carried out using a microwave network analyzer (E8364C-85070E; Keysight Technologies, Santa Rosa, CA, USA) with a reinforced internal air circulation system at a constant low power of −10 dBm over the measured frequency range and an open-end coaxial flat surface probe (high-temperature type, 19 mm diameter, electric length ~45 mm; Keysight Technologies) immersed in a conically shaped glass cell (total volume of 3.2 mL) containing a sample solution that had been degassed before loading. The cell was held at 10.00 or 20.00 ± 0.01 °C by a temperature-controlled circulator (NESLAB RTE-17, Thermo Fisher Scientific, Waltham, MA, USA). The whole measuring system was placed in an air-conditioned box maintained at 20 ± 1 °C. Probe calibration was performed by three separate runs: open-circuited to the air, short-circuited with mercury, and in contact with pure water. DR spectra were recorded over a frequency range of 0.2–26 GHz (301 frequency points in log scale) as previously described (Miyazaki et al. 2008). Briefly, DR spectra of water reference ($\varepsilon_w^*(f)$) and sample solution ($\varepsilon_{ap}^*(f)$) were measured alternately four to eight times to obtain a difference spectrum ($\Delta\varepsilon^*(f)$ by Eq. 3.1) with a reduced machine drift noise over a period of 2 h for a given solution.

3.2.3 Dielectric Spectroscopy: Data Analysis

DR spectra of $MgCl_2$ solution and water are shown in Fig. 3.1. For all solutions, the difference between a sample and water reference spectra was small but systematic. The difference DR spectrum for each sequential pair was calculated using Eq. 3.1.

$$\Delta\varepsilon^*(f) = \Delta\varepsilon'(f) - i\Delta\varepsilon''(f) = \varepsilon_{ap}^*(f) - \varepsilon_w^*(f) + i\frac{\sigma}{2\pi f \varepsilon_0} \tag{3.1}$$

where σ is the difference in static electrical conductivity between a given sample at a specified salt concentration and pure water. The difference spectra obtained from Eq. 3.1 were then averaged and smoothed to eliminate resonance noise induced by the probe over its whole length at a frequency range of 2–6 GHz. The smoothed spectral curves were determined using the third- to fifth-order polynomial functions

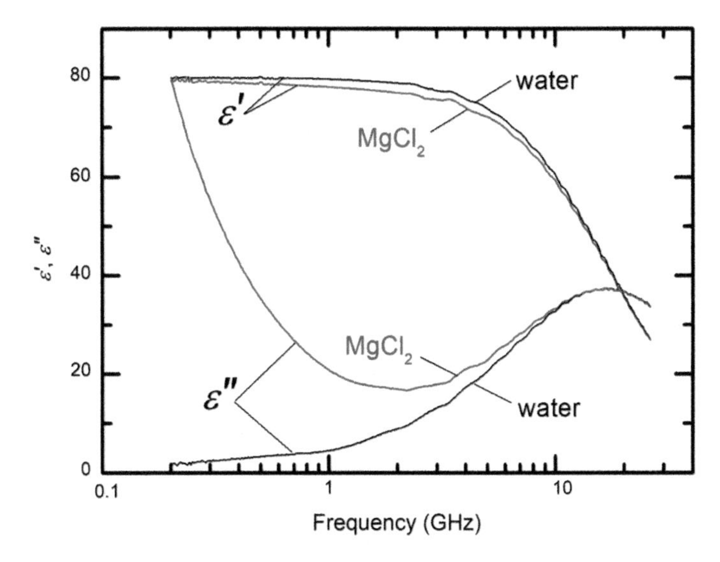

Fig. 3.1 Raw DR spectra of $MgCl_2$ aqueous solution (ε^*_{ap}) of 0.02 M and reference water (ε^*_w)

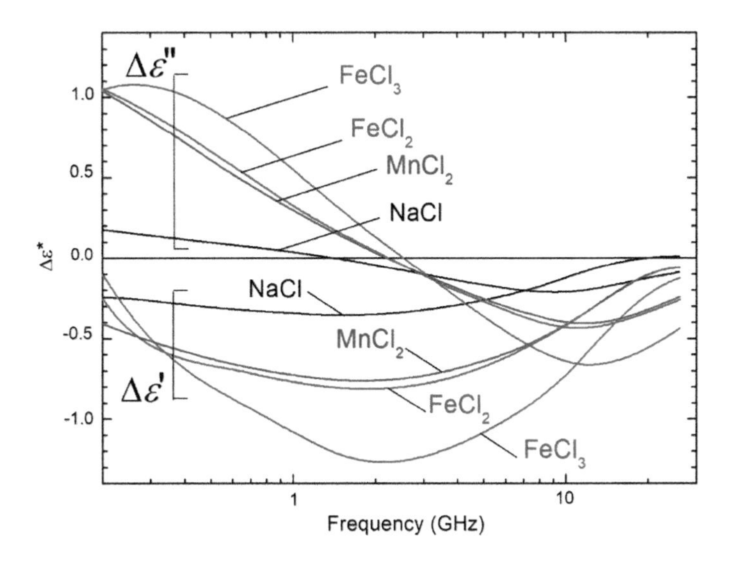

Fig. 3.2 Real and imaginary parts of the smoothed difference spectra ($\Delta\varepsilon^*$) of 0.02 M at 10 °C

of log(*f*), as shown in Fig. 3.2. The standard errors of $\Delta\varepsilon'(f)$ and $\Delta\varepsilon''(f)$ over four to eight separate experiments were 0.03 or less over 1–26 GHz. Smoothed spectra $\varepsilon^*_{ap}(f)$ were rebuilt by summing $\Delta\varepsilon^*(f)$ and $\varepsilon^*_w(f)$. The value of σ was determined by iterative parameter fitting using Eqs. 3.1–3.4, as described previously (Wazawa et al. 2010; Mogami et al. 2011, 2013).

In a previous study (Mogami et al. 2013), a dielectric response in the diluted system of the alkali halide was successfully represented with an ellipsoidal dielectric model. Next, using Asami mixture theory (Asami et al. 1980), the dielectric spectrum $\varepsilon^*_{ap}(f)$ of a sample solution was mathematically separated into two components at an arbitrary volume fraction ϕ, bulk water $\varepsilon^*_w(f)$, and solute with a hydration layer $\varepsilon^*_q(f)$ which is the average spectrum of an ellipsoidal region, as given by Eq. 3.2.

$$\begin{aligned} \varepsilon^*_{\mathrm{q}} &= \varepsilon^*_{\mathrm{w}} - \varepsilon^*_{\mathrm{w}} \frac{\{X^*(a_{0x}+a_{0y})-3\}-\sqrt{Y^*}}{2(a_{0x}a_{0y}X^*-2a_{0x}-a_{0y})} \\ X^* &= \frac{9}{\phi}\frac{\varepsilon^*_{\mathrm{ap}}-\varepsilon^*_{\mathrm{w}}}{\varepsilon^*_{\mathrm{ap}}+2\varepsilon^*_{\mathrm{w}}} \\ Y^* &= \{X^*(a_{0x}+a_{0y})-3\}^2-4(a_{0x}a_{0y}X^*-2a_{0x}-a_{0y})X^* \\ a_{0x} &= -(q^2-1)^{-1}+q(q^2-1)^{-3/2}\log\{q+(q^2-1)^{1/2}\} \\ a_{0y} &= a_{0z} = (1-a_{0x})/2 \\ & (v<\phi\ll 1) \end{aligned} \tag{3.2}$$

where v is the volume fraction of given solutes calculated by $v = cM_w v_B/1000$, where M_w, c, and v_B are the molar mass in g/mol, molar concentration of solute molecule in M (mol/L), and partial specific volume of solute in mL/g, respectively. Note that this ellipsoid model analysis at the limit of $q = 1$ precisely agrees with the sphere model analysis using the Wagner equation (Wagner 1914).

Solute with a hydration layer $\varepsilon^*_q(f)$ can be decomposed into a series of Debye functions and the bulk water component ($f_{cw} \approx 12.6$ GHz; $\delta_w \approx 78.4$ at 10 °C) by Eq. 3.3:

$$\varepsilon^*_{\mathrm{q}}(f) \cong \varepsilon^*_{\mathrm{q,sim}}(f) \equiv \varepsilon_{\mathrm{q},\infty} + \alpha(\varepsilon^*_{\mathrm{w}}(f)-\varepsilon_{\mathrm{w},\infty}) + \sum_{j=1}^{m}\frac{\delta_j}{1+i(f/f_j)} \tag{3.3}$$

where $\varepsilon_{q,\infty}$ and $\varepsilon_{w,\infty}$ are the dielectric constant of the solute in the high-frequency limit and that of water, respectively. α represents the pure-water fraction. f_j and δ_j are DR frequency and DR amplitude of the jth Debye component, respectively. Using the value $\varepsilon_{w,\infty} = 5.6$ at 10 °C, $\varepsilon_{w,\infty} = 5.4$ at 20 °C (Kaatze 1989), and the static dielectric permittivity of solutes derived to be 2.5–4.8 from the electronic polarizabilities (Szigeti 1949), the value of $\varepsilon_{q,\infty}$ is calculated to lie between 5.0 and 5.6 based on Wagner theory (Wagner 1914) for the tested samples.

In sub-GHz region, Eqs. 3.2 and 3.3 could not simulate the experimental spectral curves; then, the following Debye function model was used for counterionic cloud response,

$$\varepsilon^*_{\text{q2, sim}}(f) = \varepsilon^*_{\text{w}}(f) + \frac{\delta_3}{1 + i(f/f_3)}. \tag{3.4}$$

HMW signal depends on both relaxation frequency and dispersion amplitude, and it is impossible to compare different frequency components as it is. Therefore, D_{HMW} was introduced to express the intensity of the HMW signal in DRS as Eq. 3.5: (Suzuki et al. 2016)

$$D_{HMW} = \frac{(f_1 - f_w)\delta_1}{f_w \delta_w}. \tag{3.5}$$

3.2.4 Hydration Properties of Salt Solutions

All DR spectra were normalized to concentration of 0.02 M and set the volume fraction ϕ in Eq. 3.2 to 0.02, enabling us to compare the hydration properties of each solution in the constant volume. Since a linearity of DR spectra against a concentration has been confirmed for monovalent ion (Miyazaki et al. 2008; Mogami et al. 2013) and for multivalent ion (Mogami et al. 2011), the error propagation by this conversion process was negligible. The hydration properties and the representative fitting curve for $MgCl_2$ are shown in Table 3.1 and Fig. 3.3, respectively. As can be seen from the value of χ^2 in Table 3.1, more beautiful fitting results were obtained than $MgCl_2$. In addition to monovalent alkali halide, only one Debye relaxation component was observed for divalent and trivalent ions in addition to the bulk water component. As described previously (Mogami et al. 2011), changing the volume fraction ϕ, δ_1, and δ_{bulk} ($= \alpha \times 78.4$ at 10 °C) in Eq. 3.3 gave and took each other; therefore, observed relaxation component was assigned to the hydration component. In all tested solutions, only HMW component was observed in present frequency region.

3.2.5 Correlation of HMW Against Water Structure

The intensity of the HMW signal D_{HMW} was calculated by Eq. 3.5, as shown in Table 3.1. Since the HMW number of alkali halide correlated with the order of the Hofmeister series (Mogami et al. 2013), it was thought that HMW showed water structure breaking effect. Therefore, the correlation between hydration entropy related to water structure and D_{HMW} was plotted in Fig. 3.4. The hydration entropy of the salt was the sum of the values evaluated for each cation and anion (Marcus 1997). The hydration entropy is larger as the valence of the ion is larger, but in the case of salt of the same valence number, D_{HMW} tends to increase as hydration

Table 3.1 Hydration properties of salt solutions

Salt	T (°C)	f_1 (GHz)	δ_1	δ_{bulk}	χ^2	D_{HMW}
NaF	10	32.9 ± 1.6	4.8 ± 0.4	56.8 ± 0.4	2.0	0.099 ± 0.001
NaCl	10	19.6 ± 0.7	22.1 ± 3.5	38.1 ± 3.6	3.7	0.156 ± 0.011
NaBr	10	18.8 ± 0.3	27.9 ± 1.3	33.2 ± 1.4	2.0	0.174 ± 0.001
NaI	10	18.6 ± 0.3	33.7 ± 1.9	26.7 ± 1.9	1.7	0.204 ± 0.002
NaF[a]	20	59.7	4.6	38.4	0.9	0.077
NaCl[a]	20	25.5	40.3	5.6	0.6	0.135
NaBr[a]	20	34.3	27.3	21.5	1.0	0.185
NaI[a]	20	29.8	35.0	17.3	0.7	0.245
KF	10	21.2 ± 2.9	13.5 ± 5.2	50.5 ± 5.1	1.2	0.105 ± 0.009
KCl	10	17.8 ± 1.3	41.0 ± 9.0	23.0 ± 8.8	1.3	0.206 ± 0.006
KBr	10	18.0 ± 1.3	41.2 ± 9.6	23.9 ± 9.7	1.4	0.217 ± 0.022
KI	10	17.3 ± 0.5	57.4 ± 6.5	8.6 ± 6.6	1.3	0.271 ± 0.017
$MgCl_2$	20	27.4 ± 2.4	20.3 ± 4.1	21.0 ± 3.8	11.7	0.160 ± 0.022
$FeCl_2$	10	20.5 ± 3.0	11.3 ± 6.3	51.6 ± 6.6	2.3	0.201 ± 0.105
$MnCl_2$	10	22.2 ± 3.6	8.5 ± 3.6	55.7 ± 4.3	3.0	0.183 ± 0.040
$CaCl_2$	20	36.6 ± 7.0	16.6 ± 4.8	26.7 ± 5.2	14.8	0.238 ± 0.038
$FeCl_3$	10	27.1 ± 5.0	7.0 ± 2.4	47.9 ± 3.2	4.3	0.238 ± 0.040

[a]For NaX at 20 °C, variation of DR spectra could not be defined

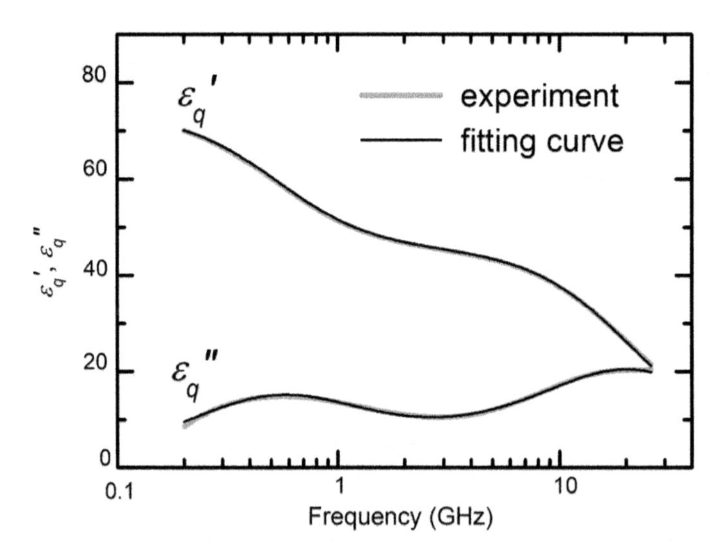

Fig. 3.3 Results of fitting with Eq. 3.2 on the spectra ε_q^* by Eq. 3.3 at $\phi = 0.02$ for $MgCl_2$. Gray line indicates the experimental curves with thickness of standard errors of $\Delta\varepsilon^*$, and black line indicates the fitting curves. Notably, for other salt solutions, it was almost the same fitting degree

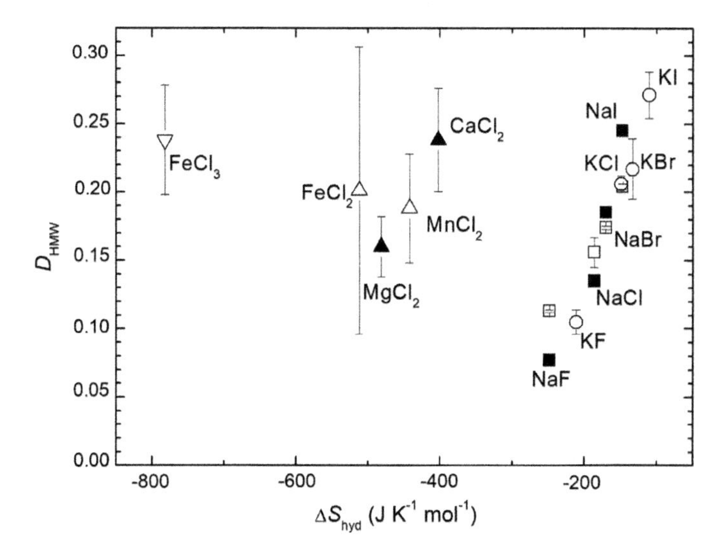

Fig. 3.4 Correlation between the hydration entropy (ΔS_{hyd}) and the signal intensity of HMW (D_{HMW}). □ NaX at 10 °C, ■ NaX at 20 °C, O KX at 10 °C, white triangle: divalent metal chloride at 10 °C, black triangle: divalent metal chloride at 20 °C, and down-pointing triangle: trivalent metal chloride at 10 °C

entropy increases. Marcus (1994) formulated molar entropy of hydration as a sum of several contributions as follows:

$$\begin{aligned} \Delta S_{hyd} &= S_{neut} + S_{el1} + S_{el2} + S_{str} \\ S_{neut} &= a + br \\ S_{el1} &= \frac{N_A e^2}{8\pi\varepsilon_0} z^2 \frac{\Delta r}{r(r+\Delta r)} \varepsilon_{el1}^{'-2} \left(\frac{\partial \varepsilon_{el1}^{'}}{\partial T} \right)_P, \\ S_{el2} &= \frac{N_A e^2}{8\pi\varepsilon_0} z^2 \frac{1}{r+\Delta r} \varepsilon_{el2}^{'-2} \left(\frac{\partial_{el2}^{'}}{\partial T} \right)_P \end{aligned} \tag{3.6}$$

where S_{neut} represents the formation of the cavity in water for the accommodation of the ion and the dispersion interactions with the surrounding water of a neutral entity of the same size r as the ion, with $a = -22$ (J K^{-1} mol^{-1}) and $b = -0.600$ (J K^{-1} mol^{-1} pm^{-1}). Parameters S_{el1} and S_{el2} represent electrostatic interactions of the ion with water in the first hydration shell and beyond, and N_{A}, e, ε_0, and z are Avogadro's number, elementary charge, dielectric constant of vacuum, and valence of ion, respectively. The value of Δr indicates the thickness of the first hydration shell (Krestov 1991) given by:

$$\Delta r = \left[\frac{45d^3|z|}{r} + r^3\right]^{\frac{1}{3}} \tag{3.7}$$

with $d = 276$ pm, the diameter of a water molecule. Next, ε'_{el1} is the relative permittivity of water in the electrostricted first hydration shell, where dielectric saturation has occurred, with $\varepsilon'_{el1} = 1.58$ and $\left(\frac{\partial \varepsilon'_{el1}}{\partial T}\right)_P = -1.0 \times 10^{-4}\mathrm{K}^{-1}$. The value of ε'_{el2}, the relative permittivity of pure water, and its isobaric temperature derivative, $\left(\frac{\partial \varepsilon'_{el2}}{\partial T}\right)_P$, are 78.4 and −0.3595 K^{-1}, respectively. The absolute molar entropy of hydration ΔS_{hyd} is experimental data and tabulated in reference (Marcus 1997), derived from the formation entropy of gas phase and aqueous phase of ion. Finally, structural entropy, S_{str}, can be calculated by subtracting each term in Eq. 3.6 and used as a scale for the water structure effects of ions. Regardless of valence, D_{HMW} and S_{str} showed a positive correlation, as shown in Fig. 3.5, suggesting that HMW reflects the effect of water structure breaking by ions and is located at the second or outer hydration layer.

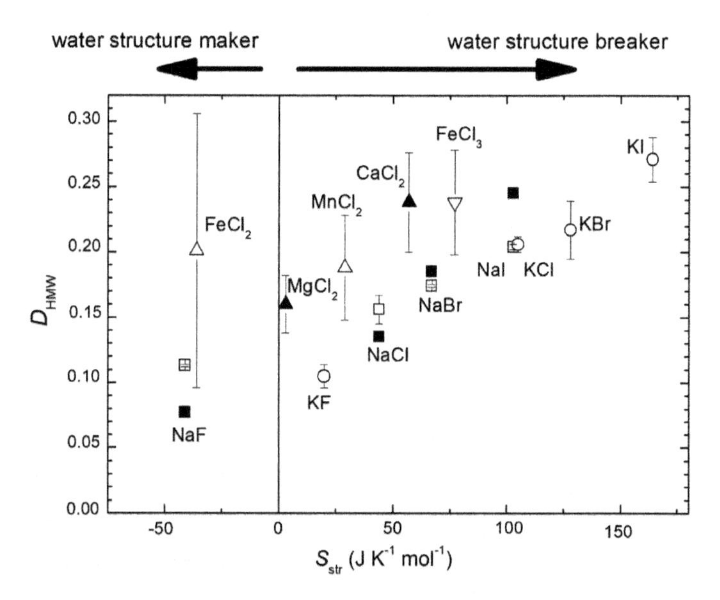

Fig. 3.5 Correlation between the structural entropy (S_{str}) and the signal intensity of HMW (D_{HMW}). □ NaX at 10 °C, ■ NaX at 20 °C, ○ KX at 10 °C, white triangle: divalent metal chloride at 10 °C, black triangle: divalent metal chloride at 20 °C, and down-pointing triangle: trivalent metal chloride at 10 °C

3.3 Hydration Energy Calculation Based on an MD Simulation

In Sect. 3.2, the hydration states of multivalence monatomic ions were measured using DRS, showing that HMW correlated with the structure entropy S_{str}. Then, in this section, to examine the role of HMW in the ionic aqueous solution, we describe the results of spatial-decomposition analysis of energetics of ionic hydration (Mogami et al. 2016).

3.3.1 Theoretical Background

Here, we omit the exact derivation of the rigorous theoretical formula but only introduce equations necessary for interpreting the results. See references (Matubayasi et al. 1994, 1998; Mogami et al. 2016) for details. The thermodynamic variable treated in the present work is the excess partial molar energy. This is the change in the total energy of the system when the solute is inserted at fixed origin. We treat an aqueous solution at infinite dilution. A single solute molecule is contained in the system and is located at the origin. Let $\rho(\mathbf{r})$ be the density of the solvent molecule (water) at position $\mathbf{r}$; $\rho(\mathbf{r})d\mathbf{r}$ is the average number of solvent molecules found in region $[\mathbf{r}, \mathbf{r} + d\mathbf{r}]$. It was then shown in reference (Matubayasi et al. 1994) that the excess partial molar energy ΔE is expressed as

$$\Delta E = \int d\mathbf{r}\rho(\mathbf{r})\left[B_u(\mathbf{r}) + \frac{1}{2}B_v(\mathbf{r}) - \frac{1}{2}B_0\right] \tag{3.8}$$

where $B_u(\mathbf{r})$ and $B_v(\mathbf{r})$ are introduced, respectively, as

$$B_u(\mathbf{r}) = \frac{\sum_i \langle u_{uv}(i)\delta(\mathbf{r}_i - \mathbf{r})\rangle}{\langle \sum_i \delta(\mathbf{r}_i - \mathbf{r})\rangle} = \frac{\langle u_{uv}(i)\delta(\mathbf{r}_i - \mathbf{r})\rangle}{\langle \delta(\mathbf{r}_i - \mathbf{r})\rangle} \tag{3.9}$$

$$B_v(\mathbf{r}) = \frac{\sum_i \langle \sum_{j \neq i} u_{vv}(i,j)\delta(\mathbf{r}_i - \mathbf{r})\rangle}{\langle \sum_i \delta(\mathbf{r}_i - \mathbf{r})\rangle} = \frac{\langle \sum_{j \neq i} u_{vv}(i,j)\delta(\mathbf{r}_i - \mathbf{r})\rangle}{\langle \delta(\mathbf{r}_i - \mathbf{r})\rangle} \tag{3.10}$$

$\mathbf{r}_i$ is the position of the ith solvent molecule, $u_{uv}(i)$ is the pair interaction energy between the solute and the ith solvent molecule, and $u_{vv}(i, j)$ is the pair interaction energy between the ith and jth solvent molecules. $B_u(\mathbf{r})$ is the average interaction between the solute and a solvent molecule (water) located at position $\mathbf{r}$. $B_v(\mathbf{r})$ is the average sum of the intermolecular interactions of a solvent molecule at $\mathbf{r}$ with the other solvent molecules in the system, and B_0 is the limiting value of $B_v(\mathbf{r})$ when the solvent molecules are separated from the solute ($\mathbf{r} \rightarrow \infty$). Note that i in the right most sides of Eqs. 3.9 and 3.10 can refer to any, single solvent molecule; there is no dependence on i actually since all the solvent molecules are equivalent. In addition, the factor 1/2 in Eq. 3.8 corrects the double counting of the solvent–solvent energy. In Eq. 3.8, the integration of $B_u(\mathbf{r})$ is the solute–solvent term in the excess partial molar energy ΔE, and the integration of $(B_v(\mathbf{r})–B_0)$ provides the solvent–solvent term (solvent reorganization term).

The extent of spatial localization of ΔE can be examined by introducing a cutoff λ into the integral of Eq. 3.8 through

$$\Delta E(\lambda) = \int_{|\mathbf{r}|<\lambda} d\mathbf{r}\rho(\mathbf{r})\left[B_u(\mathbf{r}) + \frac{1}{2}B_v(\mathbf{r}) - \frac{1}{2}B_0\right] \tag{3.11}$$

and examining the λ dependence of $\Delta E(\lambda)$. The exact value is recovered in the limit of $\lambda \rightarrow \infty$. When $\Delta E(\lambda)$ reaches the limiting value at λ corresponding to the first layer, for example, the thermodynamic quantity ΔE is spatially localized in the first layer. Similarly, the cutoff λ can be implemented to the solute–solvent and solvent–solvent (solvent reorganization) terms, respectively, through

$$\Delta E_u(\lambda) = \int_{|\mathbf{r}|<\lambda} d\mathbf{r}\rho(\mathbf{r})B_u(\mathbf{r}) \tag{3.12}$$

$$\Delta E_v(\lambda) = \frac{1}{2}\int_{|\mathbf{r}|<\lambda} d\mathbf{r}\rho(\mathbf{r})[B_v(\mathbf{r}) - B_0] \tag{3.13}$$

and the extents of spatial localization of those terms can be assessed from the λ dependencies of Eqs. 3.12 and 3.13.

3.3.2 Computational Procedure

The ionic solutes examined in the present work were Li^+, Na^+, K^+, Rb^+, and Cs^+ as monovalent cations, Cl^- and I^- as monovalent anions, and Mg^{2+}, Ca^{2+}, Mn^{2+}, and Fe^{2+} as divalent cations. The water molecule and methane were also employed as solutes for the purpose of comparison. For each solute, a single solute particle and 5000 water molecules were located in a cubic unit cell with the periodic boundary condition, and an MD simulation was conducted over 10 ns at a sampling interval of 100 fs in the NPT (isothermal–isobaric) ensemble at 20 °C and 1 bar using the MD program package NAMD2 (Philips et al. 2005). The TIP3P model (Jorgensen et al. 1983) was used for water, and the solutes were described with CHARMM22 (Brooks et al. 1983; MacKerell et al. 1998) except for Mn^{2+}, Fe^{2+}, and I^-. The force fields of Mn^{2+} and Fe^{2+} were taken from reference (Li et al. 2013), and the I^- potential employed was the one optimized against a Drude model of water (Lamoureux and Roux 2006). Although a mixed use of force fields is generally not recommended, the parameters for Mn^{2+}, Fe^{2+}, and I^- were taken from refs (Li et al. 2013; Lamoureux and Roux 2006) due to the availability. To see the effect of the choice of the potential function of the ion, MD was also carried out in TIP3P using the Na^+ model which is similarly optimized with the Drude water (Lamoureux and Roux 2006).

The electrostatic interaction was handled by the particle mesh Ewald (PME) method (Essmann et al. 1995) with a real-space cutoff of 12 Å, a spline order of 4, an inverse decay length of 0.258 $Å^{-1}$, a reciprocal space mesh size of 64 for each of the x, y, and z directions. The Lennard-Jones (LJ) interaction was truncated by applying the switching function (Brooks et al. 1983) with the switching range of 10–12 Å. The Langevin dynamics was employed for temperature control at a damping coefficient of 5 ps^{-1} with the Brünger–Brooks–Karplus algorithm at a time step of 2 fs (Van and Berendsen 1988; Brünger et al. 1984). The pressure was maintained by the Langevin piston Nosé–Hoover method with barostat oscillation and damping time constants of 200 and 100 fs, respectively (Martyna et al. 1994; Feller et al. 1995). The water molecules were kept rigid with SETTLE (Miyamoto and Kollman 1992).

3.3.3 Radial Distribution Function

Figure 3.6 shows the radial distribution function (RDF) of the oxygen atom of water around the solute ion. For all ions, RDF almost converged within ~10 Å. The RDF peaks of water molecules in the first layer corresponding to the ion radius were observed. The number of water molecules in the first layer N_{I} and the sum from the first and second layers N_{II} were obtained by integrating the RDF up to the first-minimum distance r_{I} and to the second-minimum distance r_{II}. The number N_{I} agreed with the literature value (Ohtaki and Radnai 1993). While r_{I} was mainly located at (ionic radius + water radius), in any ion r_{II} was to be ~5–6 Å and N_{II} were 23–37 for ions calculated in the present work. The number of HMW of alkali halide by DRS was estimated to be 19–37 (Mogami et al. 2013); therefore, HMW can exist in the second or outer layer.

3.3.4 Solute–Solvent and Solvent–Solvent Interaction Energies

The solute–solvent interaction energy $B_{\mathrm{u}}(r)$ and the solvent–solvent interaction energy $B_{\mathrm{v}}(r)$ were calculated by Eqs. 3.9 and 3.10 as shown in Fig. 3.7, where $B_{\mathrm{v}}(r)$ was evaluated as a shift of the solvent–solvent interaction from the bulk value $\Delta B_{\mathrm{v}}(r)$. For all the systems examined in the present work, $B_{\mathrm{u}}(r)$ and

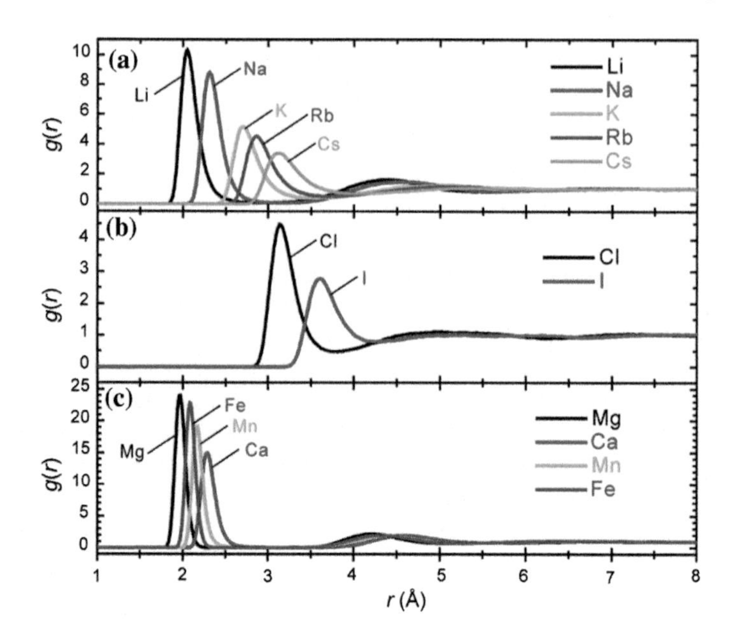

Fig. 3.6 Radial distribution function $g(r)$ of the oxygen atom of the water molecule around the solute **a** for the alkali metal ions, **b** for the halide ions, and **c** for the divalent metal ions. Modified with the permission from reference (Mogami et al. 2016). Copyright 2016 American Chemical Society

$\Delta B_v(r)$ represent negative (favorable) and positive (unfavorable) signals over the entire space, respectively, and both signals keep a significant value up to ~15 Å. However, the sum of those values ($B_u(r) + \Delta B_v(r)$), the average of the total interaction energy of a water molecule located at distance r from the solute, decays to zero in $r \geq 5$ Å, indicating the compensation between solute–solvent $B_u(r)$ and solvent–solvent $\Delta B_v(r)$ interactions. Considering from the picture of ($B_u(r) + B_v(r)$), the specificity of ion is observed only at ion–water contact region or near the first layer, while $B_u(r)$ and $\Delta B_v(r)$ for alkali halide also show the ionic specificity in the second layer. Taking into account the facts that HMW correlates the ion-specific effect and would be in the second of outer layer, the formation of HMW is thought to be derived mainly from $B_u(r)$ and $\Delta B_v(r)$. In addition, HMW was also observed in divalent ion solutions, in which there are only small differences in $B_u(r)$ and $\Delta B_v(r)$ for $r \geq 3$ Å among the divalent cation, resulting in the idea that ion-specific behavior at ion–water contact region affects the formation of HMW located second layer away from a surface of ion. Then, considering all the various factors together, at least there are two contributions to the formation of HMW; one is due to the solute–solvent and the solvent–solvent interaction at second or third layer, and the other is due to the solute–solvent interaction at contact region. The effect of a coupling of these two interactions on the HMW formation would be a profound and fascinating key term.

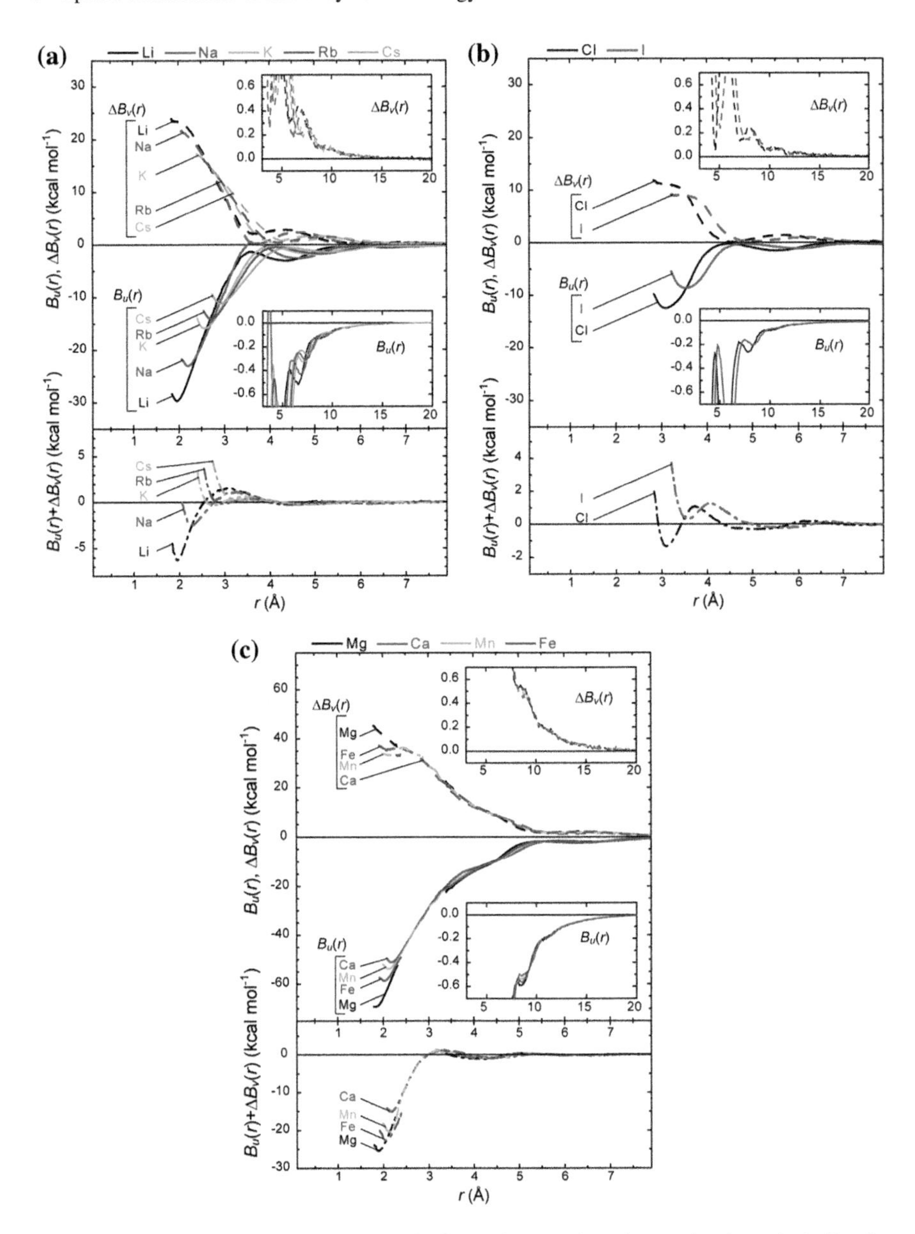

Fig. 3.7 Solute–solvent interaction B_u, shift of the solvent–solvent interaction from the bulk value ΔB_v, and their sum ($B_u + \Delta B_v$) as functions of the solute–solvent distance r **a** for the alkali metal ions, **b** for the halide ions, and **c** for the divalent metal ions. Modified with the permission from reference (Mogami et al. 2016). Copyright 2016 American Chemical Society

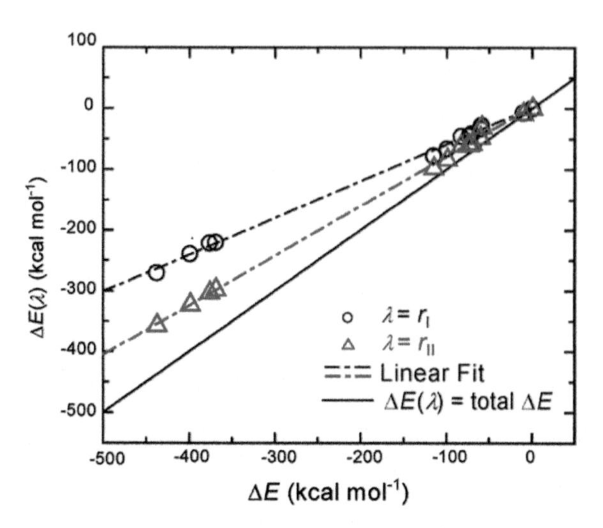

Fig. 3.8 Correlations of the total ΔE against $\Delta E(r_{\mathrm{I}})$ and $\Delta E(r_{\mathrm{II}})$. Reprinted with the permission from reference (Mogami et al. 2016). Copyright 2016 American Chemical Society

3.3.5 Extent of Spatial Localization of ΔE

As noted in Sect. 3.3.1, the extents of spatial localization of the excess partial molar energy ΔE and its solute–solvent and solvent–solvent (solvent reorganization) terms can be assessed by introducing a cutoff λ through Eqs. 3.11–3.13 and examining the convergence behaviors with respect to λ. As a representative value, $\Delta E(r_{\mathrm{I}})$ (black circle) and $\Delta E(r_{\mathrm{II}})$ (red triangle), which are localized ΔE up to the first-minimum distance r_{I} and the second-minimum distance r_{II} in RDF, respectively, are plotted in Fig. 3.8. Black solid line indicates the whole excess partial molar energy ΔE. Surprisingly, regardless of valence and ion species, the ratios of localization in the first layer and the second layer are almost constant to be ~50–70% and ~5–25%, respectively. A few tens of % of the total ΔE come from outer regions beyond the second layer.

3.3.6 Correlation Between Electric Field and Spatial Distribution

It was reported that around actin filament the number of water molecules in the region of $2 \times 10^7 - 1 \times 10^8$ V/m corresponded to the number of HMW (Suzuki et al. 2016). Then, we examined the correlation between the electric field by ion and HMW. In Fig. 3.9, the electric field made by monovalent ion and RDF of Na^+ and K^+ was plotted. The intensity of electric field in the first layer and the second layer is $\sim 2 \times 10^8 - 4 \times 10^8$ V/m and $\sim 5 \times 10^7 - 1 \times 10^8$ V/m, respectively, agreeing the previous results (Suzuki et al. 2016). Electric field by ion E_{I} and E_{II} at the first-minimum distance r_{I} and the second-minimum distance r_{II} in RDF,

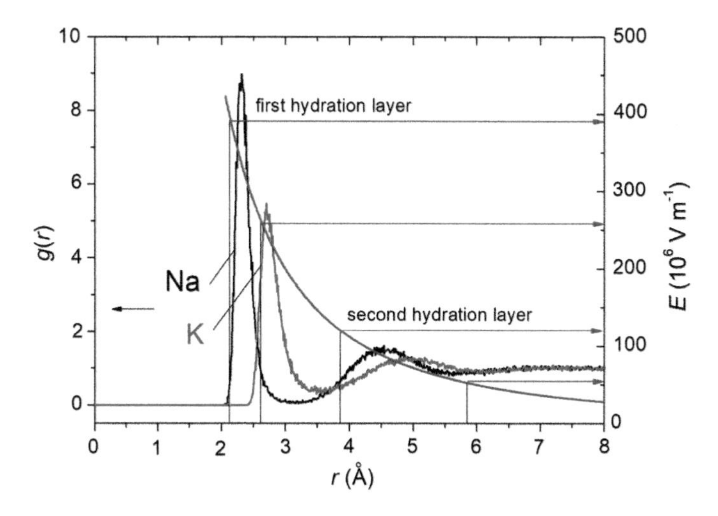

Fig. 3.9 Correlations of the electric field by monovalent ion against the RDF for Na^+ and K^+ ions

Table 3.2 Electric field by ion E_{I} and E_{II} at the first-minimum distance r_{I} and the second-minimum distance r_{II} in RDF, respectively

Solute	Li^+	Na^+	K^+	Rb^+	Cs^+	Cl^-	I^-	Mg^{2+}	Ca^{2+}	Mn^{2+}	Fe^{2+}
E_{I} (10^6 V/m)	199	175	138	131	112	124	102	530	374	426	530
E_{II} (10^6 V/m)	57	55	52	45	44	47	57	132	114	123	127

respectively, was calculated as shown in Table 3.2. According to the results, in the second layer the intensity of electric field is between E_{I} and E_{II}, and consistent with the intensity thought to be necessary to form HMW. However, in the case of Fe^{3+}, even though stronger electric field was generated in the second or outer layer, the strong HMW signal was observed in DRS measurements (Table 3.1). Where is HMW around ions with high charge density such as Fe^{3+}? In the future, it is expected that we can get to the core of the problem about HMW by elucidating the relation between energy distribution of hydration and the electric field around ion.

3.4 Conclusion

We reviewed about the spatial distribution of ionic hydration energy and hyper-mobile water (HMW). First, we described our dielectric relaxation spectroscopy (DRS) analyses of alkali halide, divalent metal chloride, and trivalent metal chloride solutions to clarify their hydration states. HMW was detected in all tested solutions correlated with the Hofmeister series. Second, we described a spatial-decomposition analysis of the energetics of hydration for a series of ionic solutes in combination with molecular dynamics (MD) simulations. As a result, it is suggested that HMW would mainly exist in the second or outer layer and the electric field generated by solute is a key to explore a mystery of Hofmeister effect including the formation mechanism of HMW.

References

Asami K, Hanai T, Koizumi N (1980) Dielectric analysis of *Escherichia coli* suspensions in the light of the theory of interfacial polarization. Biophys J 31:215–228

Bakker HJ (2008) Structural dynamics of aqueous salt solutions. Chem Rev 108:1456–1473

Brooks BR, Bruccoleri RE, Olafson BD, States DJ, Swaminathan S, Karplus M (1983) CHARMM: a program for macromolecular energy, minimization, and dynamics calculations. J Comput Chem 4:187–217

Brünger A, Brooks IIICL, Karplus M (1984) Stochastic boundary conditions for molecular dynamics simulations of ST2 water. Chem Phys Lett 105:495–500

Buchner R (2008) What can be learnt from dielectric relaxation spectroscopy about ion solvation and association? Pure Appl Chem 80:1239–1252

Essmann U, Perera L, Berkowitz ML, Darden T, Lee H, Pedersen LG (1995) A smooth particle mesh ewald method. J Chem Phys 103:8577–8593

Feller SE, Zhang Y, Pastor RW, Brooks BR (1995) Constant pressure molecular dynamics simulation: the Langevin piston method. J Chem Phys 103:4613–4621

Gurney RW (1953) Ionic processes in solution. McGraw-Hill, New York

Jorgensen WL, Chandrasekhar J, Madura JD, Impey RW, Klein ML (1983) Comparison of simple potential functions for simulating liquid water. J Chem Phys 79:926–935

Jungwirth P, Tobias DJ (2006) Specific ion effects at the air/water interface. Chem Rev 106:1259–1281

Kaatze U (1989) Complex permittivity of water as a function of frequency and temperature. J Chem Eng Data 34:371–374

Kaatze U (2013) Measuring the dielectric properties of materials. Ninety-year development from low-frequency techniques to broadband spectroscopy and high-frequency imaging. Meas Sci Technol 24:012005

Kaatze U, Feldman Y (2006) Broadband dielectric spectrometry of liquids and biosystems. Meas Sci Technol 17:R17–R35

Kabir SR, Yokoyama K, Mihashi K, Kodama T, Suzuki M (2003) Hyper-mobile water is induced around actin filaments. Biophys J 85:3154–3161

Krestov GA (1991) Thermodynamics of solvation. Ellis Horwood, New York, pp 172–177

Kubota Y, Yoshimori A, Matubayasi N, Suzuki M, Akiyama R (2012) Molecular dynamics study of fast dielectric relaxation of water around a molecular-sized ion. J Chem Phys 137 (224502):1–4

Laliberté M (2009) A model for calculating the heat capacity of aqueous solutions, with updated density and viscosity data. J Chem Eng Data 54(6):1725–1760

Lamoureux G, Roux B (2006) Absolute hydration free energy scale for alkali and halide ions established with a polarizable force field. J Phys Chem B 110:3308–3322

Li P, Roberts BP, Chakravorty DK, Merz KM Jr (2013) Rational design of particle mesh ewald compatible Lennard-Jones parameters for +2 metal cations in explicit solvent. J Chem Theory Comput 9:2733–2748

Lo Nostro P, Ninham BW (2012) Hofmeister phenomena: an apdate on ion specificity in biology. Chem Rev 112:2286–2322

Marcus Y (1994) Viscosity *B*-coefficients, structural entropies and heat capacities, and the effects of ions on the structure of water. J Sol Chem 23(7):831–848

Marcus Y (1997) Ion properties. Marcel Dekker, New York

Marcus Y (2009) Effect of ions on the structure of water: structure making and breaking. Chem Rev 109:1346–1370
Martyna GJ, Tobias DJ, Klein ML (1994) Constant pressure molecular dynamics algorithms. J Chem Phys 101:4177–4189
Matubayasi N, Levy RM (1996) Thermodynamics of the hydration shell. 2. Excess volume and compressibility of a hydrophobi solute. J Phys Chem 100:2681–2688
Matubayasi N, Gallicchio E, Levy RM (1998) On the local and nonlocal components of solvation thermodynamics and their relation to solvation shell models. J Chem Phys 109:4864–4872
Matubayasi N, Reed LH, Levy RM (1994) Thermodynamics of the hydration shell. 1. Excess energy of a hydrophobic solute. J Phys Chem 98:10640–10649
MacKerell AD Jr, Bashford D, Bellott M, JrRL Dunbrack, Evanseck JD, Field MJ, Fischer S, Gao J, Guo H, Ha S, Joseph-McCarthy D, Kuchnir L, Kuczera K, Lau FTK, Mattos C, Michnick S, Ngo T, Nguyen DT, Prodhom B, Reiher IIIWE, Roux B, Schlenkrich M, Smith JC, Stote R, Straub J, Watanabe M, Wiorkiewicz-Kuczera J, Yin D, Karplus M (1998) All-atom empirical potential for molecular modeling and dynamics studies of proteins. J Phys Chem B 102:3586–3616
Miyamoto S, Kollman PA (1992) Settle: An analytical version of the SHAKE and RATTLE algorithm for rigid water models. J Comput Chem 13:952–962
Miyazaki T, Mogami G, Wazawa T, Kodama T, Suzuki M (2008) Measurement of the dielectric relaxation property of water-ion loose complex in aqueous solutions of salt at low concentrations. J Phys Chem A 112:10801–10806
Mogami G, Wazawa T, Morimoto N, Kodama T, Suzuki M (2011) Hydration properties of adenosine phosphate series as studied by microwave dielectric spectroscopy. Biophys Chem 154:1–7
Mogami G, Miyazaki T, Wazawa T, Matubayasi N, Suzuki M (2013) Anion-dependence of fast relaxation component in Na-, K-halide solutions at low concentrations measured by high-resolution microwave dielectric spectroscopy. J Phys Chem A 117:4851–4862
Mogami G, Suzuki M, Matubayasi N (2016) Spatial-decomposition analysis of energetics of energetics of ionic hydration. J Phys Chem B 120:1813–1821
Ohtaki H, Radnai T (1993) Structure and dynamics of hydrated ions. Chem Rev 93:1157–1204
Philip G, Robert JW, Mendel T, Clara W, William D, Linda R, William R, Fahd S, Barry L (1970) “Squiggle-H_2O”. An enquiry into the importance of solvation effects in phosphate ester and anhydride reactions, 223(1):1–15
Philips JC, Braun R, Wang W Gumbart J, Tajkhorshid E, Villa E, Chipot C, Skeel RD, Kale L, Schulten K (2005) Scalable molecular dynamics with NAMD. J Comput Chem 26:1781–1802
Robinson RA, Stokes RH (1959) Electrolyte solutions, 2nd revised ed. Dover publications, Mineola, New York
Suzuki M, Kabir SR, Siddique MSP, Nazia US, Miyazaki T, Kodama T (2004) Myosin-induced volume increase of the hyper-mobile water surrounding actin filaments. Biochem Biophys Res Commun 322:340–346
Suzuki M, Imao A, Mogami G, Chishima R, Watanabe T, Yamaguchi T, Morimoto N, Wazawa T (2016) Strong dependence of hydration state of F-actin on the bound Mg^{2+}/Ca^{2+} ions. J Phys Chem B 120:6917–6928
Suzuki M, Mogami G, Ohsugi H, Watanabe T, Matubayasi N (2017) Physical driving force of actomyosin motility based on the hydration effect. Cytoskeleton. https://doi.org/10.1002/cm.21417
Szigeti B (1949) Polarisability and dielectric constant of ionic crystals. Trans Faraday Soc 45:155–166
Tu KM, Ishizuka R, Matubayasi N (2014a) Spatial-decomposition analysis of electrical conductivity in concentrated electrolyte solution. J Chem Phys 141:044126
Tu KM, Ishizuka R, Matubayasi N (2014b) Spatial-decomposition analysis of electrical conductivity in ionic liquid. J Chem Phys 141:244507

Van Gunsteren WF, Berendsen HJC (1988) A leap-frog algorithm for stochastic dynamics. Mol Simul 1:173–185

Wagner KW (1914) Erklärung der Dielectrischen Nachwirkungsvorgänge auf Grund Maxwellscher Vorstellungen. Archiv für Elektrotechnik 2:371–387

Wazawa T, Miyazaki T, Sambongi Y, Suzuki M (2010) Hydration analysis of *Pseudomonas Aeruginosa* cytochrome *c*551 upon acid unfolding by dielectric relaxation spectroscopy. Biophys Chem 151:160–169

Chapter 4
Theoretical Studies of Strong Attractive Interaction Between Macro-anions Mediated by Multivalent Metal Cations and Related Association Behavior: Effective Interaction Between ATP-Binding Proteins Can Be Regulated by Hydrolysis

Ryo Akiyama

Abstract In this chapter, calculated effective interactions between macro-anions are introduced. The macro-anions are effectively attracted to each other under certain conditions, and the aggregation behavior of macro-anions is discussed on the basis of the calculated effective interactions. The success of models that simulate the behavior of such systems indicates that theoretical discussions are important to understanding the observed aggregation of acidic proteins in solution of multivalent cations. The hydrolysis of ATP regulates the effective interaction between ATP-binding proteins, such as actin monomers. The regulation of effective interaction is discussed from the viewpoint of the calculated effective interaction between macro-anions.

Keywords Actin • Acidic protein • Effective attraction between like-charges
ATP hydrolysis • Dynamical ordering

4.1 Introduction

A motile cell has pseudopodia (Alberts et al. 2007). In the pseudopodia, filaments grow in the direction of progress. The filaments do the mechanical work on the cell membrane through the elongation of filament. The filament elongation is an "engine" of the molecular linear motors. The "engine" works with both "tire" and "clutch" molecules. In axon outgrowth, the "tire" molecule is L1-CAM (Kamiguchi

---

R. Akiyama (✉)
Department of Chemistry, Kyushu University, Fukuoka, Japan
e-mail: rakiyama@chem.kyushu-univ.jp

© Springer Nature Singapore Pte Ltd. 2018

M. Suzuki (ed.), *The Role of Water in ATP Hydrolysis Energy Transduction by Protein Machinery*, https://doi.org/10.1007/978-981-10-8459-1_4

et al. 1998), and it is the cell adhesion molecule. Shootin 1 is an important molecule in axon outgrowth (Toriyama et al. 2006, 2010; Inagaki et al. 2011; Sapir et al. 2013), and it works as the clutch molecule (Shimada et al. 2008). Shootin 1–cortactin interaction mediates signal force transduction in the system (Kubo et al. 2015). This linear molecular motor is driven by ATP hydrolysis.

The "engine" is an actin filament (F-actin), which is composed of actin monomers (G-actins). The G-actin is a negatively charged ATP-binding protein in biological systems. F-actins polymerize and elongate on the leading edge of the cell and push the cell when the tire and clutch molecules are active. Figure 4.1 shows the polymerization and the shape changes of cells. The actins (shown as blue spheres) attach to the end near the leading edge, and the cells elongate in the direction of the blue arrows. On the other hand, the retrograde flow of F-actin is induced by depolymerization (Pollard and Borisy 2003; Le Clainche and Carlier 2008). F-actins are depolymerized at the side opposite to the leading edge. The depolymerization supplies the monomer for the polymerization at the leading edge. So the G-actins are recycled in the living system.

Polymerization and depolymerization are regulated by ATP hydrolysis (Alberts et al. 2007). Actin is an ATP-binding protein that catalyzes the hydrolysis of ATP under usual living conditions. The ATP-binding actin is strongly bound at the end of the filament, and the association is more stable than that of the ADP-binding actin. These phenomena have two interesting points. First, although both ATP- and ADP-binding actins are negatively charged, the stability of association shows that there is an effective attraction between the negatively charged proteins. Second, as the amount of negative charge on the actin increases, the effective attraction becomes stronger as indicated by the difference between the association stabilities. These points appear to be counterintuitive with respect to the expected Coulomb interactions in a vacuum. However, we cannot reasonably discuss the interaction between such proteins in a vacuum; they are immersed in a biological fluid, namely an electrolyte solution. This must be a key point to understanding the phenomena.

In this chapter, the above points are discussed with reference to the effective interactions between macro-anions calculated by using integral equation theory (Akiyama et al. 2007; Akiyama and Sakata 2011; Fujihara and Akiyama 2014).

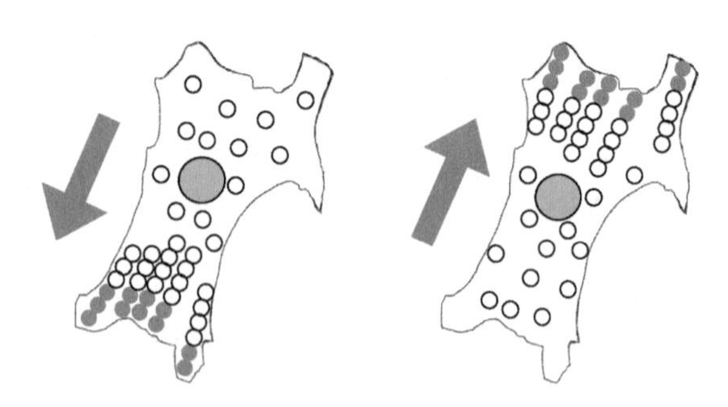

Fig. 4.1 Motile cells. In the cells, actin monomers are attached at the end of F-actin (lines of white circles) around the leading edge. The attached monomers are drawn as blue circles. The left and right cells are moving downward and upward, respectively

The dependence of the interactions on the electrolyte concentration and on the charges of the macro-anions is also investigated. We also discuss the effect of the cations in the electrolyte.

4.2 Condensation of Acidic Proteins by Multivalent Metal Cations

Before describing our theoretical studies, an important series of experimental studies on the condensation of acidic proteins by multivalent metal cations, such as Y^{3+}, should be highlighted. Zhang et al. studied such condensation and found some interesting reentrant behavior (Zhang et al. 2008, 2010, 2011). A typical phase diagram is shown in Fig. 4.2 (Zhang et al. 2008), in which the protein is bovine serum albumin (BSA) and the electrolyte is YCl_3. Two boundary conditions, namely c* and c**, can be seen, between which the proteins aggregate. Thus, if we fix the protein concentration at 10 mg/mL, when the electrolyte concentration is lower than about 0.6 mM, the proteins are dispersed in the solution. The condensed phases are observed when the electrolyte concentration ranges from 0.6 to 9.0 mM. However, the proteins are again dispersed when the electrolyte concentration is higher than about 9.0 mM. This reentrant behavior is observed for various acidic proteins, including β-lactoglobulin. Similar effective attraction between like-charges (Gosule and Schellman 1976; Bloomfield 1997; Widom and Baldwin 1980; Yoshikawa et al. 1996) and reentrant behavior has been observed in DNA solutions (Murayama et al. 2003; Besteman et al. 2007).

The isoelectric point of BSA is 4.6, and the protein is negatively charged in media with pH values above this. The condensation behavior described above shows that the negatively charged proteins attract each other under certain

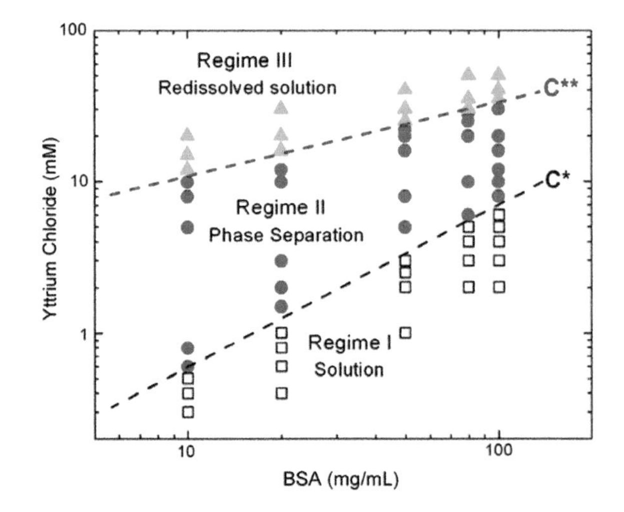

Fig. 4.2 Phase diagram of bovine serum albumin (BSA) as a function of protein and YCl_3 concentrations on the log scale. Solution and reentrant solution phases are indicated by white squares and green triangles, respectively. Red circles indicate conditions under which the solutions separate into two phases. Reprinted with permission from reference (Zhang et al. 2008). Copyright 2008 by American Physical Society

conditions. Such behavior seems counterintuitive from the viewpoint of species in a vacuum. However, this "strange" behavior is often observed in electrolyte solution. The effective attractions are mediated by multivalent metal cations, and the crystallization of these acidic proteins is controlled through ligandation of the multivalent metal cations. Structure analyses of protein crystals indicate that the multivalent metal cations are located at the attraction mediate site (Zhang et al. 2011). Figure 4.3 shows schematic diagrams based on the results of X-ray structure analysis in which the multivalent metal cations are observed between acidic residues. Each metal cation is coordinated by the acidic residues, and they seem to be connecting the proteins.

Similar conformations of the effective attraction mediators are also observed in cryo-electron microscopic images of F-actin in the presence of phosphate; in this case, the multivalent cation is Mg^{2+} (Murakami 2010). An actin association does not necessarily require Mg^{2+} ions; therefore, the effective attractions are not caused by the mediate cations only. However, it seems that the effects of the mediate cations can regulate, to some extent, the effective interaction between proteins in the assembly. A Mg^{2+} cation is surrounded by F-actin coordinating residues D286 and D288 from one actin, and E207 and Q59 from a neighbor actin (the longitudinally located second actin). If the attraction between D286, D288, and E207 is reduced or changed to repulsion, the stability of the filament is changed. Therefore, this coordination site can become a "switch" to connect G-actins.

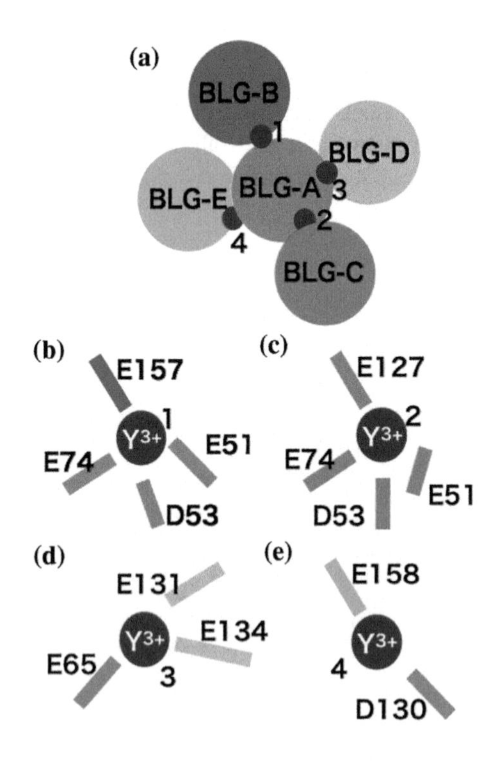

Fig. 4.3 Schematic diagrams illustrating β-lactoglobulin (BLG) in crystal. **a** A β-lactoglobulin, BLG-A, is surrounded by four β-lactoglobulins, namely BLG-B, BLG-C, BLG-D, and BLG-E. The linking yttrium sites are represented by magenta spheres. **b**–**e** Expansion of areas of (**a**) around the linking site. Each yttrium cations are coordinated by acidic residues (Zhang et al. 2011)

4.3 Integral Equation Theory for Liquids

Proteins are negatively charged in media with pH above their isoelectric point. Acidic proteins are thus macro-anions in typical biological media, which tend to be almost neutral fluids. Acidic proteins are therefore modeled with negatively charged hard spheres with diameters of 16.8 Å. Simple electrolyte solutions are also prepared. The solvent molecules, anions, and cations are modeled with uncharged, negatively charged, and positively charged hard spheres, respectively (Fig. 4.4), with diameters of 2.8 Å. The packing fraction is 0.383, which is the packing fraction of water, and this is maintained during the calculations for the electrolyte concentration dependence. In the present model, the Coulomb potential is divided by a factor of 78.5, which is the dielectric constant of water. The usual mixing rules for the excluded distance for different particles are adopted. The temperature is maintained at 298 K.

In Fig. 4.2, very dilute electrolyte solutions were also examined to study the concentration dependence. Molecular simulations of such dilute solutions are challenging because it is difficult to obtain ensembles that adequately represent the huge ratio of the number of solvent molecules and ions. In these cases, effective interactions between macro-anions in an electrolyte solution are calculated by using an Ornstein–Zernike equation with hypernetted-chain closure (HNC-OZ) theory, which is a type of integral equation theory (Hansen and McDonald 2006). HNC closure is known to be suitable for the calculation of charged particle systems. Therefore, dependence of the effective interaction on the electrolyte concentration is examined by using this theory.

In the bulk electrolyte solution, which consists of anions (A), cations (C), and solvent molecules (V), the correlation functions between particles are calculated by using the OZ equation,

$$h_{ij}(r) = c_{ij}(r) + \sum_{l=A,C,V} \int c_{il}\left(\left|\mathbf{r}-\mathbf{r}'\right|\right) \rho_l h_{lj}(r')\, d\mathbf{r}',$$

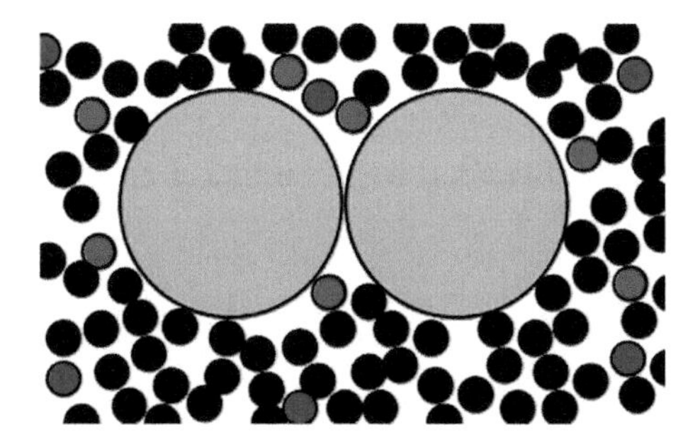

Fig. 4.4 System is modeled with uncharged and charged hard spheres. Uncharged hard spheres are solvent molecules (black circles), and charged hard spheres are cations, anions, and macro-anions (red and small blue circles and large blue circles). Although this figure is drawn in two dimensions, the three-dimensional systems were calculated

coupled with the HNC closure,

$$h_{ij}(r) = \exp\left[-\beta u_{ij}(r) + h_{ij}(r) + c_{ij}(r)\right] - 1,$$

where $h_{ij}(r)$ is the total correlation function, $c_{ij}(r)$ is the direct correlation function, ρ_{i} is the number density, and $u_{ij}(r)$ is the direct interaction between particles i and j (i, $j = A$, C, V), and β is $1/k_{\mathrm{B}}T$ (T and k_{B} are temperature and the Boltzmann constant, respectively). r is the distance between particles, and

$$r = |\mathbf{r}|.$$

After calculating the correlation functions for the bulk electrolyte solution, the correlation functions between a macro-anion (M) and bulk particles are again calculated by using the HNC-OZ theory. However, the macro-anion density goes to 0 because of the dilute limit. So the OZ equation

$$h_{Mj}(r) = c_{Mj}(r) + \sum_{l=A,C,V} \int c_{Ml}(|\mathbf{r} - \mathbf{r}'|)\, \rho_l h_{lj}(r')\, d\mathbf{r}',$$

with the HNC closure

$$h_{Mj}(r) = \exp\left[-\beta u_{Mj}(r) + h_{Mj}(r) + c_{Mj}(r)\right] - 1,$$

is solved. The potential of mean force (PMF) between two macro-anions $\Phi(r)$ is obtained as

$$\Phi(r) = u_{MM}(r) - \mathrm{k_B} T \eta_{MM}(r),$$

where η_{MM} is a sum of convolution integral

$$\eta_{MM}(r) = \sum_{l=A,C,V} \int c_{Ml}(|\mathbf{r} - \mathbf{r}'|)\, \rho_l h_{lM}(r')\, d\mathbf{r}'.$$

These equations are solved based on a successive substitution method; however, the treatment is complicated by the presence of Coulomb interactions in the present system. The details are described in earlier papers (Kinoshita and Harada 1991, 1993, 1994; Kinoshita and Berard 1996; Kinoshita et al. 1996, 1998).

4.4 Potential of Mean Force (PMF) Between Macro-anions

In this section, the PMF $\Phi(r)$, namely the effective interaction between macro-anions immersed in an electrolyte solution, is shown and discussed. The PMF is calculated by using the integral equation theory (Akiyama et al. 2007;

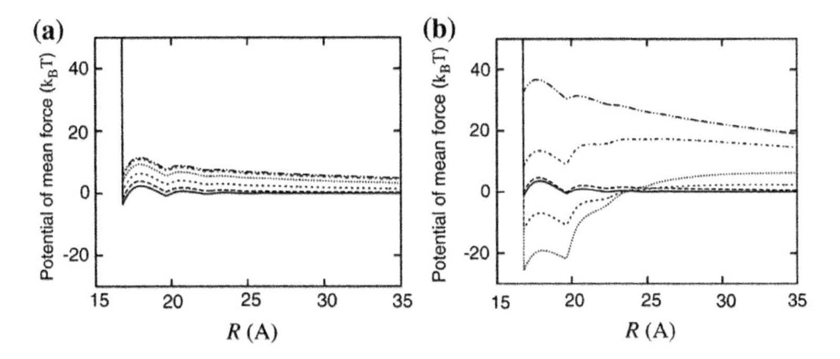

Fig. 4.5 Potential of mean force (PMF) between two macro-anions immersed in an electrolyte solution. **a** The charge of the macro-anion is −5e. **b** The charge of the macro-anion is −10e. The electrolyte concentration dependences are 1 M (solid line), 1×10^{-1} M (long dashed line), 1×10^{-2} M (short dashed line), 1×10^{-3} M (dotted line), 1×10^{-4} M (dash-dot line), and 1×10^{-5} M (dash-dot-dash line) (Akiyama and Sakata 2011)

Akiyama and Sakata 2011; Fujihara and Akiyama 2014) and is summarized in Fig. 4.5.

The PMF depends on the electrolyte concentration, and there are two typical concentration dependences. One is the Coulomb screening behavior shown in Fig. 4.5a, and the other is the reentrant behavior shown in Fig. 4.5b (Akiyama and Sakata 2011). When the charge on the macro-anions is sufficiently small, the Coulomb interactions between charged particles are weaker than the thermal energy of the system. In the case of Fig. 4.5a, the charge of the macro-anion (−5e) is small. Under these conditions, the repulsive Coulomb interaction between two macro-anions is screened by the electrolyte in solution. The PMF is similar to direct Coulomb repulsive interaction between macro-anions when the electrolyte concentration is less than 1×10^{-6}M. However, the screening of the direct repulsion becomes stronger, and the PMF shifts downward as the electrolyte concentration increases. These systems have been studied based on the linearized Poisson–Boltzmann equation. Figure 4.6 shows the logarithmic plot of Fig. 4.5a for larger separations, which

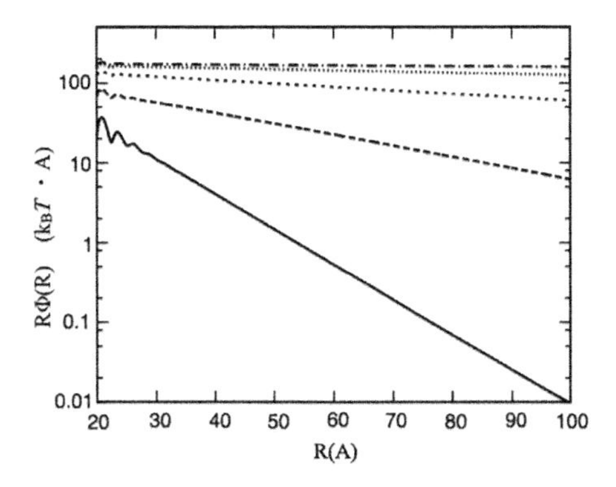

Fig. 4.6 Distance between macro-anions R times the potential of mean force (PMF) $\Phi(R)$. The logarithmic plot of $R \cdot \Phi(R)$ for Fig. 4.5a

asymptotically approach a straight line. Then, the PMFs asymptotically approach to decreasing functions that are proportional to $\exp(-\alpha r)/r$ because the vertical axis is $r \cdot \Phi(r)$. Similar results have been shown for other systems by using similar methods (Akiyama et al. 2007). Therefore, the screening picture is acceptable under these conditions.

On the other hand, strong attraction between macro-anions and a reentrant behavior is apparent in Fig. 4.5b, in which the PMFs are plotted for systems containing macro-anions with charges of −10e. The concentration dependence is completely different from that in Fig. 4.5a. In this case, the charge of macro-anions is sufficiently large that the Coulomb interactions between charged particles are stronger than the thermal energy of system. In the very dilute electrolyte solution, the PMF is also similar to direct Coulomb repulsive interactions between macro-anions. The PMF shifts downward as the electrolyte concentration increases to about 1×10^{-3} M, at which point, strong effective attraction appears, and the contact dimer of macro-anions is about 25 $k_{\mathrm{B}}T$. The strong attraction between like-charged particles disappears as the concentration increases further. This reentrant behavior is similar to experimental results obtained for the condensation behavior of acidic proteins (Zhang et al. 2008, 2010, 2011) and for the coil–globule transition of DNA (Murayama et al. 2003; Besteman et al. 2007).

The strong effective attraction between macro-anions is mediated by the cation. Figure 4.7 shows a cross section of a contact macro-anion dimer that illustrates a mediate mechanism by cations. In this system, the mediate sites make a ring-like region. When cations are located at mediate sites A_1 and A_2, the left macro-anion is pulled to the right by the mediate cations, and the right macro-anion is pulled in the opposite direction. Consequently, if the probability of locating cations at the correct sites is sufficiently high and the Coulomb attraction is sufficiently strong, then the macro-anions appear to attract each other.

To verify the above mechanism, the electrolyte concentration dependence of the free energy of dimerization of the macro-anions was compared with those with a

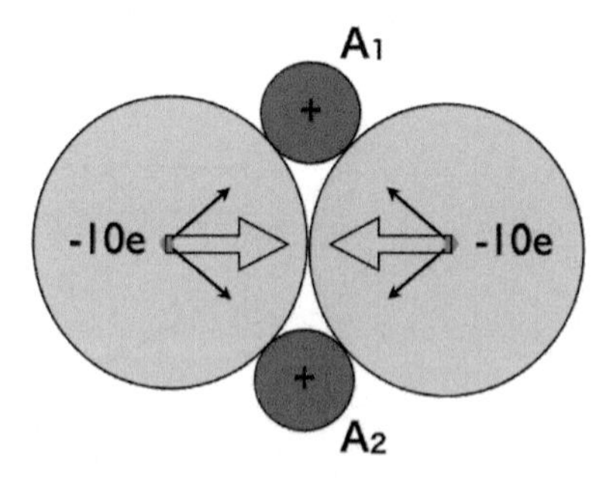

Fig. 4.7 Mediate cations are located at sites A_1 and A_2. Thin arrows on the macro-anions indicate attraction to the cations located at the attraction mediate sites. The sum of thin arrows on the left (right) macro-anion is indicated by the large right (left) arrow. The macro-anions effectively attract each other when the attraction indicated by the large arrows is stronger than the direct Coulomb repulsion between the macro-anions

local density of cations at the mediate sites. The electrolyte concentration dependence of the free energy of dimerization is shown in Fig. 4.8a (Akiyama and Sakata 2011). When the charges on the macro-anions are less than −8e, the plots decrease monotonically and large negative values do not appear. The negative gradient indicates screening of the direct Coulomb repulsive interaction between macro-anions caused by the formation of double layers. The values indicate that strongly stabilized dimers do not appear in the case of weakly charged macro-anions. By contrast, when the macro-anion charges are higher than −9e, the dimerization free energy becomes significantly negative, with a minimum reached at a cation concentration of ca. 1.0 mM. These results indicate that a strongly stabilized dimer is formed and reentrant behavior is exhibited in the case of highly charged macro-anions. As the charges on the macro-anions increase further, the dimer becomes more stable in this regime.

The concentration dependence of the local density of cations at the mediate sites is shown in Fig. 4.8b. The dependence is strongly correlated with the dimerization free energy. When a stable dimer appears, the local cation density is large. Here, we

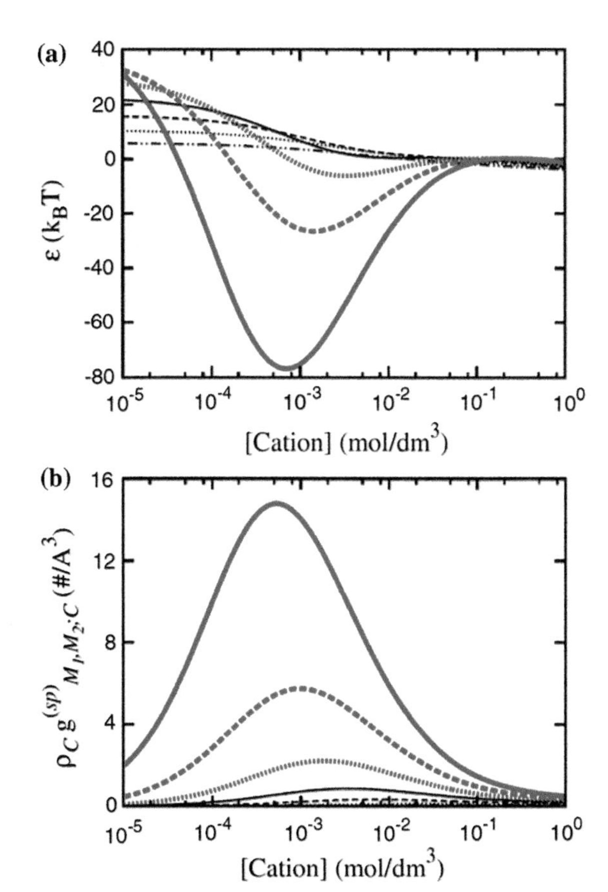

Fig. 4.8 Electrolyte concentration dependences are examined for macro-anions with various charges Q_M. $Q_M = -5e$ (dash-dot-dot line), −6e (dotted line), −7e (dashed line), −8e (solid line), −9e (red thick dotted line), −10e (red thick dashed line), and −11e (red thick solid line). **a** Dependence of free energy of dimerization of macro-anions on cation concentration. **b** Dependence of the local density of cations at the attraction mediation sites on cation concentration (Akiyama and Sakata 2011)

focus on the stability of the dimer of −11e macro-anions. As the dimer becomes more stable, the local cation density increases, and there is a minimum for the stability curve around 1 mM. The local density curve also has a maximum at almost the same concentration required to form the most stable dimer. Above this concentration, the stability decreases and the local density decreases as the concentration increases. The concentration required for the maximum local cation density and minimum dimerization free energy becomes larger as the charges on the macro-anions decrease. We can therefore summarize that the stability of the dimer correlates with the local cation concentration, and thus conclude that the mechanism shown in Fig. 4.7 is reasonable.

The reasons behind the disappearance of the attraction under reentrant conditions are not yet clear. A mechanism with which to explain the experimental results reported by Zhang's group has been proposed (Roosen-Runge et al. 2014), whereby several specific sites for mediate cations on the protein are assumed to function as attractive patches. The results are discussed with reference to the occupation of these sites by mediate cations. Each specific site can accept only one cation in their hypothesis, and high occupancy of specific sites causes the disappearance of the effective attraction under high electrolyte concentration conditions because the two occupied specific sites do not attract each other. Their idea can be used to explain the reentrant behavior that is shown in their phase diagram.

However, the reentrant behavior also appears in our calculated results, although there is no specific site on the macro-anions. Therefore, we must reevaluate our ideas to explain our calculated results, especially with respect to the disappearance of the effective attraction under high electrolyte concentration conditions. It seems that the mediate cation is pulled out into the bulk region by small anions, because the increase in the electrolyte concentration results in an increase in the number of small anions in the system. Indeed, the local density of cations at the mediate sites decreases in the reentrant concentration region. This behavior differs from the prediction based on the specific adsorption site mechanism (Roosen-Runge et al. 2014). Our group is studying the mechanism by which the decrease in the number of mediate cation occurs by using a decomposition method based on an integral equation theory. The results could lead to the proposal of a new picture for the reentrant behavior.

Our calculations indicate that monovalent cations can both mediate the effective attraction and cause the reentrant behavior. However, in practice, only multivalent cations can mediate the strong effective attraction and cause reentrant behavior. It seems that this discrepancy arises from quantitative rather than qualitative considerations. Actually, divalent cations were also examined in our study (Fujihara and Akiyama 2014). Figure 4.9 shows the dependence of dimer stability on the charge of the cation. The above results measured with 1:1 electrolyte solution show that the macro-anions do not strongly attract each other when each macro-anion has −6e charge. However, the effective attraction increases as the charge of the macro-anion increases. For a solution containing divalent cations, the stability of the dimer is about 90 $k_{\mathrm{B}}T$ at the most stable concentration. The reentrant behavior also becomes clear as the cation charge increases. These results imply that the

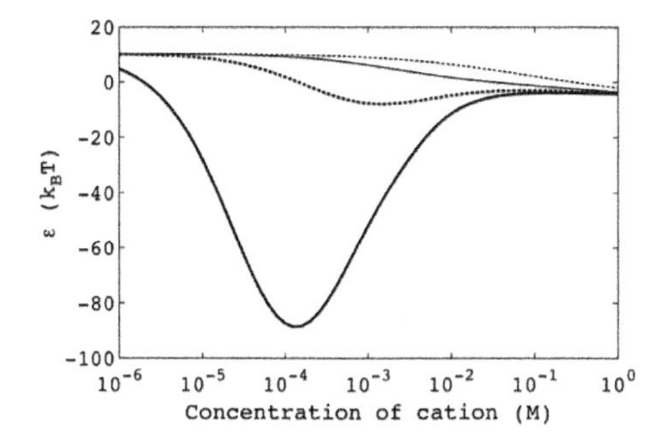

Fig. 4.9 Free energy of dimerization of −6e macro-anions in an electrolyte solution. Dependence of dimerization on the concentration of cations with valences +2.0e (thick solid line), +1.5e (thick dashed line), +1.0e (solid line), and +0.5e (dashed line). Reprinted from reference (Fujihara and Akiyama 2014), Copyright 2014, with permission from Elsevier

discrepancy between the experimental and the calculated results is eliminated if an adequate charge density is used. In the case of acidic proteins, negative charges are localized on the acidic residues, especially on oxygen atoms. The boundary between the simple screening behavior and the reentrant behavior should lie between solutions of monovalent and multivalent cations. The study is also in progress. Preliminary calculations indicate that the dependence of dimer stability on cation charge is also reasonable. It seems that the effective attraction can only be mediated by multivalent cations in the case of acidic proteins.

4.5 Association and Dissociation Model Mechanism for the Engine of a Protein Linear Motor

In this section, we discuss a model mechanism for the association and dissociation of engine proteins such as G-actin. The model is currently speculation, but it may constitute a foundation on which to base a molecular linear motor. In Sect. 4.1, fibril formation of actin is introduced as an "engine" of the linear motor in a motile cell. To push the motile cell, G-actins associate at the end of F-actin around the leading edge. It was also noted that in addition to the association of G-actin, dissociation also occurs within the engine to supply the monomer to the leading edge. The association state of ATP-binding actin (T-G-actin) is more stable than that of ADP-binding actin (D-G-actin). This difference in stability is regulated by ATP hydrolysis.

It is difficult to identify the correct charges of ATP and ADP on actin, but it is reasonable to assume that the negative charge of ATP is larger than that of ADP. Here, we assume the charge difference. If the hypothesis on the charge differences is reasonable, the stability between T-G-actin and T-G-actin, that between T-G-actin and D-G-actin, and that between D-G-actin and D-G-actin will be different (where T and D indicate ATP and ADP, respectively).

If most of the attractive interaction between the proteins is driven by the translational motion of solvent molecules (Amano et al. 2010), the effective interaction change between acidic residues can become the switch between the association and the dissociation. Figure 4.10 illustrates schematic diagrams for the switch. Figure 4.8a indicates that the effective attraction between highly charged macro-anions is stronger than that between weakly charged macro-anions, and this difference is caused by the difference in the number of mediate cations. Similar

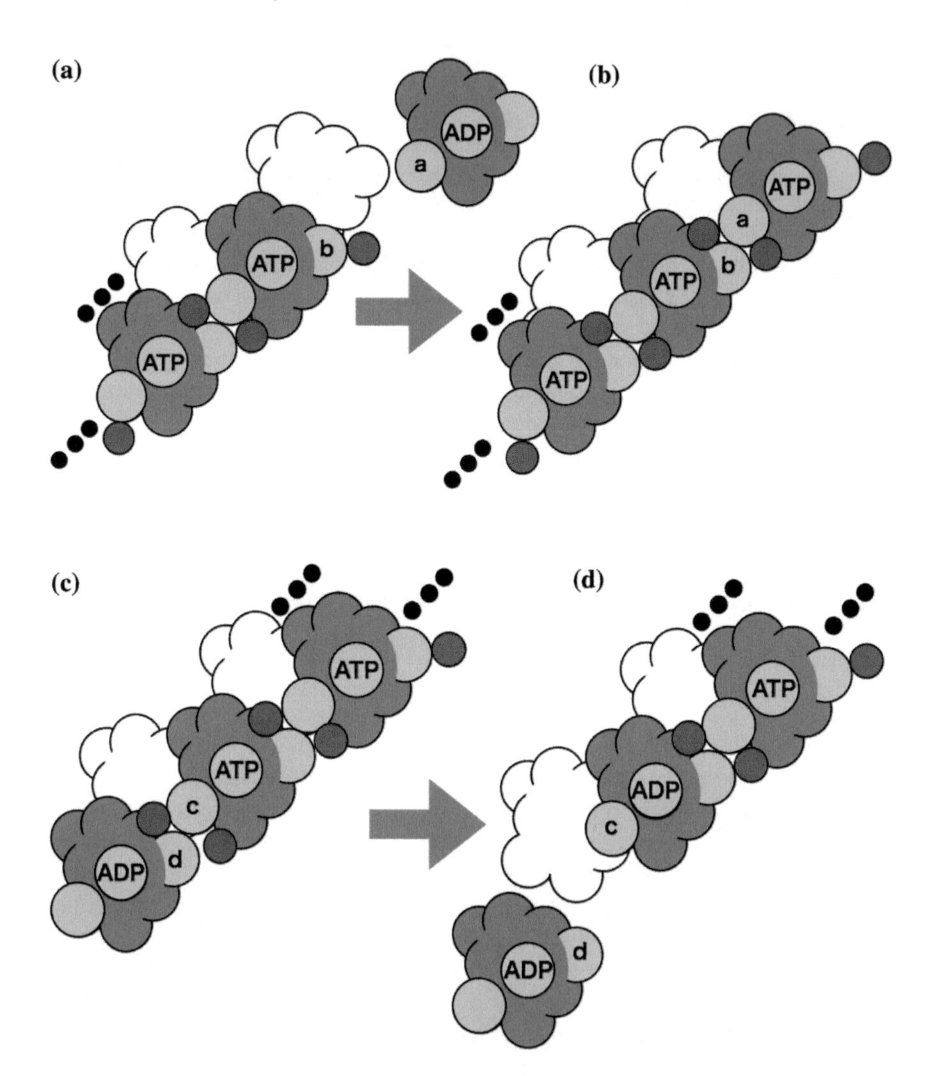

Fig. 4.10 Schematic diagrams illustrating association process (from **a** to **b**) and dissociation (from **c** to **d**) in a linear protein motor. Proteins have negatively charged sites of acidic residues on the surface (green circles). The mediate cations (red circles), the negatively charged ATP sites (blue circles with ATP), and the ADP sites (green circles with ADP) are also drawn. The spiral conformation of F-actin is simplified, with actins on the back of the spiral drawn in white

differences can be found between T-G-actin and D-G-actin. The amount of mediate cations at the acidic residues on a T-G-actin is more than that on a D-G-actin because the charge of ADP is smaller than that of ATP. Then, the affinity between F-actin at site "a" and T-G-actin at site "b" can be strong (Fig. 4.10b), whereas the affinity between F-actin at site "a" and D-G-actin at site "b" is not so strong (Fig. 4.10a). Therefore, the replacement of ADP by ATP promotes the polymerization. On the other hand, ATP hydrolysis promotes depolymerization because the amount of mediate cations at site "c" on a D-G-actin is less than that on a T-G-actin. Therefore, the effective attraction between sites "c" and "d" is reduced upon ATP hydrolysis. Here, actin is an enzyme to reduce the activation free energy for ATP hydrolysis. Associated T-G-actin changes to D-G-actin because of the hydrolysis in the F-actin, and the D-G-actin becomes easy to remove from the fibril. If the concentration and the rate of ATP hydrolysis are adequate, treadmilling likely occurs.

4.6 Summary

Effective interaction between macro-anions immersed in an electrolyte solution was studied by using HNC-OZ theory (Akiyama et al. 2007; Akiyama and Sakata 2011; Fujihara and Akiyama 2014). The results indicate that a strong attraction can exist between macro-anions and that the effective attraction is mediated by the cations. The effective attraction exhibited a reentrant behavior. Although the effective interaction is fundamentally repulsive, it becomes strongly attractive in certain concentration regimes. The concentration region was around 1.0 mM. Such interesting reentrant behavior has also been observed in various acidic protein solutions (Zhang et al. 2008, 2010, 2011). Preliminary calculations also showed that the effective attraction was also dependent on the charge on the cation. It seems that the effective attraction can be mediated only by multivalent cations in the case of acidic proteins.

A mechanism has been proposed based on the specific adsorption sites for the mediate cations on the acidic protein (Roosen-Runge et al. 2014). However, our calculations show that both strong attraction and reentrant behavior are also manifested in our simple models that do not have specific adsorption sites. Therefore, although the calculated results do not exclude the proposed mechanism, we must develop a new mechanism to explain the reentrant behavior.

A new model for dynamic (or kinetic) ordering behavior was proposed based on the calculated strong attraction. Calculated results show that the effective attraction becomes stronger as the negative charges on the macro-anions increase. This behavior corresponds to the actin polymerization behavior considering that the negative charge of ATP is larger than that of ADP. In this scenario, the attraction between G-actins should be driven by the cation-mediated attraction between acidic residues because the attraction can be switched through the replacement of ADP with ATP and through hydrolysis of ATP. It may be possible to apply this idea to construct engine parts for use in molecular linear motors.

References

Akiyama R, Fujino N, Kaneda K, Kinoshita M (2007) Interaction between like-charged colloidal particles in aqueous electrolyte solution: attractive component arising from solvent granularity. Condens Matter Phys 10:587–596

Akiyama R, Sakata R (2011) An integral equation study of reentrant behavior in attractive interactions between like-charged macroions immersed in an electrolyte solution. J Phys Soc Jpn 80:123602-1–123602-4

Alberts B, Johnson A, Lewis J, Raff M, Roberts K, Walter P (2007) Molecular biology of the cell 5th ed. Garland Science

Amano K, Yoshidome T, Iwaki M, Suzuki M, Kinoshita M (2010) Entropic potential field formed for a linear-motor protein near a filament: statistical-mechanical analyses using simple models. J. Chem. Phys. 133:045103–1–045103–11

Besteman K, Van Eijk K, Lemay SG (2007) Charge inversion accompanies DNA condensation by multivalent ions. Nat Phys 3:641–644

Bloomfield VA (1997) DNA condensation by multivalent cations. Biopolymers 44:269–282

Fujihara S, Akiyama R (2014) Attractive interaction between macro-anions mediated by multivalent cations in biological fluids. J Mol Liq 200:89–94

Gosule LC, Schellman JA (1976) Compact form of DNA induced by spermidine. Nature 259:333–335

Inagaki N, Toriyama M, Sakumura Y (2011) Systems biology of symmetry breaking during neuronal polarity formation. Dev. Neurobiol. 71:584–593

Hansen JP, McDonald IR (2006) Theory of simple liquids, 3rd edn. Academic Press, London

Kamiguchi H, Hlavin ML, Yamasaki M, Lemmon V (1998) Adhesion molecules and inherited diseases of the human nervous system. Annu Rev Neurosci 21:97–125

Kinoshita M, Berard DR (1996) Analysis of the bulk and surface-induced structure of electrolyte solutions using integral equation theories. J Comput Phys 124:230–241

Kinoshita M, Harada M (1991) Numerical solution of the HNC equation for fluids of non-spherical particles. An efficient method with application to dipolar hard spheres. Mol Phys 74:443–464

Kinoshita M, Harada M (1993) Numerical solution of the RHNC theory for fluids of non-spherical particles near a uniform planar wall: a study with dipolar hard spheres as an example system. Mol Phys 79:145–167

Kinoshita M, Harada M (1994) Numerical solution of the RHNC theory for water-like fluids near a macroparticle and a planar wall. Mol Phys 81:1473–1488

Kinoshita M, Iba S, Kuwamoto K, Harada M (1996) Interaction between macroparticles in Lennard-Jones fluids or in hard sphere mixtures. J Chem Phys 105:7177–7183

Kinoshita M, Okamoto Y, Hirata F (1998) Calculation of solvation free energy using RISM theory for peptide in salt solution. J Comput Chem 19:1724–1735

Kubo Y, Baba K, Toriyama M, Minegishi T, Sugiura T, Kozawa S, Ikeda K, Inagaki N (2015) Shootin 1-cortactin interaction mediates signal-force transduction for axon outgrowth. J Cell Biol 210:663–676

Le Clainche C, Carlier MF (2008) Regulation of actin assembly associated with protrusion and adhesion in cell migration. Physiol Rev 88:489–513

Murakami K, Yasunaga T, Noguchi TQP, Gomibuchi Y, Ngo KX, Uyeda TQP, Wakabayashi T (2010) Structural basis for actin assembly, activation of ATP hydrolysis and delayed phosphate release, cell 143:275–287

Murayama Y, Sakamaki Y, Sano M (2003) Elastic response of single DNA molecules exhibits a reentrant collapsing transition. Phys Rev Lett 90:018102-1–018102-4
Pollard TD, Borisy GG (2003) Cellular motility driven by assembly and disassembly of actin filaments. Cell 112:453–465
Roosen-Runge F, Zhang F, Schreiber F, Roth R (2014) Ion-activated attractive patches as a mechanism for controlled protein interactions. Sci Rep 4:07016-1-5
Sapir T, Levy T, Sakakibara A, Rabinkov A, Miyata T, Reiner O (2013) Shootin 1 acts in concert with KIF20B to promote polarization of migrating neurons. J Neurosci 33:11932–11948
Shimada T, Toriyama M, Uemura K, Kamiguchi H, Sugiura T, Watanabe N, Inagaki N (2008) Shootin 1 interacts with actin retrograde flow and L1-CAM to promote axon outgrowth. J Cell Biol 181:817–829
Toriyama M, Sakumura Y, Shimada T, Ishii S, Inagaki N (2010) A diffusion-based neurite length-sensing mechanism involved in neuronal symmetry breaking. Mol Syst Biol 6:394
Toriyama M, Shimada T, Kim KB, Mitsuba M, Nomura E, Katsuta K, Sakumura Y, Roepstorff P, Inagaki N (2006) Shootin 1: a protein involved in the organization of an asymmetric signal for neuronal polarization. J Cell Biol 175:147–157
Widom J, Baldwin RL (1980) Cation-induced toroidal condensation of DNA. J Mol Biol 144:431–453
Yoshikawa K, Takahashi M, Vasilevskaya VV, Khokhlov AR (1996) Large discrete transition in a single DNA molecule appears continuous in the ensemble. Phys Rev Lett 76:3029–3031
Zhang F, Skoda MWA, Jacobs RMJ, Zorn S, Martin RA, Martin CM, Clark GF, Weggler S, Hildebrandt A, Kohlbacher O, Schreiber F (2008) Reentrant condensation of proteins in solution induced by multivalent counterions. Phys Rev Lett 101:148101-1–148101-4
Zhang F, Weggler S, Ziller MJ, Ianeselli L, Heck BS, Hildebrandt A, Kohlbacher O, Skoda MWA, Jacobs RMJ, Schreiber F (2010) Universality of protein reentrant condensation in solution induced by multivalent metal ions. Proteins 78:3450–3457
Zhang F, Zocher G, Sauter A, Stehle T, Schreiber F (2011) Novel approach to controlled protein crystallization through ligandation of yttrium cations. J Appl Cryst 44:755–762

Chapter 5
Statistical Mechanical Integral Equation Approach to Reveal the Solvation Effect on Hydrolysis Free Energy of ATP and Its Analogue

Norio Yoshida and Fumio Hirata

Abstract Investigations of the hydrolysis reactions of adenosine triphosphate (ATP) and pyrophosphate in solution by the three-dimensional reference interaction site model self-consistent field (3D-RISM-SCF) theory are briefly reviewed. The theory is applied to the four different charged states of pyrophosphate, for which experimental data of hydrolysis free energies are available. The results of the reaction free energy for all the four charged states are almost constant, ~−8 kcal/mol, in accord with the experimental results, but in marked contrast to the conventional view, or the *high-energy P–O bond hypothesis*. The theory is also applied to the hydrolysis reaction of ATP to clarify the molecular origin of the energy produced by the ATP hydrolysis.

Keywords 3D-RISM-SCF · ATP hydrolysis · Pyrophosphate

5.1 Introduction

All biological functions are driven by the energy produced by the adenosine triphosphate (ATP) hydrolysis reaction that converts ATP into adenosine diphosphate (ADP) (Meyerhof and Lohmann 1932). It has long been believed that the large energy originates mainly from the backbone P–O bond in ATP, the so-called *high-energy P-O bond* (Boyd and Lipscomb 1969; Hammond et al. 1992; Hill and

---

N. Yoshida (✉)
Graduate School of Science, Department of Chemistry, Kyushu University, Fukuoka, Japan
e-mail: noriwo@chem.kyushu-univ.jp

F. Hirata
Toyota Physical & Chemical Research Institute, Nagakute, Aichi, Japan
e-mail: hirata@toyotariken.jp

© Springer Nature Singapore Pte Ltd. 2018

M. Suzuki (ed.), *The Role of Water in ATP Hydrolysis Energy Transduction by Protein Machinery*, https://doi.org/10.1007/978-981-10-8459-1_5

Morales 1951; Lipmann 1941). The high-energy bond hypothesis has been challenged by some scientists who have proposed another idea in which not only the high-energy P–O bond in ATP, but the difference in *hydration free energies* between ATP and ADP also plays a critical role in determining the reaction energy (Akola and Jones 2003; Colvin et al. 1995; George et al. 1970; Hofmann and Zundel 1974). Thus, there has not been a clear consensus regarding the origin of the ATP hydrolysis energy. Many experimental and theoretical efforts have been devoted to establishing a consensus regarding the origin of the ATP hydrolysis energy (Colvin et al. 1995; Grigorenko et al. 2006; Hong et al. 2012; Kamerlin and Warshel 2009; Klaehn et al. 2006; Takahashi et al. 2017; Wang et al. 2015; Yamamoto 2010).

For a theoretical analysis of the chemical reaction taking place in aqueous solutions, two elements of chemical physics are required, the quantum mechanics electronic structure theory and the theory of solvation. There have been many theories proposed so far that can be categorized into several groups depending on their method for treating *solvation*: the *explicit solvent* model, the *implicit solvent* model, and the integral equation theory of liquids. The explicit solvent model is the most straightforward approach for handling solvent molecules. One of these methods is the combined quantum and molecular mechanics (QM/MM) method (Aqvist and Warshel 1993; Field et al. 1990; Gao and Xia 1992). In the QM/MM method, solvent molecules are treated by classical molecular mechanics, and the solvation structure is evaluated by sampling the position and orientation of solvent molecules based on molecular simulation. A standard method of evaluating the solvation free energy in a QM/MM framework, such as the free energy perturbation method, requires calculations of the intermediate states connecting an initial un-solvated and a final solvated state, and therefore, the computational cost becomes huge. One way of avoiding this difficulty is the energy representation (ER) method proposed by Takahashi and Matubayasi, which is reviewed in Chap. 1 (Takahashi et al. 2004, 2005, 2017).

The implicit solvent model forms another category of solvation models, such as the polarizable continuum model (PCM), and the conductor-like screening model (COSMO) (Sinnecker et al. 2006; Tomasi et al. 2005; Tomasi and Persico 1994). In these methods, the solvent environment is regarded as a continuum media characterized by the dielectric constant. These methods have been widely applied to a variety of chemical processes in solution, including the hydrolysis reaction of the ATP analogue (Colvin et al. 1995; Yamamoto 2010).

The statistical mechanics integral equation theories of molecular liquids, based on the Ornstein–Zernike (OZ) theory, have noteworthy features compared with the other solvation methods (Hansen and McDonald 2006; Hirata 2003). These theories allow the treatment of the microscopic intermolecular interaction of solute and solvent molecules, and solvation thermodynamics, to be described with a complete ensemble average in thermodynamic limits. The three-dimensional reference interaction site model (3D-RISM) theory is one of the integral equation theories based on the OZ theory; it is the most versatile method and widely used for various chemical and biological phenomena in solutions (Beglov and Roux 1996, 1997;

Kovalenko and Hirata 1998; Yoshida et al. 2009). The 3D-RISM theory is combined with the Kohn–Sham density functional theory (KS-DFT) by Kovalenko and Hirata to solve problems related to the electronic structure of a macromolecule or electrode in contact with solutions. The theory is called KS-DFT/3D-RISM (Kovalenko and Hirata 1999). The KS-DFT part of the theory is replaced by the ab initio molecular orbital theory by Sato, Kovalenko, and Hirata, called the three-dimensional reference interaction site model self-consistent field (3D-RISM-SCF) theory (Sato et al. 2000). The KS-DFT/3D-RISM or 3D-RISM-SCF theory has been successfully applied to a variety of chemical reactions in solution (Sato 2013).

In this chapter, we review the application of the KS-DFT/3D-RISM or 3D-RISM-SCF theory to investigate the hydrolysis reaction of ATP and pyrophosphate, an ATP analogue, in aqueous solutions. The 3D-RISM-SCF theory has been applied to four different charged states of pyrophosphate for which experimental data for hydrolysis free energies are available (Hong et al. 2012). The calculations elucidate the physics concerning the energy produced by the hydrolysis of pyrophosphate, which is completely different from the conventional picture of the *high-energy P–O bond hypothesis* based on quantum mechanical calculations in a vacuum (Boyd and Lipscomb 1969; Hammond et al. 1992; Hill and Morales 1951; Lipmann 1941). The theory is also applied to the hydrolysis reaction of ATP and clarifies the molecular origin of the energy produced by the ATP hydrolysis reaction (Yoshida et al. Unpublished work).

5.2 The 3D-RISM Theory

In this section, the formalism of the 3D-RISM theory is introduced briefly. The details of the formalism can be found in the literature (Hirata 2003).

The 3D-RISM theory is derived from the molecular OZ (MOZ) equation theory, which is an extension of the OZ integral equation theory to molecular liquids. These theories describe the structure and thermodynamics of solvation based on pair correlation functions or pair density distribution functions (DFs). The pair density DF of molecules is defined as:

$$\rho(\boldsymbol{r}_1,\boldsymbol{r}_2,\boldsymbol{\Omega}_1,\boldsymbol{\Omega}_2)=\langle\sum_i\sum_{j\neq i}\delta(\boldsymbol{r}_1-\boldsymbol{r}_i)\delta(\boldsymbol{r}_2-\boldsymbol{r}_j)\delta(\boldsymbol{\Omega}_1-\boldsymbol{\Omega}_j)\delta(\boldsymbol{\Omega}_2-\boldsymbol{\Omega}_j)\rangle \tag{5.1}$$

If there is no external field, the translational invariance of the pair density DF can be applied and the pair DF is derived as:

$$g(\boldsymbol{r}_{12},\boldsymbol{\Omega}_1,\boldsymbol{\Omega}_2)=\left(\frac{\Omega}{\rho}\right)^2\rho(\boldsymbol{r}_{12},\boldsymbol{\Omega}_1,\boldsymbol{\Omega}_2), \tag{5.2}$$

where $\boldsymbol{r}_{12} = \boldsymbol{r}_2 - \boldsymbol{r}_1$ and $\Omega \equiv \int d\boldsymbol{\Omega}$. The MOZ equation is expressed using this function:

$$h(\boldsymbol{r}_{12}, \boldsymbol{\Omega}_1, \boldsymbol{\Omega}_2) = c(\boldsymbol{r}_{12}, \boldsymbol{\Omega}_1, \boldsymbol{\Omega}_2) + \frac{\rho}{\Omega} \int c(\boldsymbol{r}_{13}, \boldsymbol{\Omega}_1, \boldsymbol{\Omega}_3) h(\boldsymbol{r}_{32}, \boldsymbol{\Omega}_3, \boldsymbol{\Omega}_2) d\boldsymbol{r}_3 d\boldsymbol{\Omega}_3, \tag{5.3}$$

where $h = g + 1$ is the total correlation function and c is the direct correlation function that is defined through the MOZ equation. The pair correlation function in MOZ formalism is a function of the orientations of two molecules and the relative coordinates, but it is difficult to solve numerically. Although methods to solve the MOZ equation have been proposed, it is still difficult to apply them to complex molecular systems such as biomolecules. One of the ideas for reducing the computational cost involved in keeping the molecular features in the pair correlation functions is to use an interaction site model originally proposed by Chandler and Andersen (Andersen and Chandler 1972; Andersen et al. 1972; Chandler and Andersen 1972). The molecular site or the 3D correlation function is derived by taking orientational averaging of molecular pair total correlation function centered specific interaction site as:

$$h_\gamma(\boldsymbol{r}) = \frac{1}{\Omega} \int h(\boldsymbol{r}_{12}, \boldsymbol{\Omega}_1, \boldsymbol{\Omega}_2) \delta(\boldsymbol{r}_{12} + l_{2\gamma}(\boldsymbol{\Omega}_2) - \boldsymbol{r}) d\boldsymbol{r}_2 d\boldsymbol{\Omega}_2, \tag{5.4}$$

where $l_{2\gamma}$ is the vector connecting the center of molecule 2 and interaction site γ. In contrast, the 3D direct correlation function is defined by applying a superposition approximation:

$$c(\boldsymbol{r}_{12}, \boldsymbol{\Omega}_1, \boldsymbol{\Omega}_2) = \sum_\gamma c_\gamma(\boldsymbol{r}). \tag{5.5}$$

By using the 3D total and direct correlation function, the 3D-RISM equation can be derived from the MOZ equation

$$h_\gamma(\boldsymbol{r}) = \sum_{\gamma'} \left[c_{\gamma'} * \chi_{\gamma'\gamma} \right], \tag{5.6}$$

where * denotes the convolution integral and $\chi_{\gamma'\gamma}$ is the solvent susceptibility, obtained by solving the RISM equation for a bulk solvent system prior to the 3D-RISM calculation. To close the 3D-RISM equation, we need another relation between the 3D total and the direct correlation function. The equation for closing the 3D-RISM equation is called a closure relation. Several closure relations have been proposed; Kovalenko and Hirata proposed a useful closure relation for applying the 3D-RISM theory to complex molecular systems, which is known as the Kovalenko–Hirata (KH) closure:

$$g_\gamma(\boldsymbol{r}) = \begin{cases} \exp\left[d_\gamma(\boldsymbol{r})\right] & d_\gamma(\boldsymbol{r}) < 0 \\ 1 + d_\gamma(\boldsymbol{r}) & d_\gamma(\boldsymbol{r}) \geq 0 \end{cases}$$
$$d_\gamma(\boldsymbol{r}) = -\beta u_\gamma(\boldsymbol{r}) + h_\gamma(\boldsymbol{r}) - c_\gamma(\boldsymbol{r}), \tag{5.7}$$

where $u_\gamma(\boldsymbol{r})$ is a solute–solvent interaction potential at position $\boldsymbol{r}$ (Kovalenko and Hirata 1999, 2001). β is thermodynamic beta. The 3D-RISM theory coupled with the KH closure, called the 3D-RISM-KH theory, yields reasonable physical properties such as the free energy of solvation. The free energy of solvation, or excess chemical potential, is given by

$$\Delta\mu = \frac{\rho}{\beta}\sum_\gamma \int \left\{ -c_\gamma(\boldsymbol{r}) + \frac{1}{2}h_\gamma(\boldsymbol{r})^2\,\Theta\left(-h_\gamma(\boldsymbol{r})\right) - \frac{1}{2}h_\gamma(\boldsymbol{r})c_\gamma(\boldsymbol{r}) \right\} d\boldsymbol{r}, \tag{5.8}$$

where Θ is the Heaviside step function.

5.3 Hybrid 3D-RISM and Electronic Structure Theories

A hybrid method of the quantum chemical electronic structure theory with an integral equation theory of molecular liquids allows us to consider the chemical reaction in solution. A series of papers on a pioneering work has been published by Ten-no, Hirata, and Kato (Ten-no et al. 1993, 1994). They proposed a hybrid theory of an ab initio molecular orbital method with the RISM theory. The work has been followed with similar methods, mixing different combinations of quantum and statistical mechanics that have been applied to various chemical reactions in solution (Kido et al. 2015; Kovalenko and Hirata 1999; Nishihara and Otani 2017; Sato et al. 2000; Yokogawa et al. 2007; Yoshida and Kato 2000). The hybrid method of the electronic structure and 3D-RISM theories, called KS-DFT/3D-RISM or 3D-RISM-SCF, is useful for investigating the chemical reaction of a highly charged system such as ATP and pyrophosphate, because the method allows treatment of the spatial distribution of the electron density of a solute molecule without any approximation.

In the 3D-RISM-SCF formalism, the free energy of the system is defined by

$$G = E_{\text{solute}} + \Delta\mu, \tag{5.9}$$

where E_{solute} is the electronic energy of a solute molecule, evaluated by the ab initio molecular orbital (MO) method, or KS-DFT:

$$E_{\text{solute}} = \langle \Psi | \hat{H}_0 | \Psi \rangle, \tag{5.10}$$

where $\hat{H}_0$ and Ψ denote the electronic Hamiltonian of an isolated solute molecule and the wave function of a solvated solute molecule. The solute wave functions are obtained by solving the Schrödinger equation with the following Hamiltonian:

$$\hat{H}_{\text{solv}} = \hat{H}_0 + \hat{V}, \tag{5.11}$$

where $\hat{V}$ is the solvent-electron interaction term given by

$$\hat{V} = \rho \sum_{\gamma} \int g_{\gamma}(\boldsymbol{r}) \left(\frac{q_{\gamma}}{|\boldsymbol{r} - \boldsymbol{r}'|} \right) d\boldsymbol{r}, \tag{5.12}$$

q_{γ} is a point charge on the solvent site γ, and $\boldsymbol{r}'$ is the coordinate of an electron.

The free energy of solvation $\Delta\mu$ is obtained by solving the 3D-RISM-KH equation under the electrostatic potential caused by the solute molecule. The solute–solvent interaction potential is given by

$$u_{\gamma}(\boldsymbol{r}) = \sum_{\alpha} 4\epsilon_{\alpha\gamma} \left[\left(\frac{\sigma_{\alpha\gamma}}{|\boldsymbol{r} - \boldsymbol{r}_{\alpha}|} \right)^{12} + \left(\frac{\sigma_{\alpha\gamma}}{|\boldsymbol{r} - \boldsymbol{r}_{\alpha}|} \right)^{6} \right] + U_{\text{elec}}(\boldsymbol{r}), \tag{5.13}$$

where α denotes a solute atom and

$$U_{\text{elec}}(\boldsymbol{r}) = \int \frac{q_{\gamma} |\Psi(\boldsymbol{r}')|^2}{|\boldsymbol{r} - \boldsymbol{r}'|} d\boldsymbol{r}', \tag{5.14}$$

where σ and ϵ are the Lennard–Jones parameters in the usual meaning. The restrained electrostatic potential (RESP) charge q_{α} for a solute atom can also be available for the 3D-RISM-SCF formalism

$$U_{\text{elec}}(\boldsymbol{r}) = \sum_{\alpha} \frac{q_{\gamma} q_{\alpha}}{|\boldsymbol{r} - \boldsymbol{r}_{\alpha}|}. \tag{5.15}$$

When the RESP charge is used, the corresponding solvent–electron interaction in the solvated Hamiltonian is given by

$$\hat{V} = \rho \sum_{\gamma} \sum_{\alpha} \hat{q}_{\alpha} \int g_{\gamma}(\boldsymbol{r}) \left(\frac{q_{\gamma}}{|\boldsymbol{r} - \boldsymbol{r}_{\alpha}|} \right) d\boldsymbol{r}, \tag{5.16}$$

where $\hat{q}_{\alpha}$ is a partial charge operator to reproduce the RESP charge on atom α.

By solving the 3D-RISM and Schrödinger equations iteratively, the electronic structure of solute molecules and solvation structure can be obtained simultaneously.

5.4 Hydrolysis of Pyrophosphate

In this section, the 3D-RISM-SCF study of the hydrolysis reaction of the pyrophosphate, ATP analogue, is reviewed (Hong et al. 2012).

Hong et al. have applied the 3D-RISM-SCF theory to the four types of the hydrolysis reaction of the pyrophosphate which have different charged states depending on the protonation of phosphate oxygen (Scheme 5.1).

The reaction free energy of hydrolysis can be defined as

$$G_{\mathrm{aq}} = G_{\mathrm{kin}} + E_{\mathrm{gas}} + \Delta E_{\mathrm{reorg}} + \Delta \mu, \tag{5.17}$$

where G_{kin}, E_{gas}, $\Delta E_{\mathrm{reorg}}$, and $\Delta \mu$ denote the kinetic free energy, gas phase electronic energy, electronic reorganization energy, and solvation free energy, respectively. The sum of E_{gas} and $\Delta E_{\mathrm{reorg}}$ corresponds to the solute electronic energy, E_{solute}, in Eq. (5.10). KS-DFT with B3LYP/6-31+G(d) are used for the electronic structure calculation, whereas the 3D-RISM-KH theory with the OPLS and RESP parameters are used for the solvation structure calculations. The details of the computational conditions can be found in the original paper (Hong et al. 2012).

The free energy profiles of the reactions are summarized in Table 5.1 (see also Scheme 5.1). In a gas phase, the hydrolysis reaction is almost thermoneutral in the case of reaction A, whereas reaction B is exothermic. Reactions C and D are endothermic and show a large negative reaction free energy. These behaviors can be roughly understood by the charge repulsion and localization of excess charges. For reaction A, all the reactant and product species are charge-neutral; therefore, there is no large energy change involved. For reaction B, the excess charge on reactant pyrophosphate, $H_3P_2O_7^-$, is localized in the smaller space of product phosphate, $H_2PO_4^-$; thus, the product state becomes unstable compared to the reactant state. For C and D, the multiple excess charges on the reactant pyrophosphate, $H_2P_2O_7^{2-}$ and $HP_2O_7^{3-}$, respectively, are separated by the reaction. This charge separation

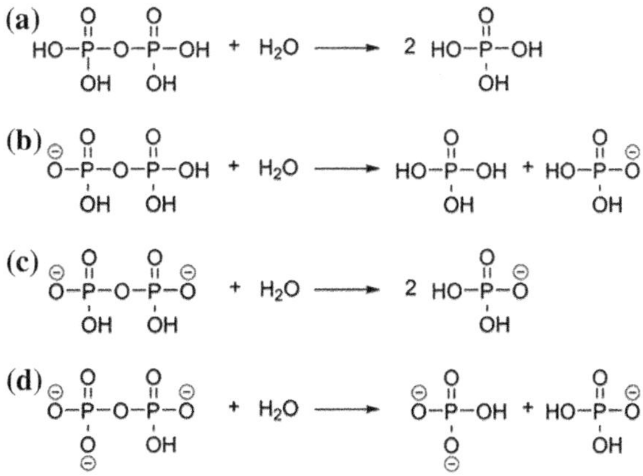

(a) HO–P(=O)(OH)–O–P(=O)(OH)–OH + H_2O ⟶ 2 HO–P(=O)(OH)–OH

(b) ⊖O–P(=O)(OH)–O–P(=O)(OH)–OH + H_2O ⟶ HO–P(=O)(OH)–OH + HO–P(=O)(OH)–O⊖

(c) ⊖O–P(=O)(OH)–O–P(=O)(OH)–O⊖ + H_2O ⟶ 2 HO–P(=O)(OH)–O⊖

(d) ⊖O–P(=O)(O⊖)–O–P(=O)(OH)–O⊖ + H_2O ⟶ ⊖O–P(=O)(O⊖)–OH + HO–P(=O)(OH)–O⊖

Scheme 5.1 Schematic description of the hydrolysis reaction of pyrophosphate for the four possible charged states. (Reprinted with permission from Hong et al. (2012). Copyright 2012 American Chemical Society.)

Table 5.1 Reaction free energies in the gas and aqueous phases computed by DFT and 3D-RISM-SCF calculations. The solvation effect means the difference between the free energies in the aqueous and gas phases. Units are in kcal/mol

Reaction	Gas phase (DFT)	Aqueous phase (3D-RISM-SCF)	Exp[a]	Solvation effect
A	−1.7	−8.9	−9.5	−7.2
B	21.3	−6.2	−7.5	−27.5
C	−56.6	−8.1	−7.7	48.5
D	−119.5	−7.7	−7.1	111.8

[a]Experimental values from George et al. (1970)

drastically reduces the Coulomb repulsion in the product state compared to the reactant state and makes these reactions exothermic associated by an unrealistically large energy production. The large energy gain, as a result of charge separation, may be the reason why the P–O bond is called a *high-energy bond*. However, such a high-energy bond may not exist in nature unless it is bound by an unrealistic chemical bond. In any case, the *high-energy P–O bond hypothesis* could not explain the experimental observation in aqueous solutions, which indicate that all four reactions are moderately exothermic, c.a. 8 kcal/mol (George et al. 1970).

In an aqueous phase, the 3D-RISM-SCF produces almost quantitative agreement for the reaction free energy with the experimental results. It clearly indicates the importance of the solvation effects as a factor for determining reactivity. In addition, it is interesting that the solvation effect on the reaction free energy is negative for A and B, but positive for C and D. This indicates that the mechanism of the reaction free energy changes because solvation varies according to the charged states. In case A, the origin of the solvation free energy change may be attributed to a change in the short-range interaction including the hydrogen bonds, as both the reactant and product species are charge-neutral. The solvation contribution to the electronic reorganization energy and the solvation free energy are summarized in Table 5.2. Here, the electronic reorganization energy is defined as the difference between the electronic energy of a solute molecule in the aqueous and gas phases. In general, the electronic reorganization, as a result of solvation, makes positive contribution to the free energy because the electronic structure is distorted by the electrostatic field of the solvent. The solvation free energy change makes a dominant contribution to the reaction free energy change (see Table 5.2). Note that the contribution from the kinetic energy change is rather small; therefore, the contribution from the kinetic energy term is omitted from the discussion. For reactions A and B, the products are stabilized by solvation more than the reactants. This converts reactions A and B from "thermoneutral" or "endothermic" in the gas phase to "exothermic" in water. On the other hand, the reactants are more stabilized, compared to the products, in reactions C and D. This changes reactions C and D from "highly exothermic" in the gas phase to "moderately exothermic" in water. Taking the solvation free energy and the electronic reorganization energy into consideration leads to a nearly quantitative agreement with the experimental data (see Table 5.1).

Table 5.2 Solvation contributions (electronic reorganization energy ΔE_{reorg} and solvation free energy $\Delta\mu$) to the reaction free energy computed by the 3D-RISM-SCF theory. Units are in kcal/mol

Reaction	Reactants		Products	
	ΔE_{reorg}	$\Delta\mu$	ΔE_{reorg}	$\Delta\mu$
A	8.2	−27.0	8.2	−32.6
B	6.7	−74.4	13.9	−113.7
C	5.8	−224.1	19.6	−194.8
D	9.7	−478.3	21.8	−385.2

To clarify the molecular origin of the solvation effects on the hydrolysis reaction, the solvation structures around the solute species were investigated. The spatial and radial distribution functions of water around the solute species are depicted in Figs. 5.1 and 5.2. As can be seen in these figures, the hydration patterns are quite different depending on the charged states of the phosphates. When the phosphate has no net charge, as in H_3PO_4 and $H_4P_2O_7$, it is primarily hydrated by water oxygen through hydrogen bonds. As stated before, reaction A is almost thermoneutral in the gas phase; therefore, the reaction free energy of −8.9 kcal/mol in the aqueous phase mainly comes from the difference in the short-range solute–solvent interactions between the reactants and products, including the energy of cavity formation, hydrogen bond, and steric effect. In contrast, in the case in which phosphates have a single negative charge, as in $H_2PO_4^-$ and $H_3P_2O_7^-$, the hydration patterns are observed to be different. In the particular case of $H_2PO_4^-$, the distributions of water hydrogen appear around the negatively charged oxygen atoms in

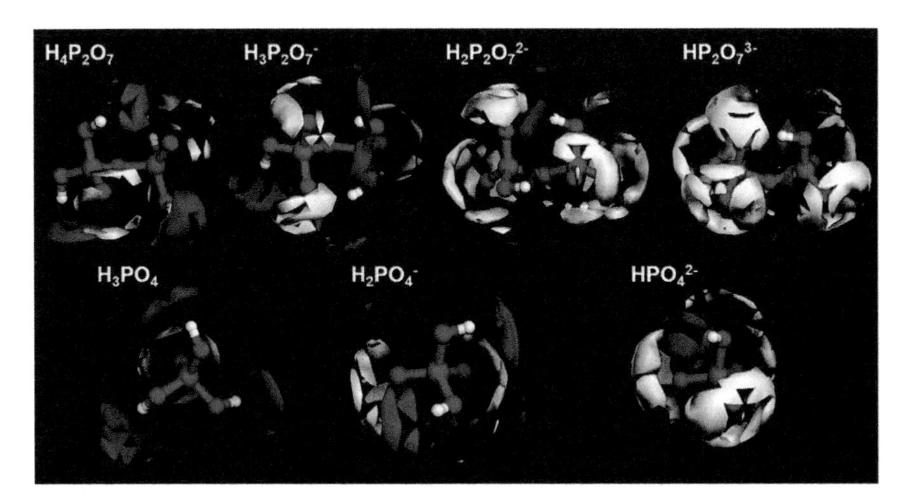

Fig. 5.1 Spatial distribution of solvent water around the solute species. The distributions of oxygen and hydrogen of water are colored red and gray, respectively. (Reprinted with permission from Hong et al. (2012). Copyright 2012 American Chemical Society.)

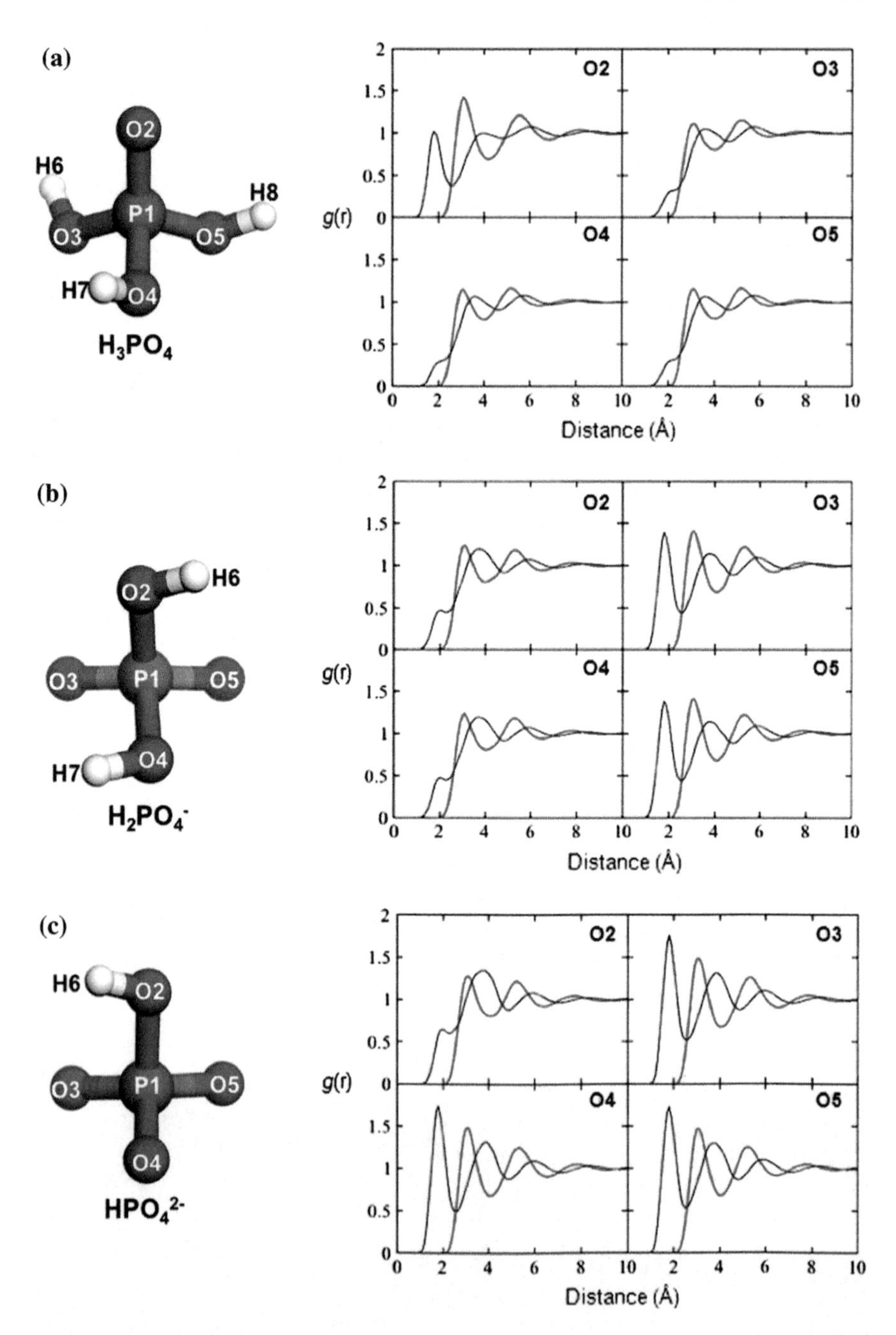

Fig. 5.2 Radial distribution function of solvent water around the phosphate oxygen of product species. The red and black curves denote the distribution functions of oxygen and hydrogen of water, respectively. (Reprinted with permission from Hong et al. (2012). Copyright 2012 American Chemical Society.)

$H_2PO_4^-$. In addition, the excess charge on $H_2PO_4^-$ is more localized than on $H_3P_2O_7^-$, making the solute–solvent electrostatic interaction for $H_2PO_4^-$ stronger than that of $H_3P_2O_7^-$. The difference in hydration patterns between the reactants and products, because of the short-range and electrostatic interactions in reaction B, converts the reaction from endothermic in the gas phase to exothermic in the aqueous phase.

When the phosphates have multiple excess charges as in HPO_4^{2-}, $H_2P_2O_7^{2-}$, and $HP_2O_7^{3-}$, the solvation structures are dramatically changed, as seen in Fig. 5.1, in which conspicuous peaks of water hydrogen are found around solute oxygen atoms. Because these phosphates have multiple excess charges, the long-range electrostatic interactions become dominant in determining their hydration free energies. This phenomenon can be explained by the simple Born model for the solvation free energy of an ion. Namely, the solvation free energy is roughly proportional to the square of the net charge of the solute molecule. Based on the model, the highly charged reactant species are stabilized by solvation more than the product state. Consequently, reactions C and D become "moderately exothermic" in water.

5.5 Hydrolysis of ATP

In this section, an application of the 3D-RISM theory to the ATP hydrolysis reaction is reviewed. In this hydrolysis reaction, the tetravalent anion ATP is assumed to be the reactant, which corresponds to the reaction in ambient condition. The reaction is depicted in Scheme 5.2. Four types of calculations were performed, namely B3LYP/6-31+G(d), B3LYP/6-31++G(d,p), M06-2X/6-31+G(d), and M06-2X/6-31++G(d,p) in aqueous phases. The geometry optimization for all reactant and product species were performed in the aqueous phase by the DFT/PCM method, whereas the free energy calculations were performed by 3D-RISM-SCF.

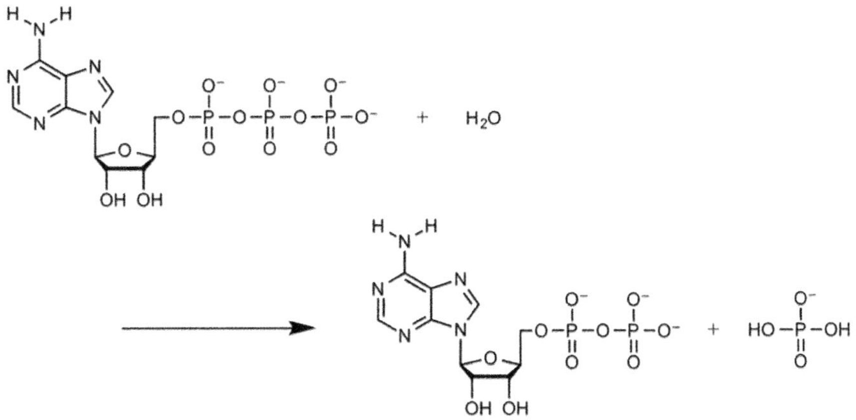

Scheme 5.2 Schematic description of the hydrolysis reaction of ATP^{4-} into ADP^{3-}

The initial structures of ATP and ADP were taken from the protein data bank (PDB) code 5LQZ (Vinothkumar et al. 2016). The GAFF parameter set was used for ATP, ADP, and phosphate; and the same solvent parameters were employed with previous calculations (Hong et al. 2012).

The reaction free energy in aqueous phase and its components are summarized in Table 5.3. The reaction free energies are evaluated by

$$\Delta G = G^{\text{product}} - G^{\text{reactant}}, \tag{5.18}$$

where G^{product} and G^{reactant} are the free energy of the product and reactant states given by Eq. (5.9), respectively. Therefore, the reaction free energy can be split into the electronic energy and solvation free energy contributions as follows.

$$\Delta G = \Delta E_{\text{solute}} + \Delta\Delta\mu \tag{5.19}$$

$$\Delta E_{\text{solute}} = E_{\text{solute}}^{\text{product}} - E_{\text{solute}}^{\text{reactant}} \tag{5.20}$$

$$\Delta\Delta\mu = \Delta\mu^{\text{product}} - \Delta\mu^{\text{reactant}} \tag{5.21}$$

Here, the kinetic energy contribution is ignored, as previous studies have shown that the value has only a minor influence on the result. All the computed values are in good agreement with the experimental value of −10.7 kcal/mol (George et al. 1970). As seen in the case of the hydrolysis reaction of pyrophosphate, ATP hydrolysis shows similar behavior; namely, the solute electronic energy makes a large negative contribution to the free energy, whereas the solvation free energy makes a large positive contribution. It can be explained by the same mechanism with the reaction of the multiple-charged pyrophosphate. Because the reactant ATP^{4-} has the multiple excess charges, the electrostatic repulsion decreased and stabilized by splitting into ADP^{3-} and $H_2PO_4^-$. On the other hand, the contribution of the solvation free energy to the reaction free energy became positive because the reactant ATP^{4-} stabilized more than the product ADP^{3-} and $H_2PO_4^-$ because of the electrostatic interactions which are roughly proportional to square of excess charges (see Table 5.4). In addition, the difference in solvation free energy between the reactants and the products is smaller than that inferred from the Born model. Such a difference may be attributed to the local solute–solvent interactions such as hydrogen bond.

The spatial distributions of water oxygen and hydrogen around the solute species are shown in Fig. 5.3. The conspicuous peaks of both oxygen and hydrogen can be

Table 5.3 Reaction free energies in the aqueous phase and the components computed by DFT/3D-RISM-SCF calculations. Units are in kcal/mol

Functional/basis set	ΔG	ΔE_{solute}	$\Delta\Delta\mu$
B3LYP/6-31+G(d)	−10.6	−167.5	156.9
B3LYP/6-31++G(d,p)	−11.0	−167.7	156.7
M06-2X/6-31+G(d)	−9.4	−167.1	157.8
M06-2X/6-31++G(d,p)	−9.6	−166.9	157.3

Table 5.4 Solvation free energy of reactant and product state. Units are in kcal/mol

Functional/basis set	$\Delta\mu^{\text{reactant}}$	$\Delta\mu^{\text{product}}$
B3LYP/6-31+G(d)	−726.0	−569.1
B3LYP/6-31++G(d,p)	−725.1	−568.5
M06-2X/6-31+G(d)	−738.4	−580.6
M06-2X/6-31++G(d,p)	−737.5	−580.2

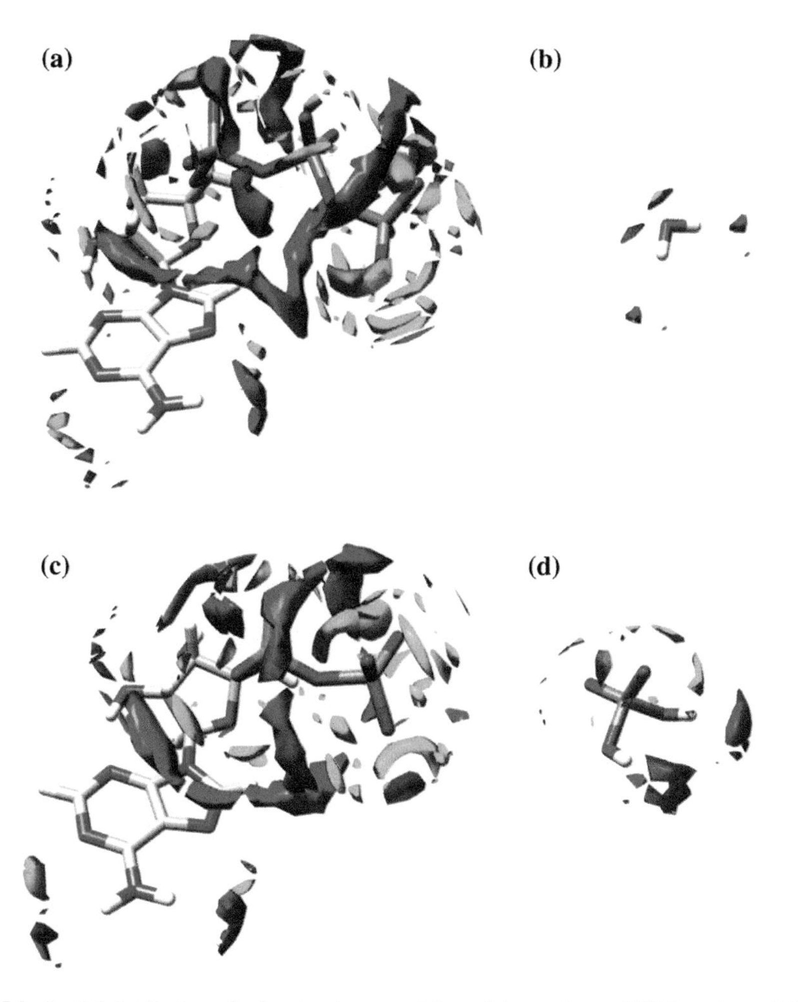

Fig. 5.3 Spatial distribution of solvent water around the solute species. **a** ATP, **b** water, **c** ADP, and **d** phosphate. Distributions of oxygen and hydrogen of water, $g(\boldsymbol{r}) > 4$, are colored red and green, respectively

found around the phosphates in ATP^{4-}, ADP^{3-}, and $H_2PO_4^-$. These peaks are attributed to the strong hydrogen bonds formed between the phosphate and water (see Fig. 5.2). In contrast, there are less conspicuous peaks around the adenine group of $ATP^{4-}$and ADP^{3-}, which may also affect the reaction free energy, but to a lesser extent.

5.6 Summary

In this chapter, applications of the 3D-RISM-SCF theory to investigate the hydrolysis reaction of ATP and pyrophosphate in aqueous solution were reviewed.

First, the hydrolysis reactions of four different dissociated states of the pyrophosphate studied experimentally by George were reviewed (George et al. 1970). In the gas phase, the results give values of reaction free energy ranging from extremely large and positive to unrealistically large and negative, depending on the dissociated states of the phosphate, which evidently contradict the experimental results. The 3D-RISM-SCF theory, on the other hand, predicts experimental results of the reaction free energy almost quantitatively for all the four dissociated states of the phosphates, indicating that the solvent plays a crucial role in determining the energy values stabilizing both the reactants and products appropriately. The results also demonstrate the importance of the hydrogen bond between the phosphate and water to make the reaction moderately exothermic, at about 8 kcal/mol.

The study of the hydrolysis reaction of ATP^{4-} into ADP^{3-} by the 3D-RISM-SCF theory was also reviewed. The computed hydration free energy shows good agreement with the experiment. The mechanism of determining the hydration free energy is essentially the same as for pyrophosphate. Therefore, the results given in this part verify the universality of the result for pyrophosphate.

The studies reviewed in the chapter provide unequivocal proof that the origin of the energy produced by the hydrolysis reaction of ATP is the balance of the hydration and electrostatic interaction and not only from the high-energy bond of the phosphate.

The theory used here, 3D-RISM-SCF, is a powerful tool to tackle the problems on chemical reactions in solution. It can produce accurate physical properties and chemical quantities at reasonable computational cost. For instance, the calculation of the free energy and solvation structure of ATP at the B3LYP/6-31++G(d,p) level can be done in about a day on a single node commodity computer that has two Xeon X5677 3.47 GHz processors. Therefore, the advanced theory based on the 3D-RISM-SCF, such as QM/MM/RISM or FMO/3D-RISM, is expected to realize the treatment of ATP hydrolysis reaction in biological environment using massively parallel super computer systems (Maruyama et al. 2014; Yoshida 2014; Yoshida et al. 2011).

References

Akola J, Jones R (2003) Atp hydrolysis in water—a density functional study. J Phys Chem B 107(42):11774–11783

Andersen H, Chandler D (1972) Optimized cluster expansions for classical fluids. 1. General theory and variational formulation of mean spherical model and hard-sphere percus-yevick equations. J Chem Phys 57(5):1918–1929

Andersen H, Chandler D, Weeks J (1972) Optimized cluster expansions for classical fluids. 3. Applications to ionic solutions and simple liquids. J Chem Phys 57(7):2626–2631

Aqvist J, Warshel A (1993) Simulation of enzyme-reactions using valence-bond force-fields and other hybrid quantum-classical approaches. Chem Rev 93(7):2523–2544

Beglov D, Roux B (1996) Solvation of complex molecules in a polar liquid: an integral equation theory. J Chem Phys 104(21):8678–8689

Beglov D, Roux B (1997) An integral equation to describe the solvation of polar molecules in liquid water. J Phys Chem B 101(39):7821–7826

Boyd DB, Lipscomb WN (1969) Electronic structures for energy-rich phosphates. J Theor Biol 25(3):403–420

Chandler D, Andersen HC (1972) Optimized cluster expansions for classical fluids. 2. Theory of molecular liquids. J Chem Phys 57(5):1930–1937

Colvin M, Evleth E, Akacem Y (1995) Quantum-chemical studies of pyrophosphate hydrolysis. J Am Chem Soc 117(15):4357–4362

Field MJ, Bash PA, Karplus M (1990) A combined quantum-mechanical and molecular mechanical potential for molecular-dynamics simulations. J Comput Chem 11(6):700–733

Gao JL, Xia XF (1992) A priori evaluation of aqueous polarization effects through monte-carlo qm-mm simulations. Science 258(5082):631–635

George P, Witonsky RJ, Trachtma M, Wu C, Dorwart W, Richman L, Richman W, Shurayh F, Lentz B (1970) Squiggle-H_2O—an enquiry into importance of solvation effects in phosphate ester and anhydride reactions. Biochim Biophys Acta 223(1):1–15

Grigorenko BL, Rogov AV, Nemukhin AV (2006) Mechanism of triphosphate hydrolysis in aqueous solution: QM/MM simulations in water clusters. J Phys Chem B 110(9):4407–4412

Hammond C, Kartenbeck J, Helenius A (1992) Effects of dithiothreitol on beta-cop distribution and golgi to er membrane traffic. Mol Biol Cell 3:A35–A35

Hansen JP, McDonald IR (2006) Theory of simple liquids. Academic Press, Amsterdam

Hill TL, Morales MF (1951) On high energy phosphate bonds of biochemical interest. J Am Chem Soc 73(4):1656–1660

Hirata F (2003) Molecular theory of solvation. Kluwer, Dordrecht

Hofmann KP, Zundel G (1974) Large hydration structure changes on hydrolyzing atp. Experientia 30(2):139–140

Hong J, Yoshida N, Chong S-H, Lee C, Ham S, Hirata F (2012) Elucidating the molecular origin of hydrolysis energy of pyrophosphate in water. J Chem Theory Comput 8:2239–2246

Kamerlin SCL, Warshel A (2009) On the energetics of atp hydrolysis in solution. J Phys Chem B 113(47):15692–15698

Kido K, Kasahara K, Yokogawa D, Sato H (2015) A hybrid framework of first principles molecular orbital calculations and a three-dimensional integral equation theory for molecular liquids: Multi-center molecular ornstein-zernike self-consistent field approach. J Chem Phys 143(1):014103

Klaehn M, Rosta E, Warshel A (2006) On the mechanism of hydrolysis of phosphate monoesters dianions in solutions and proteins. J Am Chem Soc 128(47):15310–15323

Kovalenko A, Hirata F (1998) Three-dimensional density profiles of water in contact with a solute of arbitrary shape: a RISM approach. Chem Phys Lett 290(1–3):237–244

Kovalenko A, Hirata F (1999) Self-consistent description of a metal-water interface by the kohn-sham density functional theory and the three-dimensional reference interaction site model. J Chem Phys 110(20):10095–10112

Kovalenko A, Hirata F (2001) First-principles realization of a van der waals-maxwell theory for water. Chem Phys Lett 349(5–6):496–502

Lipmann F (1941) Metabolic generation and utilization of phosphate bond energy. Adv Enzymol Rel S Bi 1:99–162

Maruyama Y, Yoshida N, Tadano H, Takahashi D, Sato M, Hirata F (2014) Massively parallel implementation of 3D-RISM calculation with volumetric 3D-FFT. J Comput Chem 35(18): 1347–1355

Meyerhof O, Lohmann K (1932) Energetic exchange connections amongst the volume of phosphoric acetic acid in muscle extracts. Biochem Z 253:431–461

Nishihara S, Otani M (2017) Hybrid solvation models for bulk, interface, and membrane: reference interaction site methods coupled with density functional theory. Phys Rev B 96(11)

Sato H (2013) A modern solvation theory: quantum chemistry and statistical chemistry. Phys Chem Chem Phys 15(20):7450–7465

Sato H, Kovalenko A, Hirata F (2000) Self-consistent field, ab initio molecular orbital and three-dimensional reference interaction site model study for solvation effect on carbon monoxide in aqueous solution. J Chem Phys 112(21):9463–9468

Sinnecker S, Rajendran A, Klamt A, Diedenhofen M, Neese F (2006) Calculation of solvent shifts on electronic g-tensors with the conductor-like screening model (COSMO) and its self-consistent generalization to real solvents (direct COSMO-RS). J Phys Chem A 110(6): 2235–2245

Takahashi H, Kawashima Y, Nitta T, Matubayasi N (2005) A novel quantum mechanical/ molecular mechanical approach to the free energy calculation for isomerization of glycine in aqueous solution. J Chem Phys 123(12):124504

Takahashi H, Matubayasi N, Nakahara M, Nitta T (2004) A quantum chemical approach to the free energy calculations in condensed systems: the QM/MM method combined with the theory of energy representation. J Chem Phys 121(9):3989–3999

Takahashi H, Umino S, Miki Y, Ishizuka R, Maeda S, Morita A, Suzuki M, Matubayasi N (2017) Drastic compensation of electronic and solvation effects on atp hydrolysis revealed through large-scale QM/MM simulations combined with a theory of solutions. J Phys Chem B 121(10): 2279–2287

Ten-no S, Hirata F, Kato S (1993) A hybrid approach for the solvent effect on the electronic-structure of a solute based on the RISM and hartree-fock equations. Chem Phys Lett 214(3–4):391–396

Ten-no S, Hirata F, Kato S (1994) Reference interaction site model self-consistent-field study for solvation effect on carbonyl-compounds in aqueous-solution. J Chem Phys 100(10):7443–7453

Tomasi J, Mennucci B, Cammi R (2005) Quantum mechanical continuum solvation models. Chem Rev 105(8):2999–3093

Tomasi J, Persico M (1994) Molecular-interactions in solution—an overview of methods based on continuous distributions of the solvent. Chem Rev 94(7):2027–2094

Vinothkumar KR, Montgomery MG, Liu S, Walker JE (2016) Structure of the mitochondrial atp synthase from pichia angusta determined by electron cryo-microscopy. Proc Nat Acad Sci USA 113(45):12709–12714

Wang C, Huang WT, Liao JL (2015) QM/MM investigation of atp hydrolysis in aqueous solution. J Phys Chem B 119(9):3720–3726

Yamamoto T (2010) Preferred dissociative mechanism of phosphate monoester hydrolysis in low dielectric environments. Chem Phys Lett 500(4–6):263–266

Yokogawa D, Sato H, Sakaki S (2007) New generation of the reference interaction site model self-consistent field method: introduction of spatial electron density distribution to the solvation theory. J Chem Phys 126(24):244504

Yoshida N (2014) Efficient implementation of the three-dimensional reference interaction site model method in the fragment molecular orbital method. J Chem Phys 140(21):214118

Yoshida N, Imai T, Phongphanphanee S, Kovalenko A, Hirata F (2009) Molecular recognition in biomolecules studied by statistical-mechanical integral-equation theory of liquids. J Phys Chem B 113(4):873–886
Yoshida N, Kato S (2000) Molecular ornstein-zernike approach to the solvent effects on solute electronic structures in solution. J Chem Phys 113(12):4974–4984
Yoshida N, Kiyota Y, Hirata F (2011) The electronic-structure theory of a large-molecular system in solution: application to the intercalation of proflavine with solvated dna. J Mol Liq 159(1): 83–92

Chapter 6
A Solvent Model of Nucleotide–Protein Interaction—Partition Coefficients of Phosphates Between Water and Organic Solvent

Hideyuki Komatsu

Abstract In attempt to experimentally evaluate hydration/solvation of phosphoric compounds in aqueous solution, partitioning of phosphoric compounds from an aqueous solution to an organic solvent has been quantified. Transfer of phosphates from an aqueous solution into octanol was greatly enhanced by addition of alkylamine as an amphiphilic extractant. This alkylamine/octanol system exhibited amine basicity, and thus the pH of solution was controlled by equilibration with buffer. Further, enthalpy changes of the transfers of ATP and ADP from water to the alkylamine/octanol were estimated from van't Hoff analysis, and these enthalpy changes depended on ionization enthalpy of buffer. This result suggests that the transfers are accompanied with protonation of phosphoric ions and deprotonation of alkylamine. Finally, the partition coefficients of ATP, ADP, AMP, and P_i were estimated under the pH-controlled condition at 25 °C. The partition coefficients depended on the pH of aqueous phase and the net charge of phosphoric compounds. Therefore, the transfer is likely to be determined by electrostatic interaction between phosphoric ion and amine. The solvent system with a nucleotide-uptake capacity may partly mimic the function of ATP-binding proteins.

Keywords Partition coefficient • Phosphoric compound • Octanol
Alkylamine • Amphiphilic extractant

6.1 Introduction

ATP hydrolysis is an important exergonic reaction providing the Gibbs free energy for thermodynamically unfavorable, endergonic reactions in living systems. In addition to the resonance stabilization of a phosphoanhydride bond and the elec-

H. Komatsu (✉)
Faculty of Computer Science and Systems Engineering, Department of Bioscience and Bioinformatics, Kyushu Institute of Technology, Fukuoka, Japan
e-mail: komatsu@bio.kyutech.ac.jp

© Springer Nature Singapore Pte Ltd. 2018

M. Suzuki (ed.), *The Role of Water in ATP Hydrolysis Energy Transduction by Protein Machinery*, https://doi.org/10.1007/978-981-10-8459-1_6

trostatic repulsions between a charged phosphoanhydride groups, the difference of solvation energies between ATP and products (ADP and P_i) has been presumed to provide the thermodynamic driving force for the ATP hydrolysis. This explanation was originally proposed by George et al. (1970) and then appeared in a biochemistry textbook (Voet et al. 2013). This presumption has been supported by experimental results by de Meis (1984), de Meis et al. (1985), Remero and de Meis (1989). They reported that the favorable free energy of pyrophosphate (PP_i) hydrolysis was greatly reduced by the decrease in the water activity by adding cosolvents (glycerol, ethylene glycol, polyethylene glycol, and dimethyl sulfoxide) to the reaction medium, indicating a large contribution of solvation for free energy of hydrolysis. In addition, the favorable Gibbs energy change for hydrolysis step of enzyme-bound PP_i is much less than that in water (Springs et al. 1981). The favorable Gibbs energy of ATP hydrolysis step on myosin is also dramatically decreased compared to that in water (Bagshaw and Trentham 1974; Kodama and Woledge 1979). These decreases of hydrolysis energy on enzyme may be attributed to the waterless environment surrounded by alkyl and aromatic side chains of the enzyme proteins.

Because transfer process of phosphoric compounds from aqueous solution to organic solvent involves the dehydration of the phosphoric compounds, the partitioning of phosphoric compounds between aqueous solution and organic solvent can be used as an indicator of the hydration energies of phosphoric compounds. An experimental assessment by partition coefficient of phosphoric compounds between aqueous solution and organic solvent was proposed by Wolfenden and Williams (1985) and was used for estimation of free energy of hydrolysis of phosphoric anhydrides in organic solvent (i.e., wet chloroform). Stockbridge and Wolfenden also estimated distribution coefficients of phosphoric compounds from water to organic solvent to investigate hydrolysis kinetics in water-saturated organic solvent (Stockbridge and Wolfenden 2009, 2010). The distribution coefficient for neopentyl phosphate and dineopentyl phosphate from water to cyclohexane in the presence of excess tetrabutylammonium was 7.0×10^{-6} and 3.9×10^{-5} at 25 °C, respectively. In addition, sodium pyrophosphate was saturated at 2×10^{-4} M in ~1% (v/v) water-contained dimethyl sulfoxide (Stockbridge and Wolfenden 2011). Thus, only slight amount of phosphoric compounds transfers from water to organic solvent, due to the high polarity of phosphate group. These low ratios of partition make it difficult to precisely estimate partition coefficients of phosphoric compounds from water and organic solvent, because extraordinarily low amount of compounds in organic solvent should be quantified.

To overcome the difficulty, amphiphilic extractants were dissolved in organic solvent, which enhance the transfer of phosphoric compounds from water to organic solvent and increase the change in concentration by partition. Phosphoric ions are probably present as their salts of amphiphilic ions in organic solvent. Such a solvent system contains the amphiphilic ions in hydrophobic alkyl chains of the organic solvent, and thus it is similar to the waterless environment on the enzyme protein. The transfer of phosphoric compounds from water to organic solvent can be regarded as a model of interaction between protein and phosphoric compounds. Such an organic solvent/amphiphilic extractant system has been developed to extract phosphoric compounds such as nucleotides from aqueous solutions by use of alkylamine as an amphiphilic extractant. In this chapter, partition of nucleotides and phosphate from water to organic solvent/amphiphilic extractant is illustrated.

6.2 Measurement of Partition Coefficients of Phosphoric Compounds Between Water and Alkylamine/Octanol

One volume of aqueous solutions containing ATP, ADP, AMP, or P_i was mixed into one volume of organic solvents and placed at 25 °C for 2–3 h. After the organic phase was removed, the concentration of solute in the aqueous phase was assayed. Concentrations of nucleotides (ATP, ADP, and AMP) were determined by absorbance at 260 nm, and concentration of P_i was determined by Malachite Green method (Kodama et al. 1986). Notably, when ATP concentration was lower than $\sim 10^{-6}$ M due to the high partition coefficient, ATP concentration in aqueous phase was determined by a luciferin–luciferase assay. Amount of solute transferred to organic solvent was estimated by subtracting the amount of solute in aqueous phase from the total amount. Partition coefficient (K_P) was estimated from ratio of concentration of solute transferred into organic solvents (c_O) to that of retained in aqueous solution (c_W) as follows.

$$K_p = \frac{c_o}{c_w} \tag{6.1}$$

6.3 Enhancement of Transfer of Phosphoric Compounds from Water to Organic Solvent by Alkylamine

6.3.1 Effect of Alkylamine on Partition of Phosphoric Compounds from Water to Organic Solvent

More than 60 years ago, alkylamine was used as extractants for solvent extraction of phosphoric compounds from aqueous solution or biological materials (Plaut et al. 1950). According to this report, the partition coefficients of pyrophosphate and ATP between water and octadecylamine-dissolved *n*-octanol were more than 40. In search of better solvent/extractant system for quantitative estimation of partition coefficients, octadecylamine and hexadecylamine were dissolved in higher alcohols (*n*-butanol, *n*-pentanol, *n*-hexanol, and *n*-octanol) and the alkylamine/alcohol solvents were tested for the transfers of ATP, ADP, and P_i under *pH-uncontrolled* condition in comparison with chloroform (Fig. 6.1). The results show that octadecylamine and hexadecylamine markedly increased the transfer of ATP, ADP, and P_i from water to the organic solvent. The increases in the transfer by hexadecylamine were higher than those by octadecylamine. Note that the water

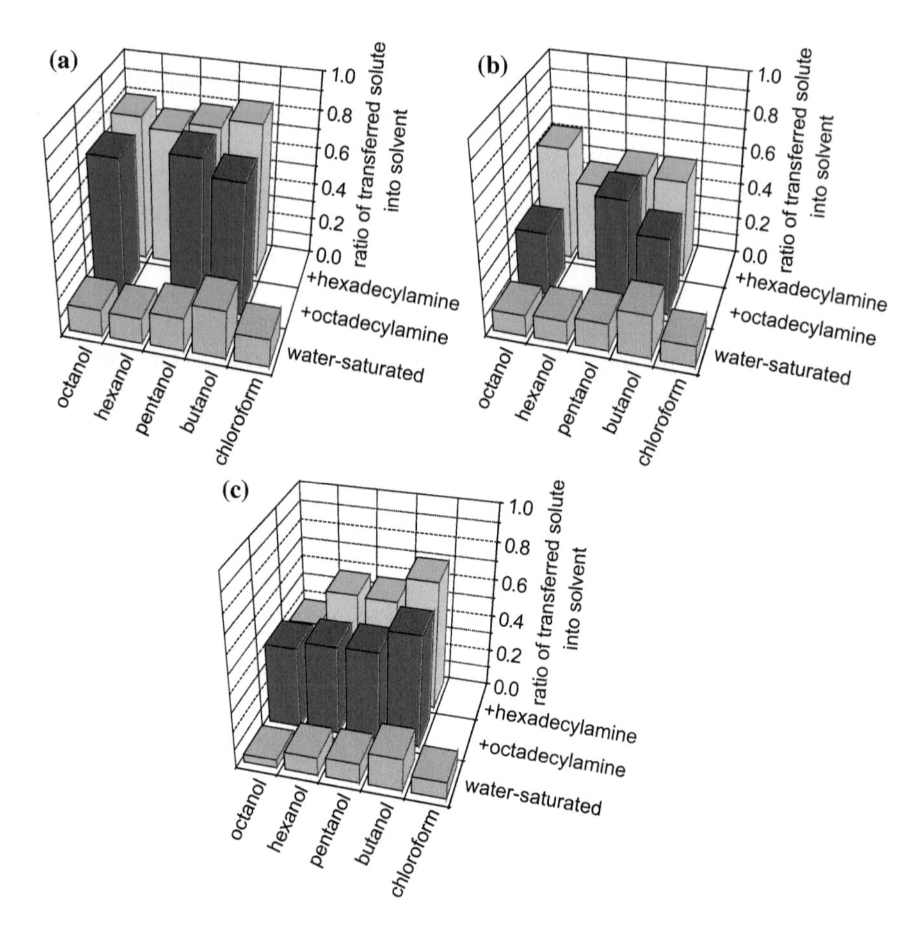

Fig. 6.1 Effect of alkylamine on transfer of phosphoric compounds (ATP, ADP, and P_i) to organic solvent, **a** ATP, **b** ADP, **c** P_i. Solvent saturated by water (*green bars*), solvent containing octadecylamine (*blue bars*), solvent containing hexadecylamine (*light blue bars*). Transfer of ATP, ADP, and P_i was tested for chloroform, butanol, pentanol, hexanol, and octanol in the absence and presence of 5% (w/v) octadecylamine or 5% (w/v) hexadecylamine. The pH of solutions was not controlled. One volume of aqueous solutions containing ATP, ADP, or P_i was mixed into one volume of organic solvents. The samples were placed at 25 °C for ca. 90 min, and the organic phases then were removed. Concentrations of ATP, ADP, and P_i in aqueous phase were determined as described in 6.2. Amount of solute transferred to organic solvent was estimated by subtracting the amount of solute in aqueous phase from the total amount

contents in water-saturated octanol were determined by using Karl Fischer moisture meters and were *ca.* 3.5% both in the absence and presence of alkylamine.

Figure 6.2 shows the effect of hexadecylamine concentration on the amounts of transferred solutes from water to hexadecylamine/octanol system at pH 8.5. The transfers of ATP and ADP apparently increased with concentration of hexadecylamine, and ATP was more transferred than ADP. Transfers of AMP and P_i were not observed in this condition, due to their low partition coefficients at pH 8.5.

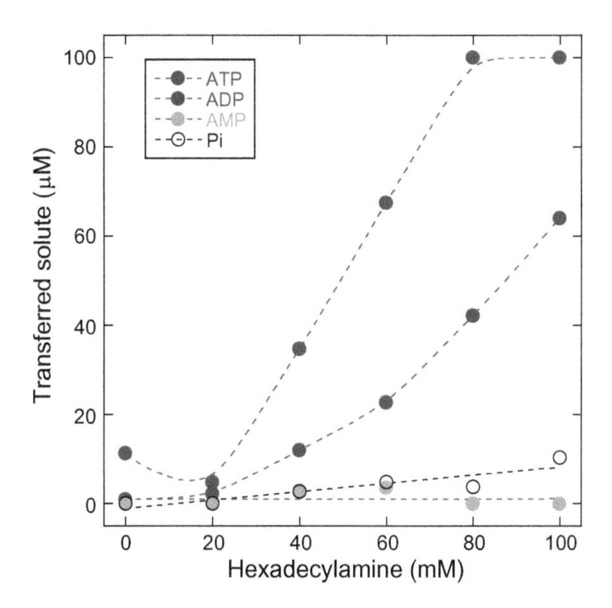

Fig. 6.2 Effect of hexadecylamine on the transfer of ATP (*red circles*), ADP (*blue circles*), AMP (*green circles*), and P_i (*open circles*) from aqueous solution (200 mM bicine–NaOH pH 8.5) to octanol. The initial concentration of each solute in aqueous solution was 100 mM

6.3.2 Homogenous Dispersion of Phosphoric Compounds in Alkylamine-Containing Organic Solvent

To determine whether phosphoric compounds adsorb on the water/octanol surface or absorb into hexadecylamine-containing octanol, fluorescent ATP analogue, 2′-(or 3′)-*O*-(*N*-methylanthraniloyl) ATP (mant-ATP), was used as a solute and its localization was visualized under UV light. As shown in Fig. 6.3, fluorescence was observed in the whole hexadecylamine-containing octanol, and therefore mant-ATP was homogeneously dispersed in the solvent.

Possibility of association of solute molecules in the organic solvent was checked by the partitioning across a wide range of solute concentrations. The concentrations of transferred phosphoric compounds to the solvent were plotted against those of remained solutes in water (Fig. 6.4). The linearities with slope of approximately

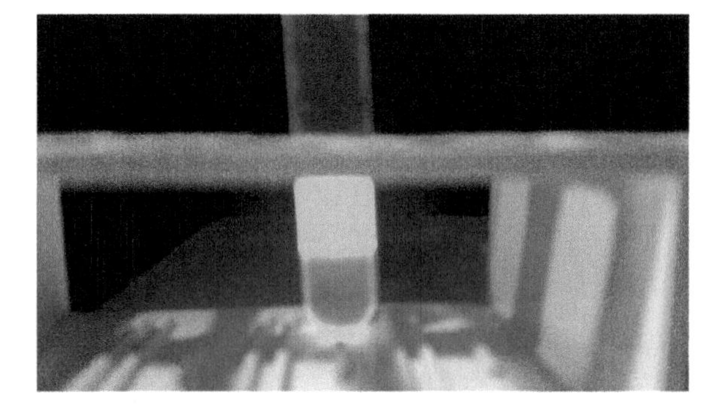

Fig. 6.3 Partitioning of fluorescent ATP analogue (mant-ATP) from aqueous to organic solvents. Mant-ATP (500 μM) was visualized under UV light

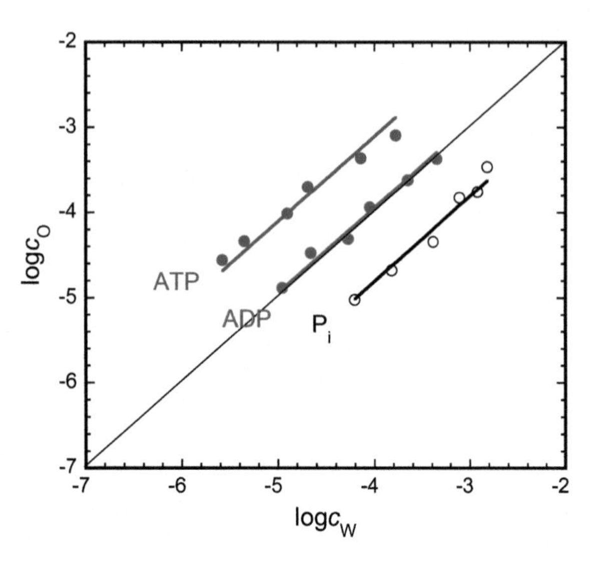

Fig. 6.4 Partitioning between water and octadecylamine/octanol of ATP (*red circles*), ADP (*blue circles*), and P_i (*open circles*) at various solute concentrations. c_O: *solute concentration transferred into organic solvents.* c_W: *solute concentration retained in aqueous solution*

1 for ATP, ADP, and P_i suggest that association of solute does not occur in the organic solvent.

6.4 Acid/Base Properties of Alkylamine-Containing Octanol Solvent System

Phosphate exists as four ionic forms depending on pH. On the other hand, alkylamine is base, and thus the pH of aqueous solution becomes basic. Therefore, pH of solution should be controlled for the accurate evaluation of the partitioning. In this section, acid/base properties of alkyl amine/octanol system are described.

6.4.1 Acid/Base Properties and pH Control of Alkylamine-Containing Octanol

The octadecylamine/octanol was titrated with HCl/isopropanol (Fig. 6.5). The titration curve indicated its basicity with a pK_a of 9.3. In fact, after mixing with alkylamine/octanol, the pH of aqueous solution (20 mM MOPS–NaOH, pH 7.0) increased to 8.5. The protonation of alkylamine appears to decrease H^+ concentration of aqueous solution.

In order to stabilize the pH of solution, alkylamine-containing octanol was equilibrated with buffer several times until pH became constant. Briefly, alkylamine

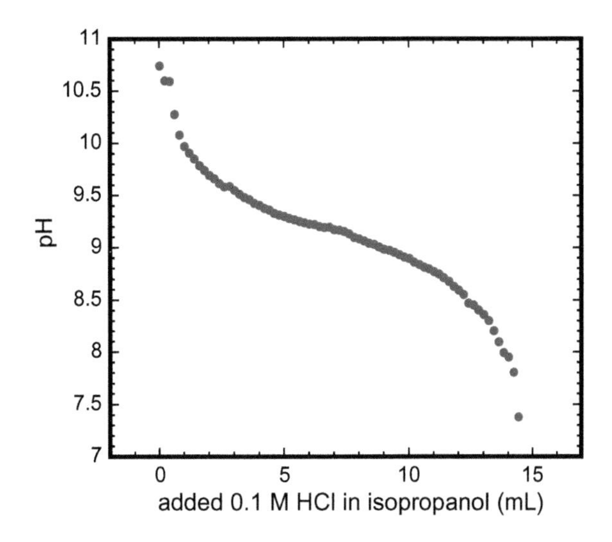

Fig. 6.5 pH titration curve of alkylamine in octanol. The octadecylamine/octanol was titrated with HCl/isopropanol. The pH was measured by using 6377-10D pH electrode (HORIBA) which is sensitive to low conductivity water and non-aqueous solvents. The pK_a was estimated as 9.3

was first dissolved in octanol at 100 mM and then mixed with equivalent volume of water. After liquid–liquid phase separation, the lower aqueous phase was discarded. This mixing–separating cycle was repeated three times to be saturated with water. Similarly, the water-saturated alkylamine/octanol was equilibrated with buffer solution by repeating mixing–separating cycle (mixing with buffer, liquid–liquid phase separation, and discarding the aqueous solution) until pH of aqueous phase was constant.

6.4.2 *Thermodynamic Analysis of the Protonation/Deprotonation Accompanying Transfer of Phosphoric Compounds Form Water to Alkylamine/Octanol*

Due to the basicity of alkylamine as described above, the transfer of phosphoric ions from water to the alkylamine/octanol may be involved in protonation of phosphoric ions and deprotonation of alkylamine. To investigate this possibility, van't Hoff enthalpies of phosphoric ion transfer were roughly estimated in the buffers with varying deprotonation enthalpy. If the transfer entails the uptake and release of proton, the observed van't Hoff enthalpy ($\Delta H_{\mathrm{VH,\ obs}}$) should be dependent on the ionization enthalpy of buffer (ΔH_{i}) (Eq. 6.2)

$$\Delta H_{\mathrm{VH,obs}} = \Delta H_0 + n\Delta H_{\mathrm{i}} \tag{6.2}$$

where ΔH_0 is the intrinsic transfer enthalpy and n is the number of proton uptake to buffer.

The K_P values of ATP and ADP from water to hexadecylamine/octanol were measured at 20–40 °C in four different kinds of buffer solution (Fig. 6.6). Significance difference of K_P values between the buffers was not observed. On the other hand, the slopes of van't Hoff plots (reciprocal temperature versus $\ln K_P$) were different in the buffers. The van't Hoff enthalpy ($\Delta H_{VH,\ obs}$) was estimated by Eq 6.3 and plotted against deprotonation enthalpy of buffer (Fig. 6.7)

$$\Delta H_{VH,\mathrm{obs}} = -R\left[\frac{\Delta \ln K_P}{\Delta(1/T)}\right] \tag{6.3}$$

where R is the gas constant and T is the temperature. The plots show that the observed van't Hoff enthalpies roughly linearly related to the buffer deprotonation

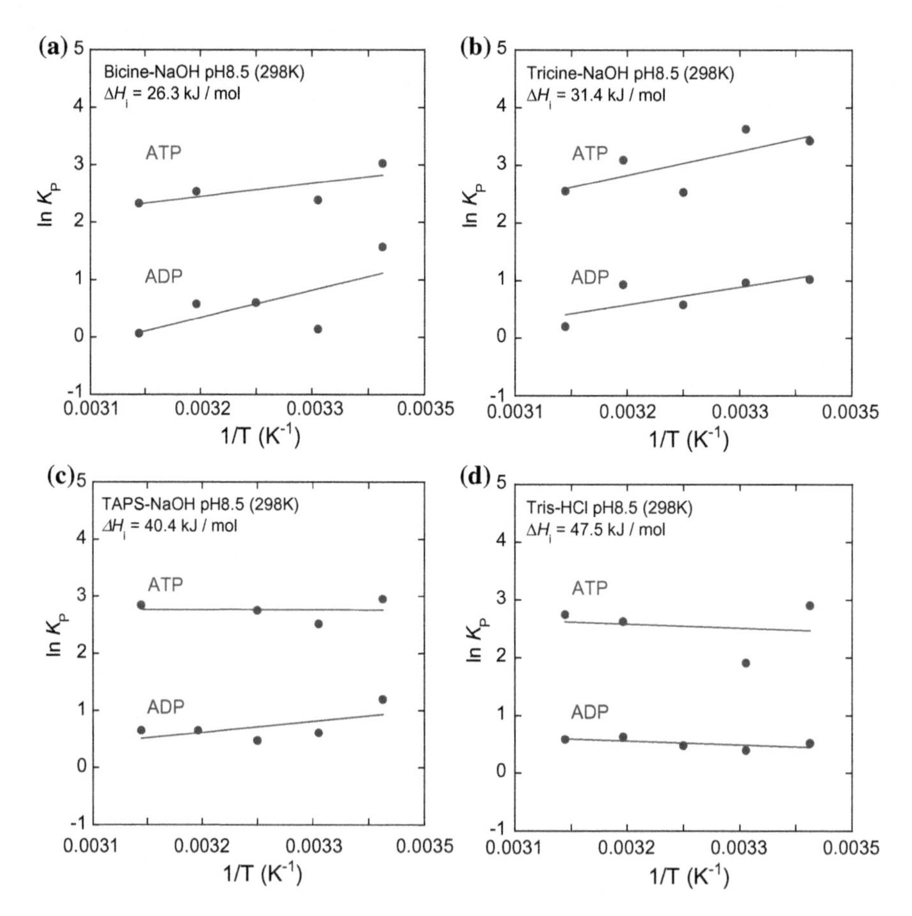

Fig. 6.6 van't Hoff analysis of the partitioning of ATP (*red circles*) and ADP (*blue circles*) from water to hexadecylamine/octanol at 20–40 °C in four different buffer solutions, **a** bicine–NaOH, **b** tricine–NaOH, **c** TAPS–NaOH, **d** Tris-HCl. The ionization enthalpies of buffer were obtained from Goldberg et al. (2002)

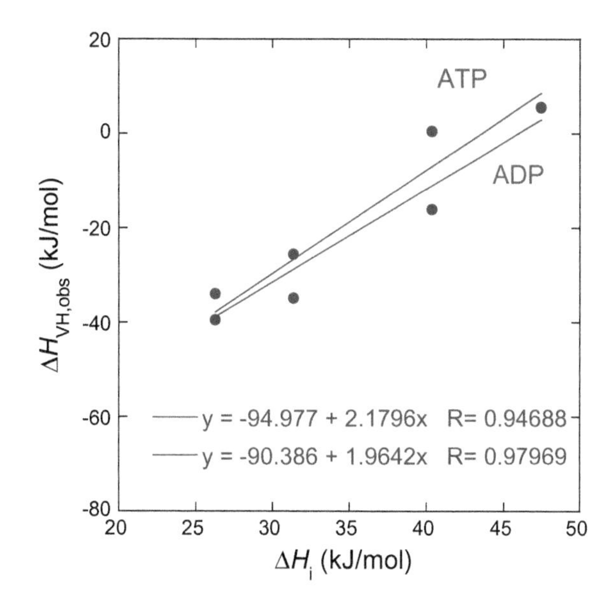

Fig. 6.7 Relationship between ionization enthalpy of buffer and van't Hoff enthalpy of the partitioning of ATP (*red circles*) and ADP (*blue circles*) from water to hexadecylamine/octanol. The data points of ATP and ADP are overlapped at 47.5 kJ/mol of ΔH_{i}

enthalpy with reasonable values of slope (n) (2.18 for ATP and 1.96 for ADP). Therefore, the transfers of ATP and ADP from water to the alkylamine/octanol are probably accompanied with protonation of phosphoric ion and deprotonation of alkylamine. In the case of ATP transfer at pH 8.3, protonated alkylamine molecules (RNH_3^+) in organic phase donate one or more protons to ATP^{4-} in aqueous phase, and ATP is transferred to organic phase as protonated ATP ($ATP \cdot nH^{(4-n)-}$)

$$ATP^{4-}(aq) + n\,RNH_3^+(or) \rightleftharpoons ATP \bullet nH^{(4-n)-}(or) + n\,RNH_2(or) \qquad (6.4)$$

The protonation and deprotonation of alkylamine and buffer were balanced at the equilibrium (see Sect. 6.4.1), and the ionization enthalpy of buffer (ΔH_{i}) appeared to come from this reaction.

$$RNH_2(or) + buffer \bullet H\ (aq) \rightleftharpoons RNH_3^+(or) + buffer^-(aq) \qquad (6.5)$$

6.5 Partition Coefficients of Phosphoric Compounds Between Aqueous Solution and Organic Solvent

Taking into account the properties of alkylamine/octanol system mentioned above, the K_{P} values of ATP, ADP, AMP, and P_{i} from aqueous to the organic phases were measured in the carefully controlled condition. These K_{P} values and the transfer energies ($\Delta G_{\mathrm{tr}} = -RT\ln K_{\mathrm{P}}$) varied by pH of solution and net charges of phosphoric compounds (Table 6.1). The K_{P} values at pH 7.2 were approximately 30- to 50-fold

Table 6.1 Partition coefficients and transfer energies of ATP, ADP, AMP, and P_i from aqueous to organic phases

pH	Solute	Net charge	K_P	ΔG_{tr}[a] kJ · mol^{-1}
8.3	ATP	−4	84.9	−11.0
	ADP	−3	13.3	−6.41
	AMP	−2	0.473	1.85
	P_i	−2	0.623	1.17
7.2	ATP	−3.5	3010	−19.8
	ADP	−2.5	440	−15.1
	AMP	−1.5	24.4	−7.91
	P_i	−1.5	30.9	−8.50

[a] $\Delta G_{tr} = -RT\ln K_P$

larger than those of pH 8.3, and the K_P value for ATP was comparable to those of submillimolar dissociation constant for protein–ligand interaction. Because the alkylamine with a pK_a of 9.3 is more protonated at lower pH, the protonation of alkylamine seems to be crucial to interact with the phosphate ions and to transfer the phosphate ions to the organic phase.

In addition, the rank order of the K_P was ATP > ADP > AMP ≈ P_i, in the order of the number of phosphate group both at pH 8.3 and 7.2. Four ionic forms of phosphate depend on pH, and their net charges are reasonably pH-dependent. Thus, the ΔG_{tr} values for each compound were plotted as a function of the net charge (Fig. 6.8). The ΔG_{tr} values were more favorable with increasing the net charge both at pH 8.3 and 7.2, indicating that the negative charge of phosphoric compounds is an important factor for the transfer.

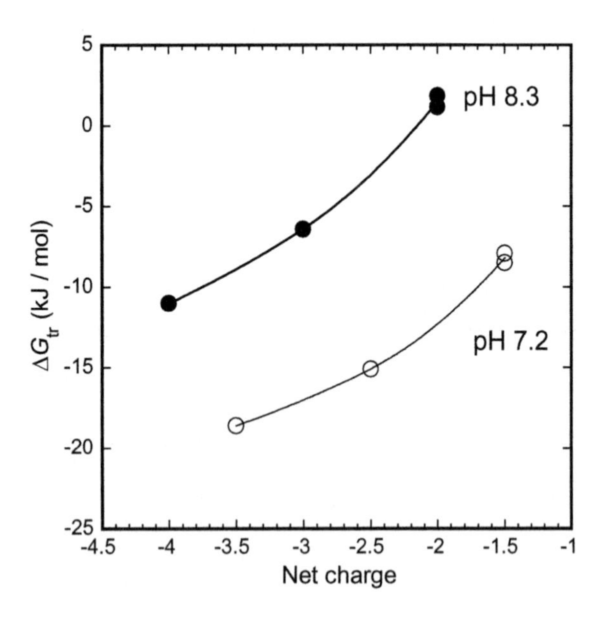

Fig. 6.8 Plot of transfer energy as function of net charge of phosphoric compound in aqueous solution at pH 7.2 (*open circles*) and pH 8.3 (*closed circles*)

Taken together, the electrostatic interaction is likely to be dominant driving force for transfer from aqueous to organic phases. The ΔG_{tr} values at pH 7.2 are substantially more favorable than those at pH 8.3. Thus, the protonation of alkylamine is probably more critical for the transfer than net charge of phosphoric compounds.

6.6 Discussion and Conclusion

Because the transfers of phosphoric compounds from water to the alkylamine/octanol are involved in the electrostatic interaction, the ΔG_{tr} values are considered to include not only solvation energy but also electrostatic energy between alkylamine and phosphate. Thus, the ΔG_{tr} values *should not* be interpreted as simple solvation energies of phosphoric compounds. Experimental separation of these two energies is difficult, and evolved strategies are needed to devise. On the other hand, in the recent QM/MM simulation of ATP and PP_i hydrolysis, the solvation energies of phosphoric compounds are distinguished from the electronic energies (Takahashi et al. 2017).

In spite of such difficulty in the interpretation of the ΔG_{tr} values, the difference of the transfer energy between ATP and its hydrolyzing products (ADP and P_i) ($\Delta\Delta G_{tr}$) was estimated (Eq. 6.6).

$$\Delta\Delta G_{tr} = (\Delta G_{tr,\,ADP} + \Delta G_{tr,\,Pi}) - \Delta G_{tr,\,ATP} \tag{6.6}$$

where $\Delta G_{tr,ATP}$, $\Delta G_{tr,ADP}$, and $\Delta G_{tr,Pi}$ are the transfer energies of ATP, ADP, and P_i, respectively. The $\Delta\Delta G_{tr}$ were estimated as $3.8\ kJ \cdot mol^{-1}$ at pH 7.2 and $-5.8\ kJ \cdot mol^{-1}$ at pH 8.3. These values seem to be considerably smaller than the Gibbs energy of ATP hydrolysis in biological standard state ($\sim -30\ kJ \cdot mol^{-1}$) (Alberty and Goldberg 1992). The free energy of ATP hydrolysis may not be explained by the difference of solubility to alkylamine-containing organic solvent between ATP and its hydrolyzing products. This interpretation appears to differ from the result that the free energy of PP_i hydrolysis was drastically decreased with deceasing water activity (concentration) by adding cosolvents (de Meis 1984; de Meis et al. 1985; Remero and de Meis 1989). However, the free energy of hydrolysis of phosphoric compounds is highly sensitive to pH, magnesium ion concentration, and ion strength (Alberty and Goldberg 1992). Therefore, the difference in these conditions between the previous literature and this study should be taken into consideration and be carefully examined.

The binding of nucleotide to myosin is known to be associated with liberation of protons in the presence of magnesium ion (Bagshaw and Trentham 1974; Chock and Eisenberg 1974; Koretz and Taylor 1975; Kodama 1981). In contrast, in the absence of divalent cations, the ADP-binding of myosin was accompanied with absorption of protons (Kardami et al. 1979). This report indicated that the protonation of alkylamine accompanied the transfers of ATP and ADP from water to alkylamine/octanol solvent system in the absence of divalent cation (see

Sect. 6.4.2). Thus, this solvent system could mimic the nucleotide-binding-induced proton absorption of myosin in the absence of divalent cation. The standard Gibbs energy of deprotonated ATP hydrolysis is positively small (3.0 $kJ \cdot mol^{-1}$) and much larger than that of protonated ATP (−36 $kJ \cdot mol^{-1}$) (Alberty and Goldberg 1992). At present, the proton of ATPase-bound nucleotide cannot be visualized in crystal structures. Thus, analysis of the protonation/deprotonation state of ATPase-bound nucleotide (e.g., a neutron diffraction analysis) is necessary to discuss energetics of ATP hydrolysis.

From a viewpoint of applied research, the alkylamine/octanol solvent system could be applied to an extraction of phosphoric compounds from polluted water and to a reaction solvent of phosphoric compound. In fact, extractions of metal ions from an aqueous solution into an organic solvent with an amphiphilic extractant are useful in applications to environmental and industrial fields (Jiang and Jia 2008; Chen et al. 2009; Ellis et al. 2012; Bu et al. 2014; Pal et al. 2015).

In conclusion, this study has successfully measured the partition coefficients of phosphoric compounds between aqueous solution and organic solvent (octanol) by employing alkylamine as amphiphilic extractants in the pH-controlled condition. The transfer of phosphoric compounds is mainly driven by the electrostatic interaction between phosphoric compounds and alkylamine. This solvent system contains the amphiphilic ions in hydrophobic alkyl chains like proteins and exhibits the high uptake capacity for phosphoric compounds and the nucleotide-transfer-induced deprotonation. Thus, the solvent may partly mimic the nucleotide-binding function of ATPase enzymes. Further improvement of solvent system may be helpful for examining the solvation energy of phosphoric compound and in future applying to environmental and/or industrial fields.

References

Alberty RA, Goldberg RN (1992) Standard thermodynamic formation properties for the adenosine 5′-triphosphate series. Biochemistry 31(43):10610–10615

Bagshaw CR, Trentham DR (1974) The characterization of myosin-product complexes and of product-release steps during the magnesium ion-dependent adenosine triphosphatase reaction. Biochem J 141(2):331–349

Bu W, Yu H, Luo G, Bera MK, Hou B, Schuman AW, Lin B, Meron M, Kuzmenko I, Antonio MR, Soderholm L, Schlossman ML (2014) Observation of a rare earth Ion−extractant complex arrested at the oil−water interface during solvent extraction. J Phys Chem B 118 (36):10662–10674

Chen YH, Liu XL, Lu Y, Zhang XY (2009) Investigation of gallium partitioning behavior in aqueous two-phase systems containing polyethylene glycol and am-monium sulfate. J Chem Eng Data 54(7):2002–2004

Chock SP, Eisenberg E (1974) Heavy meromyosin Mg-ATPase: presteady-state and steady-state H^+ release. Proc Nat Acad Sci USA 71(12):4915–4919

de Meis L (1984) Pyrophosphate of high and low energy. Contributions of pH, Ca^{2+}, Mg^{2+}, and water to free energy of hydrolysis. J Biol Chem 259(10):6090–6097

de Meis L, Behrens MI, Petretski JH (1985) Contribution of water to free energy of hydrolysis of pyrophosphate. Biochemistry 24(26):7783–7789

Ellis RJ, Audras M, Antonio MR (2012) Mesoscopic aspects of phase transitions in a solvent extraction system. Langmuir 28(44):15498–15504

George P, Witonsky RJ, Trachtman M, Wu C, Dorwart W, Richman L, Richman W, Shurayh F, Lentz B (1970) "Squiggle-H2O". An enquiry into the importance of solvation effects in phosphate ester and anhydride reactions. Biochim Biophys Acta 223(1):1–15

Goldberg RN, Kishore N, Lennen RM (2002) Thermodynamics quantities for the ionization reactions of buffers. J Phys Chem Ref Data 31(2):231–370

Jiang H, Jia J (2008) Complete reversible phase transfer of luminescent CdTe nanocrystals mediated by hexadecylamine. J Mater Chem 18(3):344–349

Kardami E, De Bruin S, Gratzer W (1979) Interaction of ADP with skeletal and cardiac myosin and their active fragments observed by proton release. Eur J Biochem 97(2):547–552

Kodama T (1981) Temperature-modulated binding of ADP and adenyl-5′-yl imidodiphosphate to myosin subfragment 1 studied by calorimetric titration. J Biol Chem 256(22):11503–11508

Kodama T, Fukui K, Kometani K (1986) The initial phosphate burst in ATP hydrolysis by myosin and subfragment-1 as studied by a modified malachite green method for determination of inorganic phosphate. J Biochem 99(5):1465–1472 (Tokyo)

Kodama T, Woledge RC (1979) Enthalpy changes for intermediate steps of the ATP hydrolysis catalyzed by myosin subfragment-1. J Biol Chem 254(14):6382–6386

Koretz JF, Taylor EW (1975) Transient state kinetic studies of proton liberation by myosin and subfragment 1. J Biol Chem 250(16):6344–6350

Pal D, Tripathi A, Shukla A, Gupta KR, Keshav A (2015) Reactive extraction of pyruvic acid using tri-n-octylamine diluted in decanol/kerosene: equilibrium and effect of temperature. J Chem Eng Data 60(3):860–869

Plaut GW, Kuby SA, Lardy HA (1950) Systems for the separation of phosphoric esters by solvent distribution. J Biol Chem 184(1):243–249

Romero PJ, de Meis L (1989) Role of water in the energy of hydrolysis of phosphoanhydride and phosphoester bonds. J Biol Chem 264(14):7869–7873

Springs B, Welsh KM, Cooperman BS (1981) Thermodynamics, kinetics, and mechanism in yeast inorganic pyrophosphatase catalysis of inorganic pyrophosphate: inorganic phosphate equilibration. Biochemistry 20(22):6384–6391

Stockbridge RB, Wolfenden R (2009) Phosphate monoester hydrolysis in cyclohexane. J Am Chem Soc 131(51):18248–18249

Stockbridge RB, Wolfenden R (2010) The hydrolysis of phosphate diesters in cyclohexane and acetone. Chem Commun 46(24):4306–4308 (Camb)

Stockbridge RB, Wolfenden R (2011) Enhancement of the rate of pyrophosphate hydrolysis by nonenzymatic catalysts and by inorganic pyrophosphatase. J Biol Chem 286(21):18538–18546

Takahashi H, Umino S, Miki Y, Ishizuka R, Maeda S, Morita A, Suzuki M, Matubayasi N (2017) Drastic compensation of electronic and solvation effects on ATP hydrolysis revealed through large-scale QM/MM simulations combined with a theory of solutions. J Phys Chem B 121 (10):2279–2287

Voet D, Voet JG, Pratt CW (2013) Fundamentals of biochemistry, 4th edn. Wiley, Hoboken

Wolfenden R, Williams R (1985) Solvent water and the biological group-transfer potential of phosphoric and carboxylic anhydrides. J Am Chem Soc 107(14):4345–4346

Part II
Basis of Protein-Ligand and Protein-Protein Interactions

Chapter 7
Energetics of Myosin ATP Hydrolysis by Calorimetry

Takao Kodama

Abstract For over ten years from the mid-1970s, ATP hydrolysis into ADP and inorganic phosphate (P_i) by myosin were investigated by reaction calorimetry using own-designed instruments. The purpose was to estimate enthalpy changes for intermediate steps of the hydrolysis cycle. The results indicated that the steps are accompanied by large enthalpy changes alternating between negative and positive values. By combining this enthalpic profile with Gibbs energy changes for the corresponding steps estimated by kinetics and equilibrium analysis, the overall energetic profile of the ATP hydrolysis cycle has been revealed. The most characteristic feature is that all the intermediates are near the same Gibbs energy levels (isoenergetic); this occurs because large enthalpy and entropy changes compensate each other in the bound ATP-hydrolyzing and subsequent P_i-releasing steps. Possible sources of the large entropy changes were investigated in the late 1990s using dielectric spectroscopy to measure changes in hydration level of myosin during the ATP hydrolysis cycle. The result indicated dehydration during the hydrolysis step and rehydration during the P_i-releasing step. The extent of these hydration changes on the myosin molecular surface was just sufficient to account for their observed entropy changes. Taken together, once myosin traps ATP, the system is brought into a low-entropy state that is maintained until hydrolysis products, P_i and ADP are released from myosin. Thus, ATP plays the role of mediator to bring negative entropy into the function of the myosin.

Keywords Muscle · Gibbs energy · Enthalpy · Entropy · Hydration and dehydration

---

T. Kodama (✉)
Computer Science and Systems Engineering, Kyushu Institute of Technology, Iizuka, Japan
e-mail: kodama@bio.kyutech.ac.jp

© Springer Nature Singapore Pte Ltd. 2018

M. Suzuki (ed.), *The Role of Water in ATP Hydrolysis Energy Transduction by Protein Machinery*, https://doi.org/10.1007/978-981-10-8459-1_7

7.1 Introduction

Myosin is the collective term of a large family of cytoskeletal motor proteins in eukaryotes that hydrolyze ATP into ADP and inorganic phosphate (P_i) and interact with actin filaments to generate mechanical force through structural changes. Muscle contraction is the most familiar manifestation of such chemo-mechanical energy transduction, and it has been a major research subject of life science and very actively studied by a wide range of physicochemical methods since the mid-1900s.

Along this trend, I started calorimetric studies of myosin ATP hydrolysis with Roger C. Woledge (Curtin and Rosenberg 2006; Barclay and Curtin 2015) at University College London, in 1974. Its kinetic characterization had been well advanced by then. Thus, it was known that the hydrolysis cycle consists of kinetically well-defined intermediates, which are distinguished from each other in their affinities for actin (Fig. 7.1). Over the next several years, most of kinetic parameters for transitions between them were accurately estimated (for a review, see Kodama 1985) and with this information, we were able to depict their Gibbs energy profiles as shown in the top panel of Fig. 7.2.

In general, for a given chemical reaction at temperature T in Kelvin, the Gibbs energy change ΔG is related with enthalpy (ΔH) and entropy (ΔS) changes by an equation:

$$\Delta G = \Delta H - T\Delta S$$

Thus, the major purpose of our calorimetric studies of myosin ATP hydrolysis is to discover the separate contributions of ΔH and ΔS to the ΔG for the intermediate

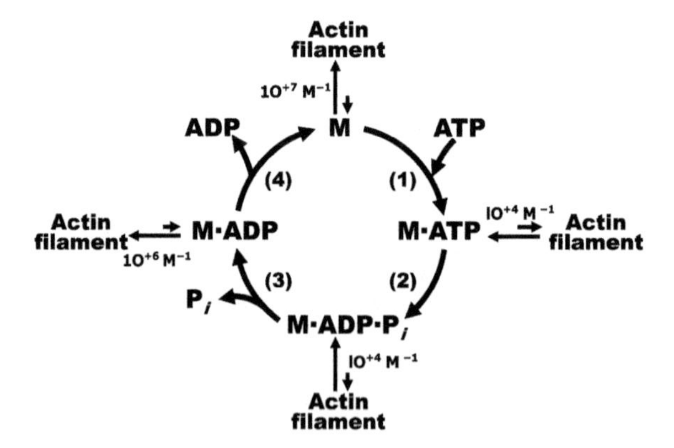

Fig. 7.1 Myosin ATP hydrolysis cycle showing intermediates with different affinities for actin filament. M denotes myosin. Each of numerical value given with a pair of arrows is binding constant between a myosin state and actin filament. Under the conditions considered in this chapter, the cycle can be divided into the rapid phase (step 1 + step 2) and the slow phase (step 3 + step 4) and the step 3 (0.03–0.04 s^{-1}) is the rate-limiting step of the cycle

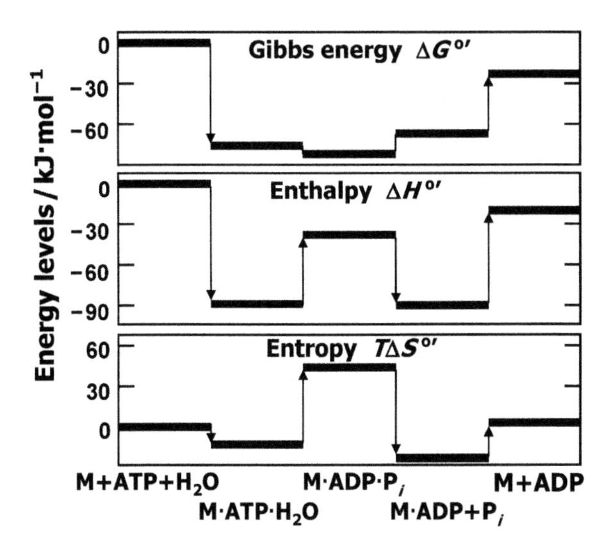

Fig. 7.2 Energy profiles of the myosin ATP hydrolysis cycle based on data from Kodama (1985)

steps in Fig. 7.1. This chapter deals with development of the instruments used in my studies, the major outcomes and their implication that led me to collaborate with Makoto Suzuki, the editor of this book, for studying hydration change of myosin accompanying the ATP hydrolysis.

Note that all thermodynamic values for myosin ATP hydrolysis are referred to conditions of 100 mM KCl, 1–5 mM $MgCl_2$, pH 7–8, and 20–25 °C (293–298 K) so that observed changes of Gibbs energy, enthalpy, and entropy are expressed as $\Delta G^{\circ\prime}$, $\Delta H^{\circ\prime}$, and $\Delta S^{\circ\prime}$, respectively, according to Alberty (1994). Myosin (abbreviated as M for describing reaction intermediates in text and Figures) denotes the head portion of myosin molecule containing sites for hydrolyzing ATP and for interacting with actin.

7.2 Instrumentation for Calorimetric Studies of Myosin ATP Hydrolysis

Reaction calorimetry is prerequisite to obtain reliable data for $\Delta H^{\circ\prime}$ values for transitions from one intermediate to another while keeping overall consistency of the reaction cycle. Although several calorimeters existed, some of which were commercially available when we started the project, we continuously used our own-designed instruments.

The first one was an LKB batch calorimeter that had been extensively modified for titration calorimetry (Woledge 1972, 1976). The device proved useful for energetic characterization of the ADP binding to myosin (the reversal of step 4 in Fig. 7.1), which provided reference data (Kodama and Woledge 1976) for our subsequent studies.

The second instrument was a rapid response calorimeter (Kodama and Woledge 1979, 1985). The wall of the cylindrical reaction vessel, diameter of 1 cm and total capacity of 2 ml, was a polyimide/Teflon film of 25-μm thickness. In close contact with the external surface of this wall was a silver-constantan thermopile consisting of 25 junctions. The calorimeter was placed in a watertight container which was immersed in a water bath, the temperature fluctuation of which was within $\pm$0.02 °C. The reaction mixture of 1 ml was stirred continuously by vertical oscillations (5 Hz with amplitude of 5 mm) of a perforated Teflon disk connected to a rigid stainless-steel tube. The reaction was started by adding a few μl of nucleotide solution over a period of 1 s from a motor-driven micrometer syringe via Teflon tubing which ran through the stainless-steel tube. The sensitivity was sufficient to measure heat changes of 0.05 mJ with a time resolution better than 1 s. Heat loss from the instrument was exponential with a time constant of 350 s, so it was suitable to follow the slow hydrolysis of ATP by myosin (k_{cat}, 0.03–0.04 s^{-1}). In addition, this instrument proved suitable for calorimetric titration (Kodama 1981). Apart from time required for thermal equilibration after loading the sample, a whole titration process was completed within several minutes. Thus, it was possible to make repeated titration experiments in a single working day. Calorimetric data obtained with this instrument were summarized in Kodama (1985).

The third instrument was a stopped-flow calorimeter made by modifying a conventional stopped-flow device. The choice of temperature sensor was of critical importance because we needed a response time faster than 50 ms and a high sensitivity. Accordingly, a thermopile for measurement of muscle heat production described by Ricchiuti and Mommaerts (1965) was used in an early version (Kodama and Kometani 1986; Kodama 1988).

However, the thermopiles for muscle heat measurement were not structurally strong enough to withstand the rapid flow of reaction mixture through the observation tube in repeated use. To overcome this difficulty, we developed a nickel/copper thermocouple array (Kodama and Kometani 1990). The array was made by essentially the same technique as for making semiconductor devices. Thus, nickel and copper patterns were made and sandwiched by layers of silicon nitride (thickness, 50 μm) on a surface of a square-shaped zirconia ceramic plate (5 × 7 × 0.05 mm). The thermocouple density is 10 mm^{-1} and the total thermal electromotive force of 1.0 mVK^{-1}. The instrument integrated with this sensor showed the dead time of 13 ms and time resolution better than 5 ms.

7.3 Energetics of Myosin ATP Hydrolysis

Using the improved instrumentation just described, $\Delta H^{\circ\prime}$ values of four reaction steps shown in Fig. 7.1 had been obtained by 1990 under essentially the same conditions as for estimation of the $\Delta G^{\circ\prime}$ values. The middle and the bottom panels of Fig. 7.2 are the relative enthalpy and entropy levels of the intermediates constituting the myosin ATP hydrolysis cycle. The outstanding features of its energy profile are:

(i) Compared to large $\Delta G^{\circ\prime}$ values for ATP-binding and ADP-releasing steps (steps 1 and 4 in Fig. 7.1, respectively), corresponding values for bound ATP-hydrolyzing and subsequent P_i-releasing steps (steps 2 and 3, respectively) are very small, so that the three reaction intermediates, $M \cdot ATP$, $M \cdot ADP \cdot P_{i,}$ and $M \cdot ADP$ are almost at the same Gibbs energy levels (isoenergetic). Hence, as pointed out earlier by White (1977), the difference between the $\Delta G^{\circ\prime}$ values for step 1 and for the reversal of step 4 is just equivalent to the $\Delta G^{\circ\prime}$ value for the overall ATP hydrolysis in aqueous solution ($ATP + H_2O \rightarrow ADP + P_i$), $\cong -30\ kJ \cdot mol^{-1}$ (Rosing and Slater 1972). It should also be noted that this difference in $\Delta G^{\circ\prime}$ values between binding of ATP and ADP to myosin can be ascribed to the difference of structure in their phosphate ester portions. In fact, AMP is known to hardly bind to myosin.
(ii) In terms of energetics, step 2, hydrolysis of bound ATP, is in marked contrast to the corresponding reaction in aqueous solution just described. Thus, its $\Delta G^{\circ\prime}$ is very small ($\cong -5\ kJ \cdot mol^{-1}$), so that the step is referred to as "near-equilibrium." This is due to a typical $\Delta H/\Delta S$ compensating effect with positive $\Delta H^{\circ\prime}$ ($\cong +50\ kJ \cdot mol^{-1}$) and negative $T\Delta S^{\circ\prime}$ ($\cong -55\ kJ \cdot mol^{-1}$). Such relations between an enzyme–substrate complex and enzyme–product complex are known to occur in other enzyme-catalyzed reactions, referred to as "on-enzyme equilibrium" (Gutfreund 1995). In addition, the endothermic nature of the step itself is noteworthy because it is quite the reverse of the exothermic hydrolysis reaction in aqueous solution ($\Delta H^{\circ\prime} \cong -20\ kJ \cdot mol^{-1}$). The $\Delta G^{\circ\prime}$ value for step 3 is also very small, which is caused by a similar $\Delta H/\Delta S$ compensating effect as in step 2, but both of $\Delta H^{\circ\prime}$ and $\Delta S^{\circ\prime}$ signs are the opposite of those for step 2.
(iii) Though not shown in Fig. 7.2, it is noteworthy that changes in the heat capacity ($\Delta C_p{}^{\circ\prime}$) of steps 2 and 3 are very large (−2.8 and +2.6 $kJ \cdot K^{-1} \cdot mol^{-1}$, respectively). These values of $\Delta S^{\circ\prime}$ and $\Delta C_p{}^{\circ\prime}$ can be compared with those accompanying protein unfolding. Considering a typical globular protein with molecular mass of 15 kDa undergoes heat denaturation to unfold with ΔS of 0.3–1 $kJ \cdot K^{-1} mol^{-1}$ and ΔC_p of 4–10 $kJ \cdot K^{-1} mol^{-1}$ (Privalov 1979; Pfeil 1986; Privalov and Dragan 2007), one can appreciate how large these thermodynamic changes in an enzyme catalytic cycle, which led us to start studies of the change in hydration state of myosin during the ATP hydrolysis cycle as described below.

7.4 Sources of Large Entropy Changes and Their Implication During Myosin ATP Hydrolysis Cycle

An obvious interpretation of large negative $\Delta G^{\circ\prime}$ for the step 1 of myosin ATP hydrolysis is that it is ascribed to the large favorable enthalpy effect ($\Delta H < 0$) due to strong interaction between ATP and amino acid residues in the ATP-binding site of myosin and accompanying conformational changes, which is inevitably accompanied by unfavorable entropy effect ($\Delta S < 0$) due to decrease in molecular motions. In general, however, observed thermodynamic quantities for a chemical reaction taking place in aqueous phase are sum of changes in thermodynamic states of all molecular species involved. Thus, it is important to consider dynamics of water molecules interacting with myosin and ATP and with the hydrolysis products (ADP + P_i) (collectively called as ligands).

Upon transfer of ATP from bulk solution to the binding site, many water molecules must be released from both ATP and the binding site. Such dehydration effects should be accompanied by large positive enthalpy and entropy changes (Ross and Subramanian 1981). Hence, the enthalpy and entropy changes for the ATP binding to myosin are estimated to be much more negative than their observed values, $\Delta H^{\circ\prime}$ and $\Delta S^{\circ\prime}$. In other words, upon ATP binding to myosin, the system is brought into very "low-entropy state" but stabilized by large enthalpy decrease.

The subsequent hydrolysis of bound ATP step is accompanied by clear $\Delta H/\Delta S$ compensating effect ($\Delta H^{\circ\prime} > 0$; $\Delta S^{\circ\prime} > 0$). Apart from the interesting issue of what chemical environment in the binding site makes "on-enzyme ATP hydrolysis" endothermic, the large value of $\Delta S^{\circ\prime}$ with a concomitant large negative $\Delta C_p^{\circ\prime}$ is suggestive of dehydration of the myosin molecular surface (Kodama 1985). In terms of entropy and heat capacity changes, the next step, P_i release, is the very opposite to the preceding step, and which suggests that the molecular surface is rehydrated.

To investigate how much hydration change occurs on myosin molecule accompanying the ATP hydrolysis, studies were made in collaboration with M Suzuki, who was then working at AIST, Tsukuba, Japan, using dielectric spectroscopy with microwaves in the frequency range between 0.2 and 20 GHz (Suzuki et al. 1997; Suzuki and Kodama 2000). The result indicated that the transition of myosin from M · ADP to M · ADP · P_i is accompanied by release of about 9% of the total of 1400 weakly restrained water molecules per myosin. These are restrained again on transition from M · ADP · P_i to M · ADP states. Since the ADP binding (the reversal of step 4) showed little water movement, the observed changes in hydration are quantitatively consistent with the accompanying large entropy and heat capacity changes described above. Hydration levels of various ATP series nucleotides were also measured (Mogami et al. 2010).

In addition, it is now possible from the Protein Data Bank (PDB) registered data to estimate the hydration state of the nucleotide binding site for different myosin/nucleotide complexes (Kimori personal communication). Taking into account of these data and other available information, Suzuki and I have been attempting to

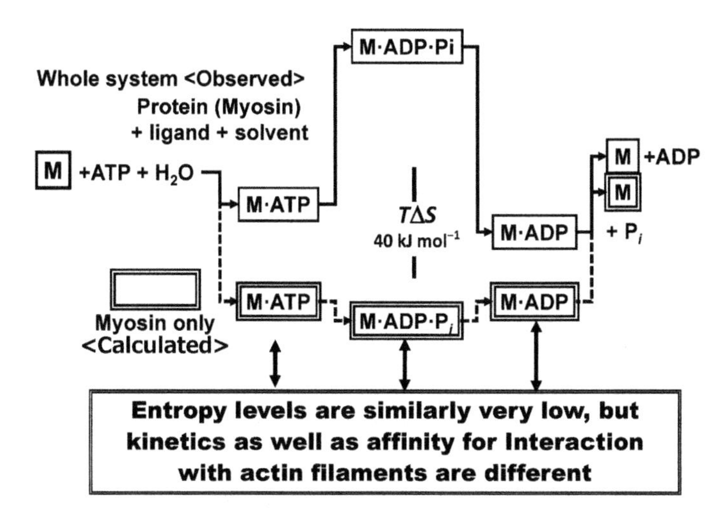

Fig. 7.3 Entropy levels of reaction intermediates of myosin ATP hydrolysis. The upper part is essentially the same as that of the bottom panel of Fig. 7.1. The lower part was drawn to show the entropy levels of the system after subtracting the entropy effect by hydration/dehydration processes estimated by dielectric spectroscopy. The difference in length of arrows below the intermediates is a rough indicator of their difference of affinity for actin filament

estimate entropy changes for transfers of ATP from bulk solution to the binding site, and P_i and ADP from the binding site to bulk solution. Figure 7.3 was drawn based on preliminary results to show the entropy state of myosin itself during the ATP hydrolysis cycle. As can be seen, once myosin traps ATP, it brings the system into a low-entropy state, which stays very low until the hydrolysis products have been released.

Correlation of such entropy changes of the system with actin/myosin interaction mode should also be noted. When the nucleotide-free myosin traps ATP, its affinity for actin is much reduced and the system enters low-entropy state. Actin in turn accelerates the release of P_i and ADP to bring the myosin back into a nucleotide-free state and the system into high entropy state. This seems the most important thermodynamic aspect obtained by calorimetry of myosin ATP hydrolysis underlying chemo-mechanical energy transduction by myosin/actin system such as muscle contraction.

Borrowing the very famous Schrödinger's phrase, "What an organism feeds upon is negative entropy" (Schrödinger 1944), we may say "myosin feeds on negative entropy through trapping and hydrolyzing ATP." It would be an interesting challenge to investigate whether ATP is the mediator to bring negative entropy into ATP-driven molecular motors in general.

Acknowledgement I am grateful to Nancy A. Curtin, Professor Emeritus of Imperial College London, for critical reading of the manuscript and for many useful suggestions. I would also like to express my sincere gratitude to Makoto Suzuki, Professor Emeritus of Tohoku University, for his collaboration with me over the past two decades. Finally, this chapter is dedicated to the memory of Roger C. Woledge.

References

Alberty RA (1994) Recommendations for nomenclature and tables in biochemical thermodynamics. Pure Appl Chem 66:1641–1666

Barclay C, Curtin N (2015) Roger C. Woledge 1938–2015. J Muscle Res Cell Motil 36:301–303

Curtin N, Rosenberg M (2006) An interview with Roger Woledge. http://www.physoc.org/sites/default/files/page/Roger_Woledge_0.pdf

Gutfreund H (1995) Thermodynamic information from kinetics. In: Kinetics for the life sciences: receptors, transmitters and catalysts. Chapter 4 kinetic analysis of complex reactions. Cambridge University Press, Cambridge, UK/New York, pp 110. ISBN: 978-0521485869

Kodama T, Woledge RC (1976) Calorimetric studies of the interaction myosin with ADP. J Biol Chem 251:7499–7503

Kodama T, Woledge RC (1979) Enthalpy changes for intermediate steps of the ATP hydrolysis catalyzed by myosin subfragment-1. J Biol Chem 254:6382–6386

Kodama T (1981) Temperature modulated binding of ADP and adenyl-5′-yl imidodiphosphate to myosin subfragment 1 studied by calorimetric titration. J Biol Chem 256:11503–11508

Kodama T (1985) Thermodynamic analysis of muscle ATPase mechanisms. Physiol Rev 65: 467–551

Kodama T, Woledge RC (1985) Rapid response microcalorimeters for biochemical experiments. Rev Sci Instrum 56:40–42

Kodama T, Kometani K (1986) A stopped-flow calorimetric study of ATP hydrolysis by myosin subfragment 1. Proc Japan Acad 62B:105–108

Kodama T (1988) Stopped-flow calorimetry of myosin ATP hydrolysis: an implication of chemomechanical energy transduction. Adv Exp Med Biol 226:671–676

Kodama T, Kometani K (1990) A stopped-flow calorimeter with a thermocouple array sensor for studies of energetic aspects of enzymatic reactions. Thermochim Acta 163:105–110

Mogami G, Wazawa T, Morimoto N, Kodama T, Suzuki M (2010) Hydration properties of adenosine phosphate series as studied by microwave dielectric spectroscopy. Biophys Chem 151:160–169

Pfeil W (1986) Unfolding of proteins. In: Hinz HJ (ed) Thermodynamic data for biochemistry and biotechnology. Springer, pp 349–376. ISBN: 978-3-642-71116-9

Privalov PL (1979) Stability of proteins: small globular proteins. Adv Protein Chem 33:167–241

Privalov PL, Dragan AI (2007) Microcalorimetry of biological macromolecules. Biophys Chem 126:16–24

Ricchiuti NV, Mommaerts WFHM (1965) Technique for myothermic measurements. Physiologist 8:259

Rosing J, Slater EC (1972) The value of ΔG° for the hydrolysis of ATP. Biochim Biophys Acta 267:275–290

Ross PD, Subramanian S (1981) Thermodynamics of protein association reactions: forces contributing to stability. Biochemistry 21:3096–3102

Schrödinger E (1944) It feeds on 'negative entropy'. In: What is life—the physical aspect of the living cell. Chapter 6 order, disorder and entropy. Cambridge University Press, Cambridge, UK, pp 68–75. http://dlab.clemson.edu/11._Erwin_Schrodinger_-_What_is_Life__1944_.pdf

Suzuki M, Shigematsu J, Fukunishi Y, Harada Y, Yanagida T, Kodama T (1997) Coupling of protein surface hydrophobicity change to ATP hydrolysis by myosin motor domain. Biophys J 72:18–23

Suzuki M, Kodama T (2000) Motor protein mechanism coupled with hydrophobic hydration/dehydration cycle. In: Osada Y, DeRossi, D (eds) Polymer sensors and actuators. Springer, pp 361–369. ISBN: 978-3-662-04068-3

White H (1977) …and biochemistry. Nature 267:754–755

Woledge RC (1972) Heat, work and phosphocreatine splitting during muscular contraction. Cold Spring Harb Symp 37:613–618

Woledge RC (1976) Calorimetric studies of muscle and muscle proteins. In: Lamprecht I, Saarschmidt R (eds) Application of calorimetry in life sciences. Proceedings of the international conference in Berlin, August 2–3, 1976. de Gruyter, Berlin, pp 183–197. ISBN: 3110069199

Chapter 8
Orchestrated Electrostatic Interactions Among Myosin, Actin, ATP, and Water

Mitsunori Takano

Abstract The electrostatic interactions are deeply involved in the force-generating function of the actomyosin molecular motor where myosin, actin, ATP, and water are interacting with each other in a orchestrated manner. In this chapter, an electrostatic perspective is presented based on our recent molecular dynamics simulation studies on the force-generation mechanisms of the actomyosin molecular motor. First, as an unusual property of the electrostatic interaction in water, thermodynamics of association between oppositely singed charges is addressed. Then, our computational results regarding the electrostatic interaction between myosin and actin are described, featuring a sawtooth-like asymmetric energy landscape on which myosin generates forces by multiple mechanisms including the Brownian ratchet-like mechanism. Then the role of ATP is discussed, with a focus on "dielectric allostery" that we found in myosin as an allosteric response to the ATP binding, which serves as weakening the actin–myosin electrostatic interaction and causes myosin to dissociate from actin. Finally, the role of water is discussed from the viewpoint of the association thermodynamics of biomolecules.

8.1 Electrostatic Interaction in Water

In this chapter, I will summarize our recent studies on the actomyosin molecular motor by molecular dynamics simulation, featuring how the electrostatic interactions are involved in the force-generating function of this molecular motor. To do this, I first address the fundamental physics of the electrostatic interaction in water that requires some caution.

From a microscopic viewpoint, biological functions in living organisms are based on well-orchestrated association–dissociation dynamics of molecules such as proteins, DNA, RNA, saccharides, lipids, and other small molecules. Since most of these biological molecules are electrolytes, the electrostatic interaction plays a vital

M. Takano (✉)
Department of Pure and Applied Physics, Waseda University, Okubo 3-4-1, Shinjuku-Ku, Tokyo 169-8555, Japan
e-mail: mtkn@waseda.jp

© Springer Nature Singapore Pte Ltd. 2018

M. Suzuki (ed.), *The Role of Water in ATP Hydrolysis Energy Transduction by Protein Machinery*, https://doi.org/10.1007/978-981-10-8459-1_8

role in their association–dissociation dynamics and thermodynamics (e.g., Sheinerman et al. 2000; Cherstvy 2009). Therefore, adequate physical understanding of electrostatics is necessary to reveal the design principle that is inherent in the highly orchestrated dynamics of those molecules in a cell. However, the electrostatic interactions among biomolecules in water are far from trivial. Firstly, this is because of the structural complexity of biomolecules, where positive and negative charges distribute heterogeneously and the repulsive and attractive interactions complicatedly coexist so that the system is prone to be highly frustrated, like spin glasses where ferromagnetic and anti-ferromagnetic interactions coexist in a complex manner. The second yet equally significant reason comes from the unusual physical properties of water as the solvent of the biomolecules.

One of the most noteworthy results that originate from the unusual physical properties of water can be seen in the thermodynamics of the association between oppositely singed charges. One may intuitively think that the association should be exothermic (i.e., the enthalpy change upon association, ΔH, should be negative). However, the association in water often becomes endothermic (i.e., $\Delta H > 0$), as was discussed theoretically by Schellman (1953) and pointed out based on the calorimetric data by Kodama (1985).

The physical reason of the above-mentioned subtlety of ΔH can be explained as follows. The electrostatic interaction energy (free energy) involves the dielectric constant ε, which is dependent on the temperature; remember that the Coulomb interaction formula contains the factor of $1/\varepsilon$, and more generally the electrostatic energy of the system is obtained by the spatial integration of the energy density, i.e.,

$$G_{\mathrm{el}} = \frac{1}{2} \int \boldsymbol{E} \cdot \boldsymbol{D} \, \mathrm{d}\boldsymbol{r} \\ = \frac{1}{2} \int \frac{1}{\varepsilon} \boldsymbol{D}^2 \, \mathrm{d}\boldsymbol{r} \, ,$$

where $\boldsymbol{D}$ and $\boldsymbol{E}$ represent the electric flux density and the electric field, respectively, and the spatial integration runs over the entire space (outside the charges). Then, the electrostatic energy change upon association is described by,

$$\Delta G_{\mathrm{el}} = \frac{A}{2\varepsilon} \, ,$$

$$A = \int \boldsymbol{D}_{\mathrm{a}}^2 \, \mathrm{d}\boldsymbol{r} - \int \boldsymbol{D}_{\mathrm{d}}^2 \, \mathrm{d}\boldsymbol{r} \, ,$$

where $\boldsymbol{D}_{\mathrm{d}}$ and $\boldsymbol{D}_{\mathrm{a}}$ represent the electric flux density before association and that after association, respectively (ε is assumed to be spatially uniform). Note that ΔG_{el} is generally negative because of the cancellation of the electric fields upon association (i.e., $A < 0$ because $\int \boldsymbol{D}_{\mathrm{a}}^2 \mathrm{d}\boldsymbol{r} < \int \boldsymbol{D}_{\mathrm{d}}^2 \mathrm{d}\boldsymbol{r}$). Therefore, the oppositely signed charges attract each other via the electrostatic interaction in water, even though the attractive interaction is weakened by the factor of $\varepsilon_0/\varepsilon$ compared to that in vacuum. This physical picture remains essentially the same when the charges are located on the surfaces of two interacting molecules even though the polarization charges that arise at the dielectric boundaries between the solute and the solvent regions need to be properly taken into account (Mizuhara et al. 2017). Then, how about the enthalpy change upon association? ΔH_{el} is given by,

$$\begin{aligned} \Delta H_{\mathrm{el}} &= \Delta G_{\mathrm{el}} + T\Delta S_{\mathrm{el}} \\ &= \frac{A}{2\varepsilon} + \frac{AT}{2\varepsilon^2}\frac{\partial \varepsilon}{\partial T} \\ &= \frac{A}{2\varepsilon^2}\frac{\partial(\varepsilon T)}{\partial T}. \end{aligned}$$

Thus, the sign of $\partial(\varepsilon T)/\partial T$ determines whether the association is endothermic or exothermic (Schellman 1953). One of the unusual physical properties of water is that the temperature dependence of ε is quite strong, which makes an appearance here. The temperature dependence is well approximated by $\varepsilon \sim T^{-b}$, with $b \sim 1.4$ for the bulk water (Fig. 8.1), whereas b is normally less than one for other solvents. Therefore, ΔH_{el} becomes positive (i.e., endothermic) for water, whereas ΔH_{el} becomes negative (i.e., exothermic) for other normal solvents. Furthermore, it should be noted that the dielectric response of water in the vicinity of the surfaces of macromolecules may be reduced compared to that for the bulk water (e.g., Despa et al. 2004; Ahmad et al. 2011; Chen et al. 2015), which leads to a reduced temperature dependence of ε and makes ΔH_{el} negative in some cases when the charges are on the surfaces of two interacting macromolecules. [Of course, structural changes of the macromolecules upon association can be one of the causes of $\Delta H_{\mathrm{el}} < 0$ (Kodama 1985; Sheinerman et al. 2000)]. We previously addressed this peculiarity of water regarding the

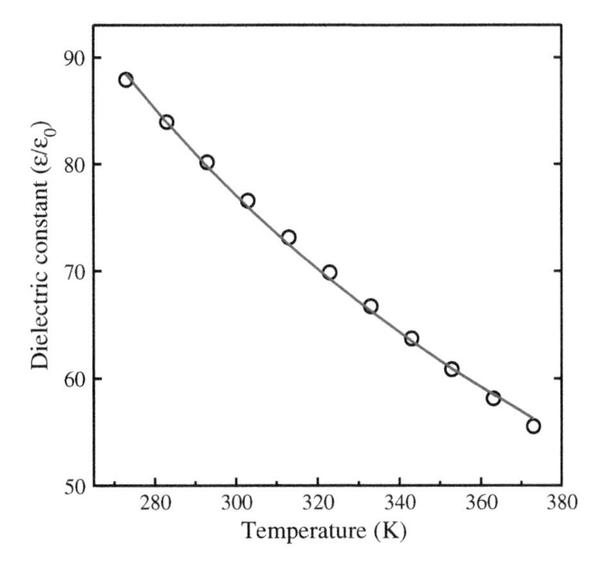

Fig. 8.1 Temperature dependence of ε for the bulk water (Archer and Wang 1990). The solid curve represents $\varepsilon \sim T^{-b}$ with the least squares fitting parameter of $b = 1.45$

temperature dependence of ε by computational study of the association between myosin and actin (Okazaki et al. 2012), where we reinterpreted the thermodynamic data (Katoh and Morita 1996) and highlighted the temperature enhancement of the electrostatic interaction as the driving force for the myosin–actin association. The characteristics of the electrostatic interaction between myosin and actin, as found by our molecular dynamics simulations, are addressed in the next section.

8.2 Electrostatic Interaction Between Myosin and Actin

The involvement of the electrostatic interaction in the association between myosin and actin was early recognized from the experimental observation that the association constant decreases as the ionic strength of the solution is increased (Tonomura et al. 1962; Highsmith 1977; Katoh and Morita 1996). Amino acid substitution studies also indicated the involvement of the charged residues in myosin and actin (regarding the mutation in actin, e.g., Sutoh et al. 1991; Johara et al. 1993).

To clarify the role of the electrostatic interactions in the actin–myosin association, and furthermore, to investigate the force-generating mechanism of the actomyosin molecular motor, we conducted molecular dynamics simulations of the actomyosin system using theoretical models where 3D structure of myosin and actin and the physical interactions between them are properly taken into account in a course-grained manner at the amino acid residue level for proteins and also using the dielectric continuum model for water (Takano et al. 2010; Okazaki et al. 2012; Nie et al. 2014).

The most important finding in our MD simulation (Takano et al. 2010) was that myosin exhibits a net unidirectionality in the Brownian motion along the actin filament. Moreover, the net unidirectional Brownian motion was found to be caused by the energy landscape for the myosin–actin interaction that exhibits a sawtooth-like asymmetric profile along the actin filament, as illustrated in Fig. 8.2 (for details, see the original article). We further showed that this characteristic profile of the energy landscape was produced by the electrostatic interaction between myosin and the actin filament, particularly, between a positively charged loop, called "loop 2", of myosin and negatively charged regions of actin (see the circles shown in Fig. 8.2). Due to the asymmetric 3D structures of myosin and the actin filament on which charged residues are neatly arranged, the interaction interface is changed gradually and asymmetrically as myosin translates along the long axis of the actin filament. Indeed, when the charge reversal mutations were computationally applied to the acidic residues in actin, the unidirectional Brownian motion and the sawtooth-like energy landscape were both diminished (Takano et al. 2010). The unidirectional Brownian motion of myosin observed by our MD simulation was remarkably similar to that observed by the sophisticated single-molecule experiment by the Yanagida group (Kitamura et al. 1999). Our MD simulation thus substantiated the existence of the Brownian ratchet-like mechanism inherent in the actomyosin molecular motor as envisioned in the single-molecule experiment by Kitamura et al. (1999).

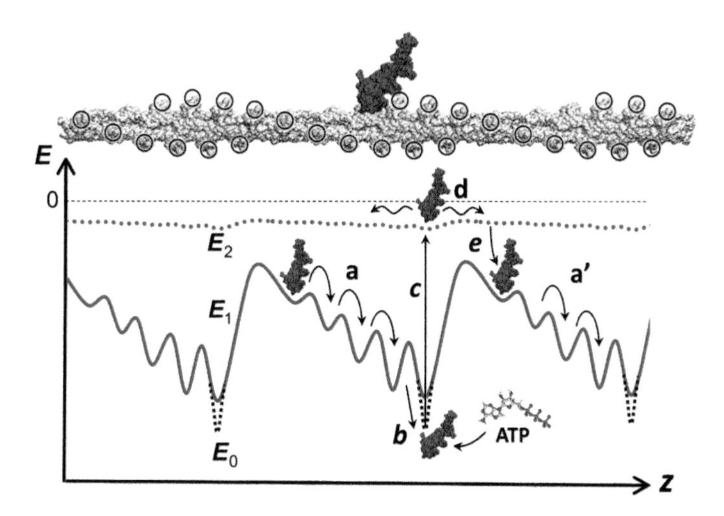

Fig. 8.2 Energy landscape between myosin (red) and actin (gray) along the long axis (z-axis) of the actin filament (consisting of 39 actin subunits). Positions of some of the acidic residues in actin are indicated by circles [Asp1/Glu2 (yellow) and Glu99/Glu100 (cyan)]. Three landscapes for the myosin–actin interaction energy (mainly electrostatic), E_0, E_1, and E_2 are schematically depicted as a function of z. E_1 represents the sawtooth-like asymmetric energy landscape on which myosin exhibits unidirectional Brownian motion toward the plus direction of z (process "a") (Takano et al. 2010). In this state, myosin is assumed to contain the hydrolysis products. E_0 expresses the energy landscape that occurs in the strong actin-binding state (see text) after the release of the hydrolysis products (process "b"). E_2 represents the state where myosin is apart from actin, which is induced by the binding of ATP to myosin (process "c"). After experiencing nearly isotropic 1D diffusion (process "d"), myosin cleaves ATP into Pi and ADP, by which the myosin–actin interaction energy becomes a bit stronger (process "e")

The sawtooth-like asymmetric energy landscape alone is not sufficient for the Brownian ratchet mechanism to work; the system must be energized and get out of equilibrium. Needless to say, it is ATP that plays this role. I will address the role of ATP in the next section. Here I mention the force-generating mechanisms, other than the Brownian ratchet one, that were deduced from our MD studies. When myosin interacts weakly with actin, the orientation of myosin relative to the actin filament [i.e., the angle formed between the long axis of myosin (viewed as a prolate) and that of the actin filament] is disordered. Then, upon strong binding of myosin to the actin filament, the orientation of myosin becomes ordered and fixed at an angle tilted toward the plus direction of z, which leads to a force generation. This force-generating mechanism was early deduced from experimental studies by EPR (Berger and Thomas 1994) and time-resolved EM (Walker et al. 1999). Our MD study indicated intrinsic disordering of the myosin orientation in the weak actin-binding state (Okazaki et al. 2012), where the electrostatic interactions played a dominant role, supporting the force-generation mechanism through the disorder-to-order transition of the orientation of myosin upon strong binding to actin (process "b" in Fig. 8.2). Then how about the lever-arm swing model, which has been the most widely recognized force-generating mechanism (e.g., Thomas et al. 2009; Sweeney and Houdusse 2010)? Here, it should be emphasized that multiple force-generating mechanisms can coexist (Fujita et al. 2012), and myosin seems to have evolved so as to use whatever it can exploit; actually, our MD study showed that the lever-arm mechanism can coexist with the Brownian ratchet mechanism (Nie et al. 2014). Interestingly, our recent MD simulation using an all-atom model showed that myosin possesses a piezoelectric property; the lever-arm swing induces the electrostatic potential change in the actin-binding region that should affect the electrostatic interaction with actin. This physical property of myosin, which we referred to as "piezoelectric allostery" (Ohnuki et al. 2016), suggests that when the negative electrostatic potential (due to actin) is applied to the actin-binding region, myosin allosterically induces the lever-arm swing via the "converse piezoelectric effect", offering another mechanism of force generation.

8.3 ATP as an Energizing Electrostatic Modulator

The force generation is coupled with the hydrolysis of ATP. In the elementary processes of the ATP hydrolysis, the binding of ATP to myosin can be considered as the most important process in the force-generating function because it is accompanied by the largest free-energy change ($\Delta G° \approx -65$ kJ/mol) (Kodama 1985). It is also noteworthy that the ATP binding to myosin is accompanied by a large negative enthalpy change ($\Delta H° \approx -90$ kJ/mol) (Kodama 1985). Multiple factors are likely to be involved in this large negative enthalpy changes, including the electrostatic interaction between the highly negatively charged triphosphate moiety of ATP and the positively charged residues in the ATP-binding region and some kind of structural changes in myosin. On the other hand, upon binding of ATP to myosin, the attractive interaction between myosin and actin is largely decreased (the association constant is reduced by $1/10^3$ (Kodama 1985)). Thus, there exists an interrelationship between the myosin–ATP and the myosin–actin bindings. From the viewpoint of the myosin–actin interaction energy landscape (Fig. 8.2), ATP can be regarded as an energizer that brings myosin into a higher myosin-actin interaction energy (i.e., out of equilibrium) state where myosin is given a potential to do work (the process "c" in Fig. 8.2). Then, what is the physical mechanism that couples the myosin–ATP binding with the myosin–actin binding? This mechanism has been investigated extensively, and a rigid body-like mechanical motion between subdomains in the actin-binding region was supposed to play the major role (Rayment et al. 1993). However, our recent MD simulation using the all-atom model (Sato et al. 2016) showed that in addition to such a rigid body-like motion, myosin exhibits a substantially large dielectric response upon ATP binding; the net negative charge of ATP induces a large-scale rearrangement of the electrostatic bond network in myosin, which reaches distant surface regions including the actin-binding region. We referred to this novel allosteric response as the "dielectric allostery" (Sato et al. 2016), which is closely related to the "piezoelectric allostery" (Ohnuki et al. 2016) that was mentioned above. We further found that loop 2 of myosin, which is highly positively charged and interacts with the negatively charged regions of actin as mentioned above, moves toward the direction away from actin in response to the surface potential change caused by the dielectric allostery, which leads to weakening of the actin–myosin electrostatic interaction and eventually dissociation of myosin from actin (see the original paper for the details).

In addition to the energizer, ATP can be also regarded as a modulator of the electrostatic interaction between myosin and actin. The modulation is forcibly done by coupling with the net free-energy decrease that accompanies the ATP hydrolysis. In addition to the ATP binding, the release of the inorganic phosphate (Pi) is accompanied by a large free-energy decrease due the low concentration of Pi under the physiological condition (Kodama 1985), by which the force-generating reaction cycle is pushed forward. In the course of ATP cleavage and subsequent ADP-Pi separation and Pi dissociation, the dielectric allostery is expected to occur and gradually induce the electrostatic potential change in the actin-binding region so as to strengthen the interaction with actin again. Note that the dielectric allostery caused by Pi and ADP releases can also induce electrostatic potential change at the inter-domain interface with the converter subdomain from which the lever-arm extends, suggesting the existence of the lever-arm swing caused by the dielectric allostery (Sato et al. 2017).

In passing, I just comment on the coupling between the chemical state of the ATP hydrolysis and the energy landscape. Since the dielectric allostery is caused by a large-scale rearrangement of the complicated electrostatic bond network within myosin, involving a large number of degrees of freedom, the coupling between the ATP-binding region and the actin-binding region is loose, exhibiting large fluctuation (Sato et al. 2016). Accordingly, the coupling between the chemical state of the ATP hydrolysis and the energy landscape for the myosin–actin electrostatic interaction as presented in Fig. 8.2 is loose as well. While we supposed that myosin contains ADP and Pi during the unidirectional Brownian motion on the energy landscape E_1 (Takano et al. 2010), it is possible that ADP-Pi separation proceeds and even Pi release may occur during the unidirectional Brownian motion, which deepens the the energy landscape (as represented by E_0). Note that the energy landscape depicted in Fig. 8.2 only considers the myosin–actin interaction energy; the free-energy change accompanying the elementary processes of the ATP hydrolysis can be considered through the (forcible) change of the energy landscapes [e.g., the change of $E_0 \rightarrow E_2$ (process “c”) can be thermodynamically realized by considering the free-energy decrease accompanying the binding of ATP to myosin].

8.4 Water as an Electrostatic Coordinator

For the myosin–actin association to be dynamical and modulated by ATP, it is necessary that the electrostatic interactions are not so strong. For example, the electrostatic interaction energy of a single electrostatic bond between oppositely charged elementary charges that are separated by 5 Å becomes as much as −300 kJ/mol in vacuum, whereas it is reduced to −3 kJ/mol in water, which comes into the realm of thermal

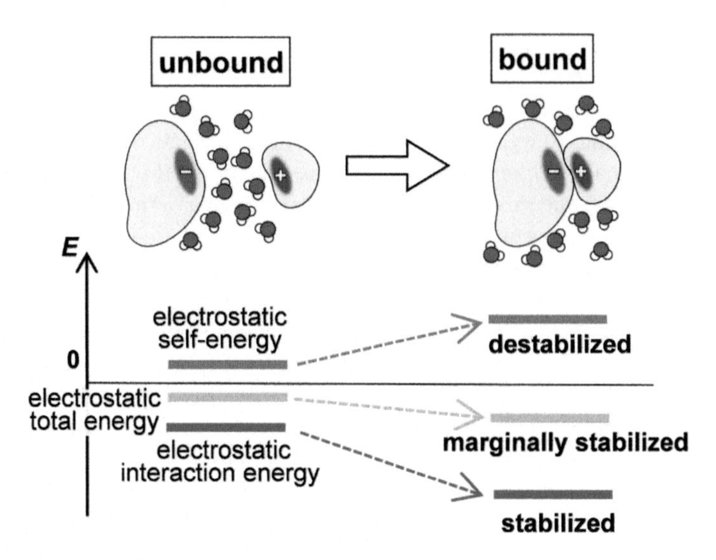

Fig. 8.3 Association of charged macromolecules in water (upper), and electrostatic energy changes that accompany the association (lower)

energy. In this sense, the fact that the electrostatic interaction is largely reduced in water due to its large dielectric constant (see Fig. 8.1) can be regarded as an important benefit. This argument remains essentially the same when the charges are on the surfaces of protein molecules as shown in Fig. 8.3; Upon the association of the protein molecules (Fig. 8.3), the effective dielectric constant around the charges at the interaction interface is decreased because of the low dielectric constant of protein molecules (Sheinerman et al. 2000). Because of this decrease of the effective dielectric constant, the electrostatic attraction between the charges is enhanced, whereas the electrostatic self-energies of the charges is destabilized [i.e., the stabilization by the electrostatic interaction with water (Born energy) is reduced]. As a result, the total electrostatic stabilization becomes marginal and comes into the range of the thermal energy (Mizuhara et al. 2017).

Inhomogeneous nature of water is also important to coordinate the electrostatic interaction between proteins. For example, the dielectric constant near the hydrophobic surface is supposed to be reduced, which results in enhancing the electrostatic interaction between the charges on the hydrophobic surface (Despa et al. 2004; Chen et al. 2015). Although the hydrophobic interaction plays a crucial role in the association of biomolecules, particularly when high stability is required to maintain the associated state against thermal agitation, it would be worth remembering that the hydrophobic surface can enhance the electrostatic interaction through the coordination of water. It is also noteworthy that the dielectric allostery that we found in myosin (Sato et al. 2016, 2017) can further extend from the surface of myosin to the surrounding water molecules. Actually, the change in the hydration waters surrounding myosin that is induced by the ATP hydrolysis was detected by the microwave dielectric spectroscopy (Suzuki et al. 1997).

In this chapter, I remarked that the actomyosin motor seems to fully exploit multiple mechanisms for the force generation. It is worth noting that the solvation energy of myosin can offer another force-generating mechanism (Suzuki et al. 2017): The electric field due to the actin filament is considered to be strong, and the strong electric field affects the orientation of the hydration water, making the hydration water "hyper-mobile" (Kubota et al. 2012), which leads to enhancing the water-stabilizing solvation energy of a protein molecule in a strong electric field (Suzuki et al. 2017). The electric field due to actin should be position dependent, so that the solvation energy of myosin varies along the filament, giving rise to another sawtooth-like asymmetric landscape due to the solvation energy [as envisaged in the original paper (Suzuki et al. 2017), the physical state variation of the actin filament can be considered as well].

In my opinion, water plays the coordinating role of the electrostatic interaction among myosin, actin, and ATP, which is indispensable for the actomyosin molecular motor to work. Without water, Coulombic force would remain too stubborn to allow emergence of life.

Acknowledgements Parts of the studies described in this chapter have been done in collaboration with M. Sasaki (Nagoya Univ), T. P. Terada (Nagoya Univ), and K. Okazaki (IMS). The efforts of the members in my laboratory, particularly, T. Sato, J. Ohnuki, Y. Mizuhara, and D. Parkin have advanced the understanding of the role of the electrostatic interaction involved in the molecular motors. I thank M. Suzuki (Tohoku Univ) for continuous encouragement and helpful comments, and the member of Innovative Scientific Research Area "Water plays the main role in ATP energy transfer" led by Prof. Suzuki for stimulating discussions, by which I really became aware of the importance and the depth of water.

References

Ahmad M, Gu W, Geyer T, Helms V (2011) Adhesive water networks facilitate binding of protein interfaces. Nat Commun 2:261

Archer DG, Wang P (1990) The dielectric constant of water and debye-hückel limiting law slopes. J Phys Chem Ref Data 19:371–411

Berger CL, Thomas DD (1994) Rotational dynamics of actin-bound intermediates of the myosin adenosine triphosphatase cycle in myofibrils. Biophys J 67:250–261

Chen S, Itoh Y, Masuda T, Shimizu S, Zhao J, Ma J, Nakamura S, Okuro K, Noguchi H, Uosaki K, Aida T (2015) Subnanoscale hydrophobic modulation of salt bridges in aqueous media. Science 348:555–559

Cherstvy AG (2009) Positively charged residues in DNA-binding domains of structural proteins follow sequence-specific positions of DNA phosphate groups. J Phys Chem B 113:4242–4247

Despa F, Fernández A, Berry RS (2004) Dielectric modulation of biological water. Phys Rev Lett 93:104–228

Fujita K, Iwaki M, Iwane AH, L M, Yanagida T, (2012) Switching of myosin-V motion between the lever-arm swing and brownian search-and-catch. Nature Commun 3:956

Highsmith S (1977) The effects of temperature and salts on myosin subfragment-1 and F-actin association. Arch Biochem Biophys 180:404–408

Johara M, Yano Toyoshima Y, Ishijima A, Kojima H, Yanagida T, Sutoh K (1993) Charge-reversion mutagenesis of Dictyostelium actin to map the surface recognized by myosin during ATP-driven sliding motion. Proc Natl Acad Sci USA 90:2127–2131

Katoh T, Morita F (1996) Binding of myosin subfragment 1 to actin. J Biochem 120:189–192

Kitamura K, Tokunaga M, Iwane A, Yanagida T (1999) A single myosin head moves along an actin filament with regular steps of 5.3 nanometres. Nature 397:129–134

Kodama T (1985) Thermodynamic analysis of muscle ATPase mechanisms. Physiol Rev 65:467–551

Kubota Y, Yoshimori A, Matubayasi N, Suzuki M, Akiyama R (2012) Molecular dynamics study of fast dielectric relaxation of water around a molecular-sized ion. J Chem Phys 137:224–502

Mizuhara Y, Parkin D, Umezawa K, Ohnuki J, Takano M (2017) Over-destabilization of protein-protein interaction in generalized born model and utility of energy density integration cutoff. J Phys Chem B 121:4669–4677

Nie QM, Togashi A, Sasaki TN, Takano M, Sasai M, Terada TP (2014) Coupling of lever arm swing and biased Brownian motion in actomyosin. PLoS Comput Biol 10:e1003552

Ohnuki J, Sato T, Takano M (2016) Piezoelectric allostery of protein. Phys Rev E 94:012406

Okazaki K, Sato T, Takano M (2012) Temperature-enhanced association of proteins due to electrostatic interaction: a coarse-grained simulation of actin-myosin binding. J Am Chem Soc 134:8918–8925

Rayment I, Holden HM, Whittaker M, Yohn CB, Lorenz M, Holmes KC, Milligan RA (1993) Structure of the actin-myosin complex and its implications for muscle contraction. Science 261:58–65

Sato T, Ohnuki J, Takano M (2017) Long-range coupling between ATP-binding and lever-arm regions in myosin via dielectric allostery. J Chem Phys 147(215):101

Sato T, Ohnuki J, Takano M (2016) Dielectric allostery of protein: response of myosin to ATP Binding. J Phys Chem B 120:13,047–13,055

Schellman JA (1953) The application of the Bjerrum ion association theory to the binding of anions by proteins. J Phys Chem 57:472–475

Sheinerman FB, Norel R, Honig B (2000) Electrostatic aspects of protein-protein interactions. Curr Opin Struct Biol 10:153–159

Sutoh K, Ando M, Sutoh K, Yano Toyoshima Y (1991) Site-directed mutations of Dictyostelium actin: Disruption of a negative charge cluster at the N terminus. Proc Natl Acad Sci USA 88:7711–7714

Suzuki M, Shigematsu J, Fukunish Y, Harada Y, Yanagida T, Kodama T (1997) Coupling of protein surface and hydrophobicity change to atp hydrolysis by myosin motor domain. Biophys J 72:18–23

Suzuki M, Mogami G, Ohsugi H, Watanabe T, Matubayasi N (2017) Physical driving force of actomyosin motility based on the hydration effect. Cytoskeleton 74:512–527

Sweeney HL, Houdusse A (2010) Structural and functional insights into the myosin motor mechanism. Annu Rev Biophys 39:539–557

Takano M, Terada TP, Sasai M (2010) Unidirectional Brownian motion observed in an in silico single molecule experiment of an actomyosin motor. Proc Natl Acad Sci USA 107:7769–7774

Thomas DD, Kast D, Korman VL (2009) Site-directed spectroscopic probes of actomyosin structural dynamics. Annu Rev Biophys 38:347–369

Tonomura Y, Tokura S, Sekiya K (1962) Binding of myosin a to F-actin. J Biol Chem 237:1074–1081

Walker M, Zhang XZ, Jiang W, Trinick J, White HD (1999) Observation of transient disorder during myosin subfragment-1 binding to actin by stopped-flow fluorescence and millisecond time resolution electron cryomicroscopy: evidence that the start of the crossbridge power stroke in muscle has variable geometry. Proc Natl Acad Sci USA 96:465–470

Chapter 9
Protonation/Deprotonation of Proteins by Neutron Diffraction Structure Analysis

Ichiro Tanaka, Katsuhiro Kusaka and Nobuo Niimura

Abstract Neutron protein crystallography can reveal nuclear position and it is very useful to find hydrogen or protonation/deprotonation of protein. It is, however, an intensity-limited experiment and requires large and good quality single protein crystal, so the user population has been so small. Recently, new intense neutron source makes ones to find several protonation states in proteins; PcyA complex (complex of Phycocyanobilin: Ferredoxin Oxidoreductase and Biliverdin IXα), cellulase and substrate complex and farnesyl pyrophosphate synthase (FPPS)-drug complex. At the same time, new techniques for neutron measurement such as high pressure freezing and dynamic nuclear polarization of protein have been also tried to be developed. Finally, a plan of new neutron facility to gain more S/N ratio is expected so that the sample crystal volume can be much small to find protonation/deprotonation.

Keywords Neutron protein crystallography · Proton · Dynamic nuclear polarization · High-pressure freezing · Spallation neutron source

9.1 Introduction

There are several techniques to analyze protein structures in the atomic level. Among them, X-ray protein crystallography (XPC) is the most popular one. On the other hand, neutron protein crystallography (NPC) is one of rare methods; however, it has a unique and powerful potential. Neutrons are much more difficult to produce

I. Tanaka (✉)
College of Engineering, Ibaraki University, Hitachi, Japan
e-mail: ichiro.tanaka.h27@vc.ibaraki.ac.jp

K. Kusaka · N. Niimura
Frontier Research Center for Applied Atomic Sciences,
Ibaraki University, Tokai, Japan
e-mail: katsuhiro.kusaka.1129@vc.ibaraki.ac.jp

N. Niimura
e-mail: nobuo.niimura.n@vc.ibaraki.ac.jp

© Springer Nature Singapore Pte Ltd. 2018

M. Suzuki (ed.), *The Role of Water in ATP Hydrolysis Energy Transduction by Protein Machinery*, https://doi.org/10.1007/978-981-10-8459-1_9

than X-rays concerning to operating cost and beam flux. Indeed, there are presently only several facilities in the world that can conduct NPC experiments. Even so, the fluxes in these facilities are about 10 orders smaller than from X-ray sources (especially synchrotron radiation and free electron laser). This makes NPC experiments to take much more time (typically, more than about 1 week) than XPC experiments (typically, less than 10 min). Therefore, NPC experiments should be used only when other methods cannot provide any information.

The statistics of the Protein Data Bank (PDB) can present clearly the results of these differences. Since the beginning of the PDB in 1970s, the number of folding structures registered in the PDB has increased exponentially year by year; it was more than 130,000 as of November 14, 2017 (Fig. 9.1). About 90% of them, more than 120,000 structures have been determined by XPC, whereas the number of structures determined by NPC is 124 as of November 14, 2017 (Fig. 9.2). It is very small and about 1/1000 of XPC.

Neutron is, however, still valuable because one can see hydrogen easily. It is strongly diffracted by hydrogen nuclei. X-rays are little diffracted by atomic nuclei but by electron, and X-rays are not strongly diffracted by hydrogen or its isotopes, deuterium, and tritium since hydrogen has only one electron. Therefore, neutrons can give ones unique information about hydrogen atoms; especially, protonation states and water orientations including two hydrogens per molecule. Generally,

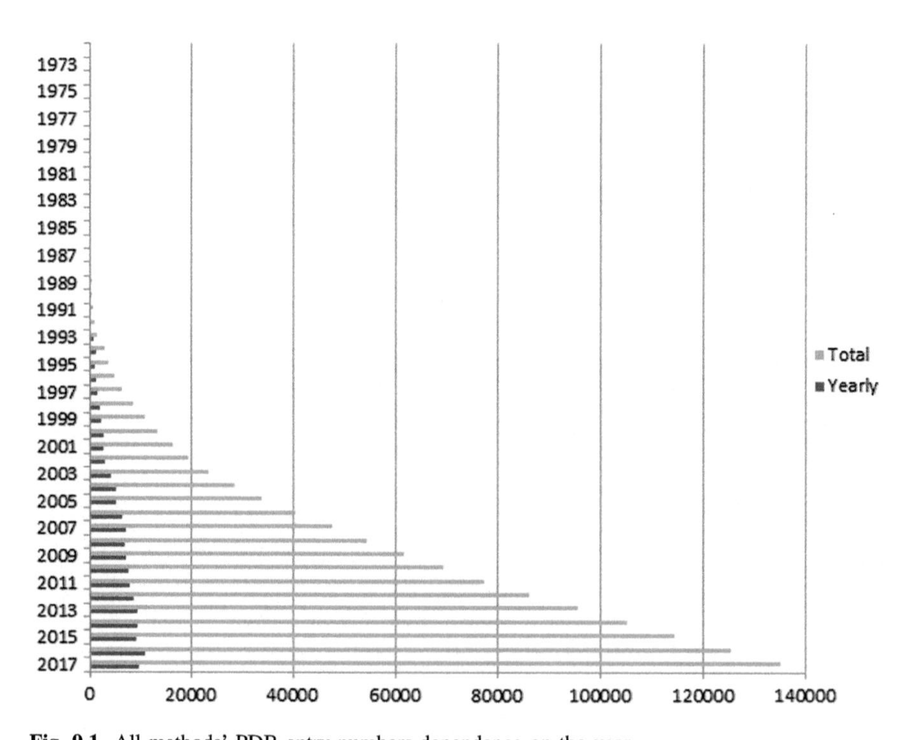

Fig. 9.1 All methods' PDB entry numbers dependence on the year

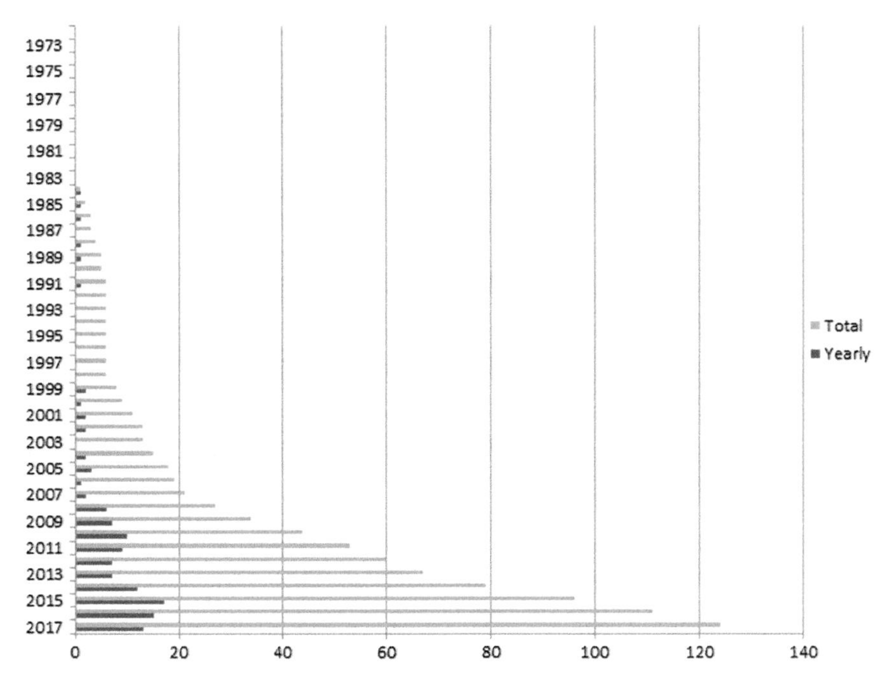

Fig. 9.2 Neutron PDB entry numbers dependence on the year

about half of the constituent atoms of a protein are hydrogen, and in principle, all the hydrogen atoms including protonation states can be identified by neutron diffraction. So, the purpose of NPC is not to determine the folding structure of biomolecules but rather the structure that is involved in chemical reactions. By identifying the hydrogen atoms in and around the biomolecule, NPC can provide the next step beyond the folding structure and give ones information about the hydrogen bonding, protonation states, and hydration configuration of the biomolecule. These are critical features to understand how it actually functions. This unique information makes NPC a powerful tool as the new structural biology.

The general subject of NPC has been reviewed by several authors (Niimura and Bau 2008; Blakeley 2009; Niimura and Podjarny 2011; Niimura et al. 2016), so here only the latest progresses and the expected advances in the near future were described.

9.2 Neutron Specific Features for Biomacromolecules

The radiation damage of protein crystals strongly depends on the relationship between energy and wavelength. A neutron is a particle with a significant mass but an X-ray is a photon particle with no mass (an electromagnetic wave). As a result, the relationship between the energy and the wavelength is very different for each "particle." The energies of an X-ray and a neutron of 1 Å in wavelength are about 12.4 keV and 82 meV, respectively, and the ratio of which is more than 10^5. Since the energy of a 1 Å neutron is less than that of hydrogen bond (about 20 kJ/mol), neutrons do not bring about radiation damage to proteins. This is why it is not necessary in NPC experiments to freeze crystals to liquid nitrogen temperatures to avoid or delay radiation-induced damage. NPC experiment can be carried out at ambient temperatures, such as 293 K. And the same crystal which has been used for a neutron diffraction experiment can be measured later in an X-ray diffraction experiment to provide the initial model for NPC. Moreover, the NPC might be useful to conduct the diffraction experiment of redox-free proteins, though the XPC may cause a redox reaction in metalloproteins or oxidoreductase by the radiation-induced radicals.

The neutron scattering lengths (NSL) and X-ray atomic scattering factors of some elements are the same dimension, those of which correspond to amplitudes scattered by each atom, but the tendencies on elements are different. The Fourier map of neutron diffraction data should be called as neutron scattering length (NSL) density map, not as nuclear density map nor neutron density map, because it has the best physical and mathematical meaning (Niimura et al. 2016).

Distinctive features of neutrons are as follows:

(1) An important technical issue is that hydrogen atoms have very large incoherent-scattering cross sections (80 barns). The nature of the incoherent scattering causes a high background in the diffraction pattern which makes weak signals to be undistinguished from the background, so it is recommended that hydrogen (^{1}H) atoms should be replaced by deuterium (^{2}H or D), because the incoherent-scattering cross sections of deuterium atoms are very small (2 barns). To replace H to D partially, hydrogenated protein crystals are soaked in heavy water (D_2O) solutions. By doing this, ones can obtain the H/D exchange ratio of exchangeable hydrogen atoms in proteins. For full replacement of H, cells are grown in D_2O with full-deuterated culture medium to obtain fully deuterated protein molecules, and then the crystallization should be carried out in D_2O solutions.

(2) The cancellation or neutralization of NSL density in Fourier map somtimes occur. Because the b_H of hydrogen atom is negative, at medium or low resolution (>2.0 Å), the Fourier map of a methylene group ($-CH_2-$) nearly cancels the carbon and two hydrogen atoms each other. This is because the sum ($b_C + 2 \times b_H$) becomes close to zero.

(3) The hydrogen atoms of the $C(NH_2)_3$ group of arginine, the NH_2 group of lysine, and the NH group of amide are easily replaced by deuterium atoms and become $C(ND_2)_3ND$ and ND after soaking crystals in D_2O. They highly become visible in Fourier map. On the other hand, the b_S of a sulfur atom is comparatively small; therefore, the –S–S– bonds are rather difficult to be observed.

9.3 Neutron Sources and Diffractometers for NPC in the World

There are not a small number of neutron sources in the world. But there are very few diffractometers installed there for neutron protein crystallography. Including diffractometers under construction or suspended operation, there are only 10 instruments all over the world (Fig. 9.3). From 1990s (in the era of the beginning of modern neutron detection), diffractometers with ^{3}He position-sensitive area detectors and neutron imaging plate (Niimura et al. 1994, 1997; Helliwell 1997) have been constructed in research reactors; BIX-3 (Tanaka et al. 2002) and BIX-4 (Kurihara et al. 2004) in JRR-3, Japan; IMAGINE (Meilleur 2013) in HFIR, US; BioDIFF (Ostermann and Schrader 2015) in FRM-II, German. On the other hand, diffractometers for pulsed neutron have been and will be constructed in spallation neutron sources; iBIX (Tanaka et al. 2009) in J-PARC, Japan; MaNDi (Coates et al. 2010) in SNS, US; NMX (2017) in ESS, Sweden. Generally, pulsed neutron source can generate higher peak neutrons than steady-state reactors (conventional neutron source), so it is expected for ones to obtain better S/N ratio data, that is, to measure sample with less time and with smaller crystal sample. Details will be discussed in Sect. 9.7.

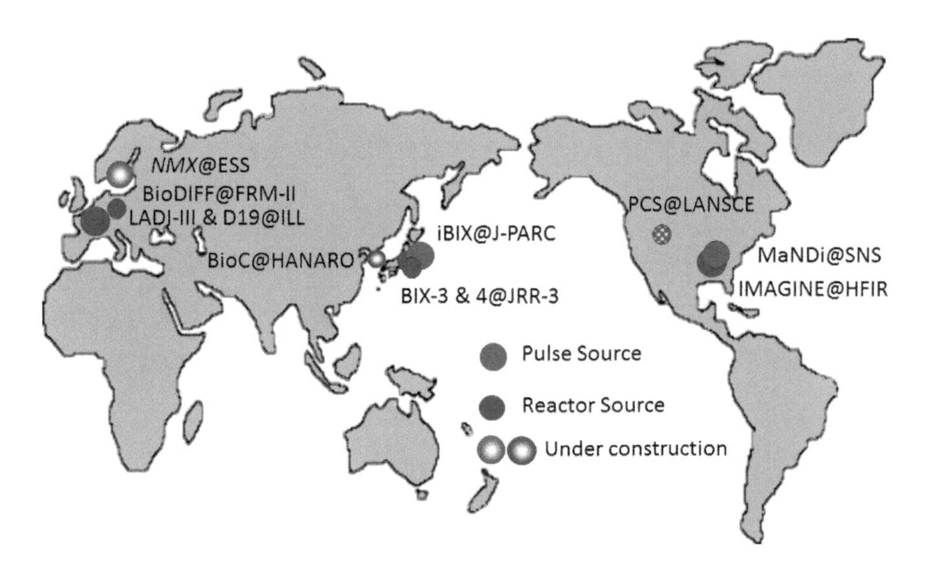

Fig. 9.3 Neutron diffractometers for protein crystallography in the world

9.4 Neutron Diffractometer for Protein Crystallography in J-PARC-iBIX

IBARAKI biological crystal diffractometer called iBIX was constructed in Japan Proton Accelerator Research Complex (J-PARC) to realize high-throughput single crystal neutron structure analysis for mainly biological macromolecules in various life processes at BL03, Material and Life Science Facility (MLF) in 2008 (Tanaka et al. 2010).

To realize high performance, authors have succeeded to develop a new photon-counting two-dimensional detector system using scintillator sheets and wavelength-shifting (WLS) fiber arrays for the X/Y axes (Hosoya et al. 2009). In 2012, authors have upgraded the 14 current detectors and add 16 new detectors for iBIX diffractometer (Fig. 9.4). The total solid angle of detectors subtended by a sample becomes two times, and the average of detector efficiency becomes three times (Kusaka et al. 2013). The total measurement efficiency of iBIX increased 10 times from the previous one with the increased accelerator power. In the end of 2012, iBIX could be started to user experiments for biological macromolecules in earnest. The final specifications of the iBIX are shown in Table 9.1.

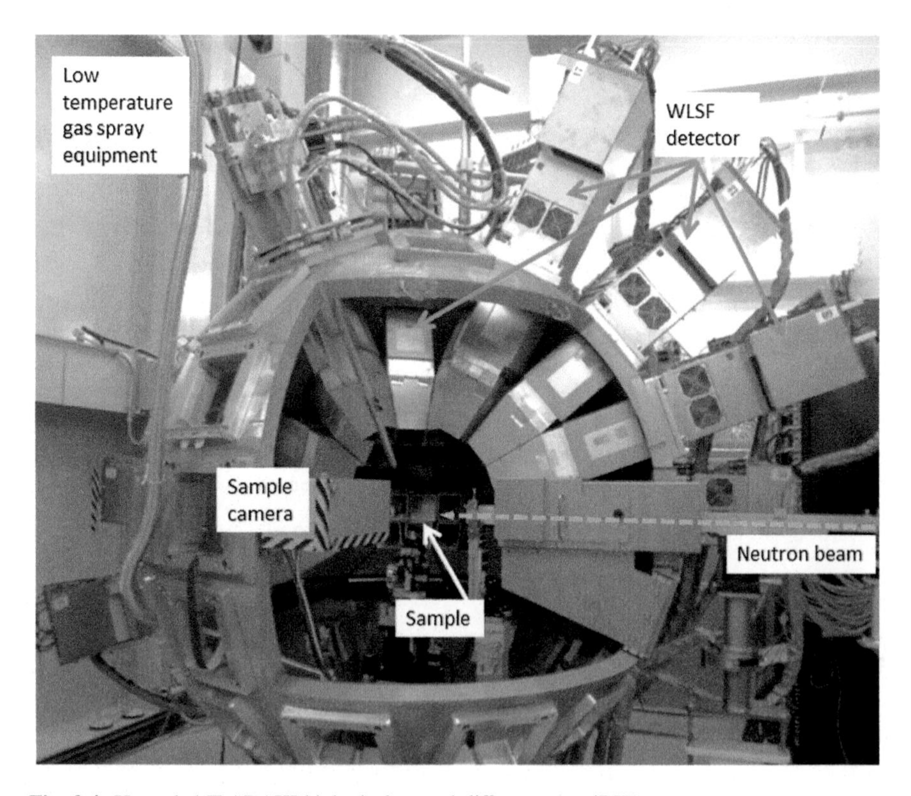

Fig. 9.4 Upgraded IBARAKI biological crystal diffractometer iBIX

Table 9.1 New specifications of iBIX

Moderator	Coupled H_2 (para) 100 × 100 mm^2
Guide tube (m)	25
L1 (m)	40
L2 (m)	0.49
Max. unit cell length ($Å^3$)	~135^3
Measurement region in d spacing (Å)	0.35 < d < 50
Range of neutron wavelength (Å)	0.5 < λ < 9.0 or more
Neutron flux (neutrons/s/cm^2)	7 × 10^7
Sample environment	Gas flow type cooling system, He: ~20 K, N_2: ~90 K
Detector	Two-dimensional, scintillator, wavelength shift fiber type
Size of sensitive area (mm^2)	133 × 133
Spatial resolution (mm)	Less than 1.0
Standard size of sample (mm^3)	1.0
Standard measurement time (days)	0.5 for Organic compounds 3 for Biological macromolecules

In 2018, the accelerator power of J-PARC will become more than 500 kW. Achievement for full data set of biological macromolecules for neutron structure analysis by using iBIX is as below. Maximum unit cell size was 110 × 110 × 70 Å. Average sample volume was about 2 mm^3, and average measurement time was about 7 days. If the accelerator power will become 1 MW, the total measurement time or the sample size will be reduced less than half.

In order to obtain precise integrated intensity of weak reflections, authors developed a profile-fitting method for the peak integration of the data reduction software STARGazer (Ohhara et al. 2009). In order to determine proper-fitting functions, four asymmetric functions were evaluated using strong intensity peaks. The Gaussian function convolved with two back-to-back exponentials was selected as the most suitable fitting function for TOF diffraction data by iBIX. A profile-fitting algorithm for the integration method was developed, and the integration component was implemented in STARGazer. The profile-fitting component was applied to the TOF diffraction data set of standard protein samples obtained by iBIX. From the results, the integrated intensities and model structure obtained by the profile-fitting method were more accurate than those of summation integration method especially at higher resolution shells (Yano et al. 2016). We have prepared the latest user manual and a distribution package of the data reduction software including the profile-fitting component.

To respond to user's needs, authors develop the new equipment of sample environments, cooling system for capillary enclosed sample. Temperature of a capillary enclosed sample is controlled around 0 °C. Both of Peltier cooling system mounted on the goniometer head and N_2 gas flow cooling system were used for sample cooling in order to prevent the condensation of dew inside the capillary.

This equipment can be mounted on the three-axis goniometer. In 2016, authors tried to test measurement with this equipment.

In the near future, the accelerator power of J-PARC will be 1 MW. Then, iBIX will be available regularly for full data set measurement of 1 mm^3 sample size. Authors will continue developing the data reduction software and beam line instruments in order to improve the accuracy of intensity data obtained from small samples in a shorter time. Furthermore, authors will soon enable ones to measure a full data set, to reduce the raw data, and to analyze the structure analysis of the sample with large unit cell (133 × 133 × 133 Å).

Authors should successfully complete to develop the utility equipments for sample environments (heating, extension system for polymer sample, pulse laser system, and cooling system for capillary enclosed sample). And then, those equipments have started to be applied for user experiment since 2016.

As to the sample preparation, it is very important to grow a large single crystal for the NPC experiment practically. Such a kind of statements has been always written in literature of NPC, but it is not so strongly expressed that the high-quality single crystal is important, not the large one. The overall B-factor obtained by modified Wilson plot is one of the best ways to estimate the quality of the crystal universally (Arai et al. 2004).

9.5 Recent Results of iBIX—Protonation/Deprotonation

From the beginning of 2015, three distinctive scientific outcomes have been published, which make the most of the merit of the neutron diffraction experiment by iBIX as follows.

9.5.1 PcyA-BV Complex (Complex of Phycocyanobilin: Ferredoxin Oxidoreductase and Biliverdin IXα)

The neutron crystal structure of the PcyA-BV complex was determined at room temperature. It was observed that approximately half of the BV bound to PcyA was BVH^+, a state in which all four pyrrole nitrogen atoms were protonated (N-protonated), and both of the lactam structures of BV were preserved in wild-type PcyA. A hydronium ion was also found near the active site, which is very rare. The results of them will provide crucial information for elucidating the unique catalytic mechanisms of PcyA (Unno et al. 2015).

9.5.2 *Cellullase and Substrate Complex*

The neutron crystal structures of inverting cellulase unliganded and libanded with cellopentaose were determined at room temperature. Those results indicate a key role of multiple tautomerizations of asparagine residues and peptide bonds, which are finally connected to the other catalytic residue via typical side-chain hydrogen bonds, in forming the "Newton's cradle" like proton relay pathway of the catalytic cycle (Nakamura et al. 2015).

9.5.3 *Farnesyl Pyrophosphate Synthase (FPPS)-Drug Complex*

FPPS, which functions actively when osteoporosis develops, catalyzes the condensation of isopentenyl pyrophosphate (IPP) and dimethylallyl pyrophosphate to FPP. This protein is known to be a molecular target of osteoporosis drugs. Risedronate (RIS), which is a nitrogen-containing bisphosphonate, is one of them. NPC at iBIX and BioDIFF reveals the protonation states and hydration structure of RIS bound to FPPS. According to the structure analysis, it was revealed that the phosphate groups of RIS were fully deprotonated with the abnormally decreased p*K*a clearly, and that the roles of Glu93, Glu168 and Asp264 are canceling the extra negative charges upon the binding of ligands, RIS and IPP, respectively (Yokoyama et al. 2015).

9.6 The Latest Progresses in Neutron Measurements for Protonation/Deprotonation

Though NPC can more easily find hydrogen in macromolecules, it might be difficult to detect hydrogen atoms involved in chemical reactions in the biological systems by conventional neutron methods because they have low occupancies. Proton polarization method, that is, dynamic nuclear polarization (DNP) of hydrogen in protein, is expected to gain hydrogen detection sensitivity about eight times larger than the normal NPC (Niimura and Podjarny 2011; Tanaka et al. 2013, 2018; Zhao et al. 2013). Neutron scattering lengths of atoms, especially, hydrogen, can vary depending on the polarization rates both of protons in sample and neutrons of incident beam as Fig. 9.5. On the contrary, other atoms, for example, carbon, nitrogen, and oxygen, are almost constant compared with hydrogen (Stuhrmann 2004). In this figure, -1 on horizontal axis presents that both spins of neutron and sample proton are antiparallel, and $+1$ presents parallel under both polarization rates are 100%. Zero on horizontal axis presents no polarization condition both of them, that is, both polarization rates are 0%. These features make ones to improve the visibility of hydrogen by subtracting up-spin neutron data from down-spin one

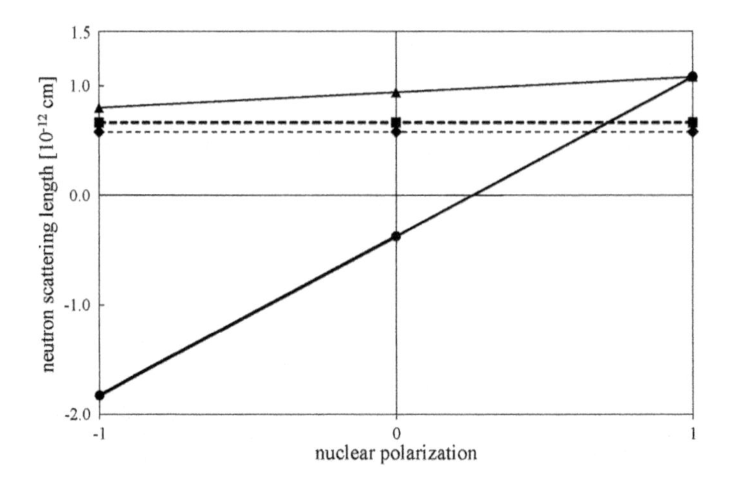

Fig. 9.5 Neutron scattering length dependence on nuclear polarization of each atom. Circle on thick line: ^{1}H, square on thick broken line: ^{12}C: diamond on thin broken line: ^{16}O, triangle on thin line: ^{14}N

under the condition of sample proton highly polarized (Niimura and Podjarny 2011; Tanaka et al. 2013). There are some trials of proton polarization with neutron experiments (Zimmer et al. 2016; Piegsa et al. 2013), NPC applications, however, have not been done yet. Practically, some technical difficulties should be overcome to realize the DNP method in NPC; freezing a large protein single crystal with high successful rate and obtaining higher proton polarization rate of protein sample doped with proper concentration of a radical molecule under low temperature (~1 K) and high magnetic field (~3 T). Even when DNP can be applied, relatively large protein single crystal will be necessary for NPC because of neutron low flux. To freeze larger crystals certainly, high-pressure freezing is one of the reliable and excellent methods for diffraction (Kim et al. 2005). When the 200 MPa pressure is applied on a protein crystal, water viscosity becomes maximum. Then, the crystal is cooled quickly (flash-cooling), depressurized and kept it in liquid nitrogen, and the water becomes vitreous or amorphous solid (not crystalline), the state of which is quasi-stable (Mishima 1996). This high-pressure freezing method does not always require cryoprotectants, which is usually used and surveyed carefully not to affect the crystal quality or structure by itself. So, this freezing method may be also useful in synchrotron radiation X-ray experiment. As tracking protein and solvent movements in chemical reaction could be realized by this cooling method (Kim et al. 2016), cryo-NPC is expected for proton pathway tracing during enzymatic reaction with the DNP technique, too. On the other hand, there was an example that a neutron cryo-crystallography of a protein could explain a reaction mechanism by freezing under normal pressure (Casadei et al. 2014).

Recently, two-key experiments to improve detection sensitivity of hydrogen in protein were carried out; high-pressure freezing as a practical experiment for diffraction, and DNP as the 1st time experiment of proton polarization in protein doped by highly soluble radical TEMPOL, so we show these results as following.

9.6.1 High-Pressure Freezing

High-Pressure Cryo Cooler HPC-201 made by ADC Inc. in 2014 for high-pressure freezing (Fig. 9.6) and EM CPC produced by Leica microsystems Ltd. (Fig. 9.7) were used. The CPC is one of the best useful cryotools because frozen samples are easy to be handled through dried and cooled transparent nitrogen gas under the well-controlled temperature condition and there is little worry to get frost on and around the sample. For various dimensions of sample crystals, two kinds of pressure tubings were prepared; thicker one was specially ordered to ADC Inc. It takes about 3 min in case of thinner tube and about 8 min in case of thicker one to arrive at the maximum pressure (200 MPa), respectively. No cryoprotectants were used in this time.

Frozen crystals were irradiated at an X-ray diffractometer in a synchrotron facility under 100 K nitrogen gas, and full data were collected. As data reduction, HKL2000 (Otwinowski and Minor 1997) or XDS (Kabsch 2010) were used.

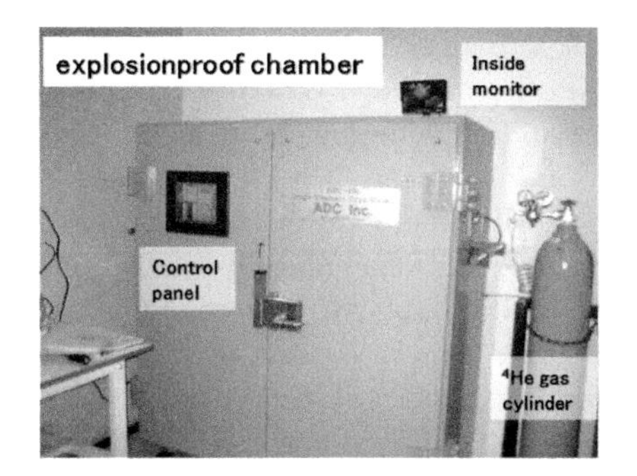

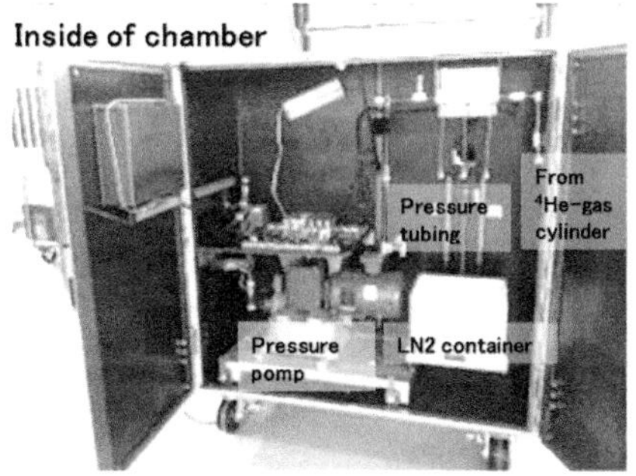

Fig. 9.6 High-pressure freezing machine, HPC-201: outside view (left) and inside view (right)

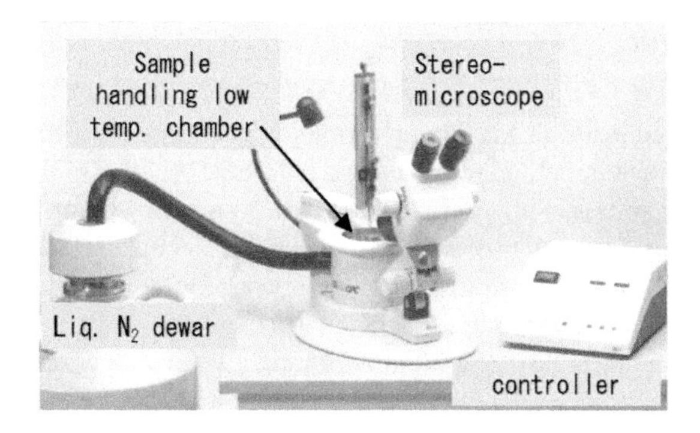

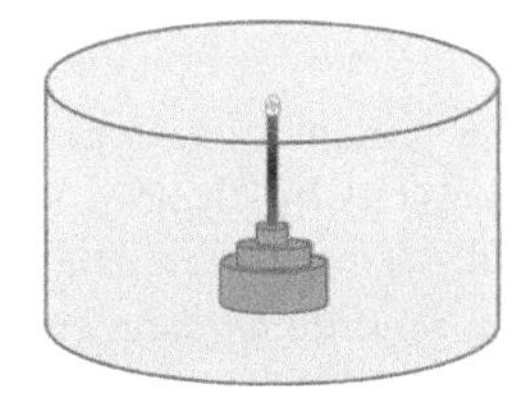

Fig. 9.7 Low-temperature sample handling tool, EM CPC: overview (left) and sample chamber schematic view (right), here one can handle cooled sample in cool nitrogen air in the chamber with stereomicroscope

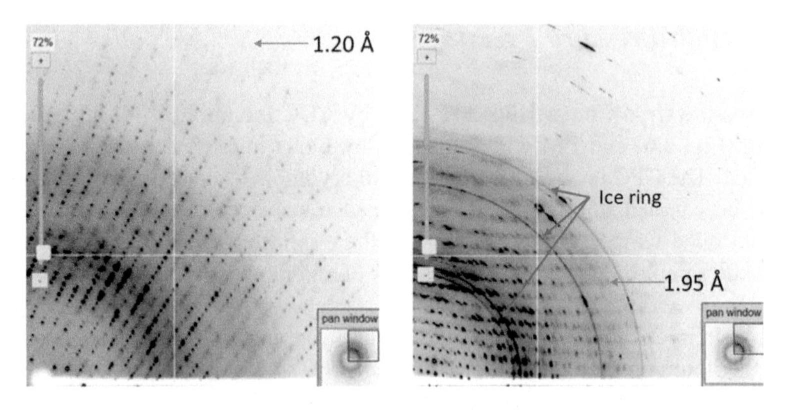

Fig. 9.8 X-ray diffraction image of high-pressure freezing (left) and normal pressure freezing (right) both without cryoprotectants

As structural analysis, Phenix (Adams et al. 2010) and Coot (Emsley and Cowtan 2004) were used.

A relatively large volume lysozyme crystal with no cryoprotectants about 1 mm cube was successfully frozen under 200 MPa by thin tube (2.11 mm diameter) without cryoprotectants, and it diffracted up to 1.25 Å resolution at an 2.5 GeV synchrotron source (Fig. 9.8 left). On the other hand, nearly the same crystal was frozen without cryoprotectant under normal pressure, and it diffracted up to only 2 Å resolution with many ice rings (Fig. 9.8 right). After data reduction and analysis of high-pressure freezing crystal, the R_{merge} is 0.092 (0.378 in outer shell), and R_{work} and R_{free} are 0.212 and 0.237, respectively, with 153 waters at 1.20 Å resolution. This result seems standard one for lysozyme at synchrotron radiation facility.

9.6.2 Electron Spin Resonance Experiments for Dynamic Nuclear Polarization

The protein single crystals doped with TEMPOL in various concentrations were measured by electron spin resonance method with an X-band JES-FA300 (JEOL Ltd.) at room temperature. One protein single crystal (0.03–0.2 mm^3 in volume) was put into a quartz capillary for X-ray after removing buffer at the crystal surface and sealed with a small amount of buffer separated from the crystal for avoiding to be dried. Then, this capillary was put into a normal ESR quartz capillary, and ESR measurements were conducted and calculated the number of radicals based upon an experiment of a certain concentration TEMPOL buffer solution with the nearly same order of volume to protein crystal (~1 mm^3).

ESR measurement was conducted for crystals grown from 0 (no TEMPOL) to 200 mM TEMPOL. TEMPOL has one unpaired electron (radical) per molecule.

Under this crystallization condition, the lysozyme protein grows crystal in a $P4_32_12$ space group and the solvent volume fraction in this crystal is nearly 0.5, so the radical number per lysozyme protein molecule dependence upon TEMPOL concentration when co-crystallization with lysozyme can be calculated. A good linearity between radical number per lysozyme protein molecule and TEMPOL concentration in co-crystallization with lysozyme was observed. As a radical density for DNP, it is said that one radical per 1000 protons is suitable and lysozyme has about 1000 hydrogen per molecule (Bunyatova 1995; Kumada et al. 2012), so crystal in 50 mM TEMPOL condition was selected for the DNP experiment.

9.6.3 Dynamic Nuclear Polarization

To prepare sample for DNP, about 300 mg of lysozyme protein polycrystals grown in TEMPOL 50 mM was collected after taking the moisture around crystal by centrifuging lightly and filled into the Teflon tube of the 10 mm diameter (Fig. 9.9). The filled protein was sandwiched by Teflon sheet with small holes deformed to fit the tube inside surface. The small holes are for liquid helium to permeate the sheet but not for crystals. The mass of sample was necessary for the NMR coil used this

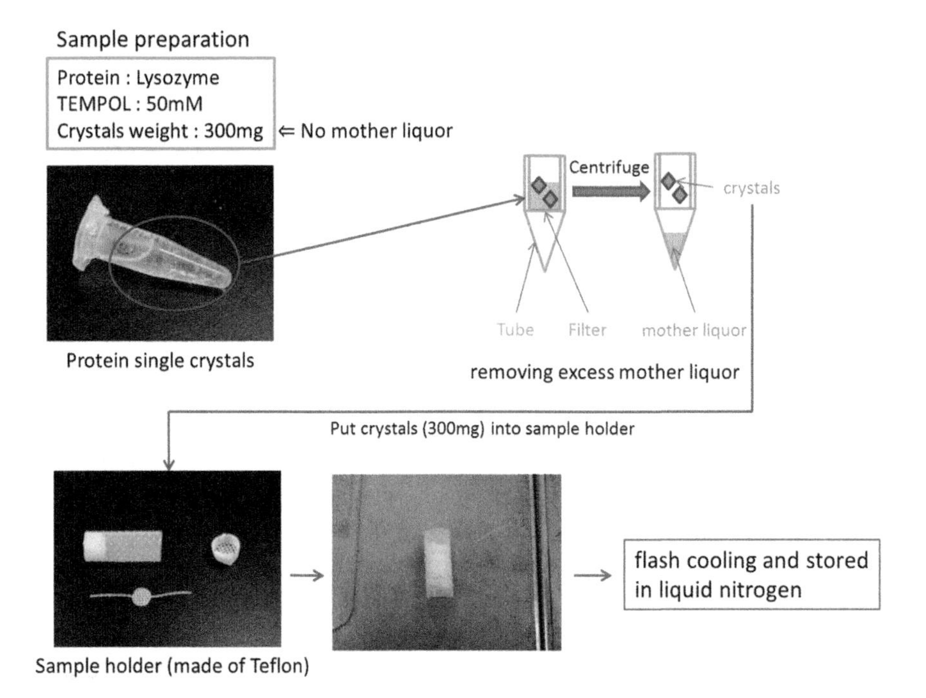

Fig. 9.9 Sample preparation scheme for DNP experiment of protein crystal

Fig. 9.10 Instruments for DNP experiment

time to detect NMR signal in DNP experiment. Then, it was immersed in liquid nitrogen and stored until the DNP measurement. When starting DNP measurement, the sample in Teflon was set to the NMR coil at the bottom of a cryostat in liquid nitrogen. Then, the cryostat was moved to the center of a magnet. In this way, the protein sample was polarized at a temperature of 0.5 K by a ^{3}He refrigerator in a 2.5 T normal conducting magnetic field with about 70 GHz microwave (Fig. 9.10).

After setting temperature at 0.5 K, magnetic field at 2.5 T and microwave frequency at 70 GHZ, it took about 90 min to the maximum proton polarization rate, 22.3 ± 0.7% (Fig. 9.11). At that time, the microwave power was 57 mW. This preliminary polarization experiment had to stop due to the shortage of ^{4}He. The polarization rate is still low for the enhancement of hydrogen visibility. For increasing the rate, the concentration of TEMPOL should be increased, and/or it

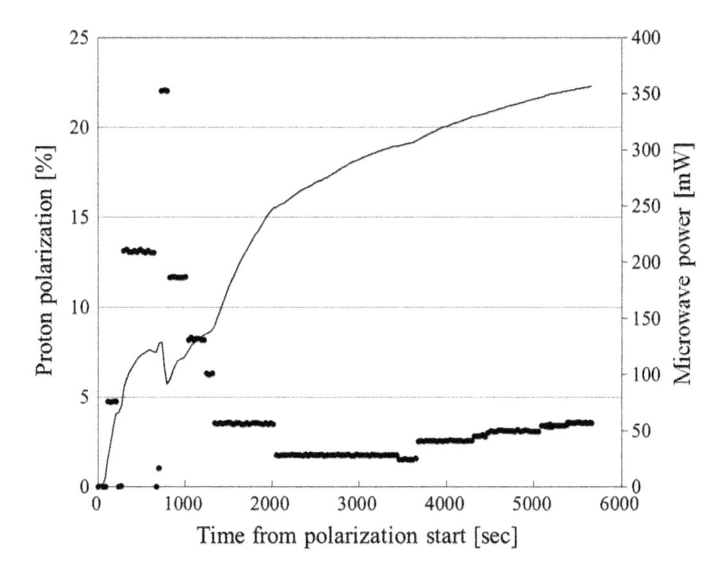

Fig. 9.11 Time dependence of polarization rate and microwave power applied to protein sample; solid curve: proton polarization in left axis, closed circle: microwave power in right axis

might be necessary to reduce dissolved oxygen in solution, because dissolved oxygen in sample to be polarized might make the polarization relaxation time T_1 shorter (de Boer 1973). T_1 at 1.1 K was 527 s at this time, and the temperature dependence of T_1 was similar to other polarized samples (de Boer 1973), so the sample preparation seemed to have no problems but T_1 should be longer to obtain higher polarization rate by anyhow. In addition, sample holding method in this condition for NPC should be also well considered.

9.7 The Future Development in Neutron Sources—Expectation to the Second Target Station in J-PARC

As described in Sect. 9.3, there are very few neutron diffractometers for NPC. In addition, NPC is an intensity-limited experiment, so many efforts have been done to obtain more neutrons or make sample crystal larger and better and so on. Among them, there is a great expectation in the developments of neutron source itself. At present, the neutron source of J-PARC has the brightest neutron beam per pulse among the world's ones. This is because the proton intensity per pulse is the largest. The time-averaged neutron intensity is comparable to reactor sources, but the high pulse peak of neutrons is advantageous to obtain high S/N ratio data in case of using time-resolved detectors which are common in time-of-flight (TOF) measurements. As a result, less volume crystal can be measurable. On the other hand, the

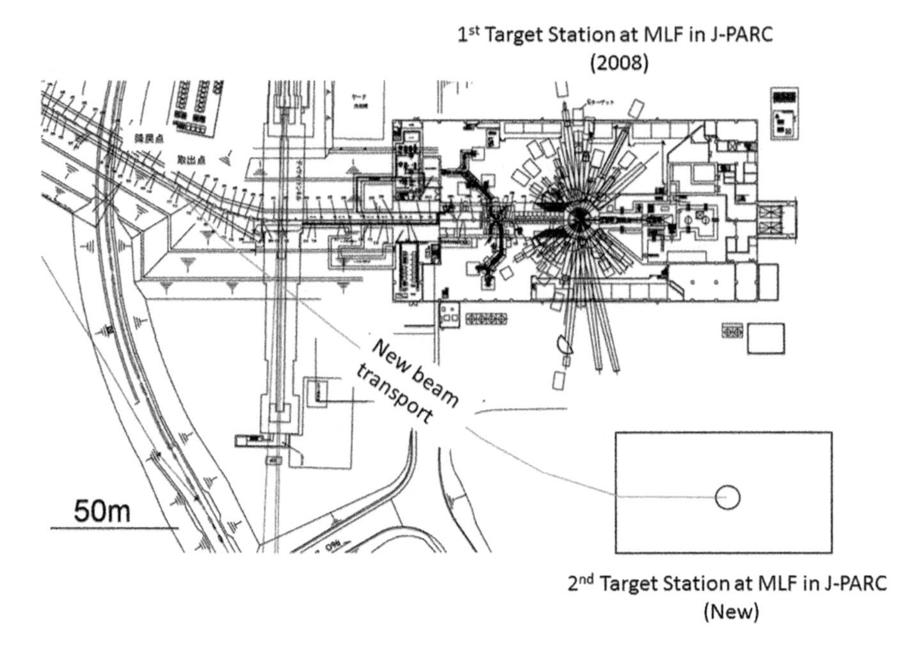

Fig. 9.12 Second target station plan at MLF in J-PARC

measurement time will depend on the time-averaged neutron flux, but it does not affect the S/N ratio (data quality). High peak neutrons available at iBIX in MLF in J-PARC have enabled ones to get the sample crystal size (about 1 mm^3) smaller than in BIX-3 or BIX-4 at JRR-3 (research reactor; about 10 mm^3) by at least one order in volume, even when the proton power in J-PARC is around 300 kW. Recently, a future plan consideration of the second target station at MLF in J-PARC has started (Fig. 9.12), whose peak intensity will be 10 times larger than the current one after arriving 1 MW power (J-PARC Center 2010). This specification is expected for ones to reduce the crystal size to about 0.1 mm^3, so that protonation/deprotonation survey with neutron will develop very much.

9.8 Conclusion

In this chapter, it was described how protonation/deprotonation of protein is studied by neutrons with its specific features, together with examples, the latest progresses in NPC and the future prospect. NPC is valuable in detecting proton easily, but for all practical purposes, is intensity-limited experiment, and one of difficult techniques. In order to overcome them, new attempts such as high-pressure freezing, nuclear polarization by DNP and development of new neutron sources have been conducted and will be planned, respectively. Obtaining direct information of hydrogen or proton experimentally still has been difficult without neutrons, so they will be found to be useful in the future, so ones should understand neutrons better and continue to develop the method and apply to wider field of life sciences and others.

Acknowledgements Authors would thank Profs. T. Iwata and Miyachi and their students in Yamagata University for conducting and discussing ESR and DNP measurements, Prof. Chatake in Kyoto University for preparation of protein crystals and discussion, Dr. T. Kumada in Japan Atomic Energy Agency (JAEA) for providing a radical concentration information, and Drs. H. Seto in High Energy Accelerator Research Organization (KEK) and Harada in JAEA for providing information of the second target station at MLF in J-PARC. Synchrotron radiation experiment was conducted at BL5A of the Photon Factory in KEK, Ibaraki, Japan [2014G650]. Finally, authors are profoundly grateful to iBIX users, Prof. T. Yamada, Dr. N. Yano, Mrs. S. Ninomiya and J. Hiroki, and all students in Tanaka laboratory in Ibaraki University.

References

Adams PD, Afonine PV et al (2010) PHENIX: a comprehensive Python-based system for macromolecular structure solution. Acta Cryst D66:213–221

Arai S et al (2004) More rapid evaluation of biomacromolecular crystals for diffraction experiments. Acta Cryst D 60:1032–1039

Blakely M (2009) Neutron macromolecular crystallography. Cryst Rev 15:157–218

Bunyatova EI (1995) New investigations of organic compounds for targets with polarized hydrogen nuclei. Nucl Instrum Methods Phys Res A 356:29–33

Casadei CM et al (2014) Neutron cryo-crystallography captures the protonation state of ferryl heme in a peroxidase. Science 345:193–197

Coates L, Stoica AD, Hoffmann C, Richards J, Cooper R (2010) The macromolecular neutron diffractometer (MaNDi) at the Spallation Neutron Source, Oak Ridge: enhanced optics design, high-resolution neutron detectors and simulated diffraction. J Appl Cryst 43:570–577

de Boer W (1973) High proton polarization in 1,2-propanediol at ^{3}He temperatures. Nucl Instrum Methods 107:99–104

Emsley P, Cowtan K (2004) Coot: model-building tools for molecular graphics. Acta Cryst D60:2126–2132

Helliwell JR (1997) Neutron Laue diffraction does it faster. Nat Struct Mol Biol 4:874–876

Hosoya T et al (2009) Development of a new detector and DAQ systems for iBIX. Nucl Instrum Methods Phys Res A 600:217–219

J-PARC Center, MLF (2010) Report of J-PARC materials and life science experimental facility future planning task (in Japanese). J-PARC 10-02

Kabsch W (2010) XDS. Acta Cryst D66:125–132

Kim CU, Kapfer R, Gruner SM (2005) High-pressure cooling of protein crystals without cryoprotectants. Acta Cryst D61:881–890

Kim CU et al (2016) Tracking solvent and protein movement during CO_2 release in carbonic anhydrase II crystals. Proc Nat Acad Sci USA 113:5257–5262

Kumada T, Noda Y, Ishikawa N (2012) Dynamic nuclear polarization of electron-beam irradiated polyethylene by pairs of alkyl free radicals. J Magn Reson 218:59–65

Kurihara K, Tanaka I et al (2004) A new neutron single crystal diffractometer dedicated for biological macromolecules (BIX-4). J Synchrotron Radiat 11:68–71

Kusaka K et al (2013) Evaluation of performance for IBARAKI biological crystal diffractometer iBIX with new detectors. J Synchrotron Radiat 20:994–998

Meilleur F et al (2013) The IMAGINE instrument: first neutron protein structure and new capabilities for neutron macromolecular crystallography. Acta Cryst D 69:2157–2160
Mishima O (1996) Relationship between melting and amorphization of ice. Nature 384:546–549
Nakamura A et al (2015) "Newton's cradle" proton relay with amide imidic acid tautomerization in inverting cellulase visualized by neutron crystallography. Sci Adv 1:e1500263 (2015)
Niimura N, Bau R (2008) Neutron protein crystallography: beyond the folding structure of biological macromolecules. Acta Cryst A64:12–22
Niimura N, Podjarny A (2011) Neutron protein crystallography: hydrogen, protons, and hydration in bio-macromolecules. In: IUCr monographs on crystallography 25. Oxford University Press, Oxford
Niimura N, Karasawa Y, Tanaka I et al (1994) An imaging plate neutron detector. Nucl Instrum Methods A349:521–525
Niimura N, Minezaki Y et al (1997) Neutron Laue diffractometry with an imaging plate provides an effective data collection regime for neutron protein crystallography. Nat Struct Mol Biol 4:909–914
Niimura N, Takimoto-Kamimura M, Tanaka I (2016) Application of neutron diffraction in studies of protein dynamics and functions. In: Encyclopedia of analytical chemistry. Wiley, New York, pp 1–30
NMX (Macromolecular Diffractometer) at ESS (European Spallation Source) (2017). https://europeanspallationsource.se/instruments/nmx. Accessed 20 Nov 2017
Ohhara T et al (2009) Development of data processing software for a new TOF single crystal neutron diffractometer at J-PARC. Nucl Instrum Methods A 600:195–197
Ostermann A, Schrader T (2015) BIODIFF: diffractometer for large unit cells. J Large-Scale Res Facil 1:A2
Otwinowski Z, Minor W (1997) Processing of X-ray diffraction data collected in oscillation mode. Methods Enzymol 276:307–326
Piegsa FM et al (2013) Polarized neutron Laue diffraction on a crystal containing dynamically polarized proton spins. J Appl Cryst 46:30–34
Stuhrmann HB (2004) Unique aspects of neutron scattering for the study of biological systems. Rep Prog Phys 67:1073–1115
Tanaka I et al (2002) A high-performance neutron diffractometer for biological crystallography (BIX-3). J Appl Cryst 35:34–40
Tanaka I et al (2010) Neutron structure analysis by IBARAKI biological crystal diffractometer (iBIX) in J-PARC. Acta Cryst D66:1194–1197
Tanaka I, Komatsuzaki N et al (2018) Cryoprotectant-free high-pressure freezing and dynamic nuclear polarization for more sensitive detection of hydrogen in neutron protein crystallography. Acta Cryst D. (submitted)
Tanaka I, Kusaka K et al (2009) Overview of a new biological neutron diffractometer (iBIX) in J-PARC. Nucl Instrum Methods A 600:161–163
Tanaka I, Kusaka K, Chatake T, Niimura N (2013) Fundamental studies for the proton polarization technique in neutron protein crystallography. J Synchrotron Radiat 20:958–961
Unno M et al (2015) Insights into the proton transfer mechanism of a bilin reductase PcyA following neutron crystallography. J Am Chem Soc 137:5452–5460
Yano N et al (2016) Application of profile fitting method to neutron time-of-flight protein single crystal diffraction data collected at the iBIX. Sci Rep 6:36628
Yokoyama T et al (2015) Protonation state and hydration of bisphosphonate bound to farnesyl pyrophosphate synthase. J Med Chem 58:7549–7556
Zhao JK, Robertson L, Herwig K, Crabb D (2013) Polarized neutron in structural biology—present and future outlook. Phys Procedia 42:39–45
Zimmer O, Jouve HM, Stuhrmann HB (2016) Polarized proton spin density images the tyrosyl radical locations in bovine liver catalase. IUCr J 3:326–340

Chapter 10
All-Atom Analysis of Free Energy of Protein Solvation Through Molecular Simulation and Solution Theory

Nobuyuki Matubayasi

Abstract Solvation affects the protein structure strongly, and its effect is quantified by the solvation free energy in statistical thermodynamics. In the present chapter, a fast and accurate method of computation is introduced for the free energy of protein solvation with explicit solvent. The application of the method is then presented for the equilibrium fluctuation in pure-water solvent and the effect of urea on the protein structure. The roles of the solvent water and the urea cosolvent are discussed from the standpoint of energetics, and it is seen that the variation of the protein structural energy is induced and compensated by the solvent water during the course of equilibrium fluctuation. The unfolding mechanism of added urea is also addressed in terms of the energetics of transfer of the protein solute from pure-water solvent to the urea–water mixed solvent, and an extended structure of protein is shown to be favored through the direct, van der Waals interaction between the protein and urea.

Keywords Solvation · Free energy · Solution theory · Molecular dynamics simulation · Cosolvent

10.1 Introduction

The structure of protein is strongly coupled to solvation. Solvent water plays a key role in structure determination of protein, indeed, and addition of urea cosolvent leads to denaturation, for example. The solvation effect is determined by the balance between the attractive and repulsive interactions of protein with the solvent.

N. Matubayasi (✉)
Division of Chemical Engineering, Graduate School of Engineering Science, Osaka University, Toyonaka, Osaka 560-8531, Japan
e-mail: nobuyuki@cheng.es.osaka-u.ac.jp

N. Matubayasi
Elements Strategy Initiative for Catalysts and Batteries, Kyoto University, Katsura, Kyoto 615-8520, Japan

© Springer Nature Singapore Pte Ltd. 2018
M. Suzuki (ed.), *The Role of Water in ATP Hydrolysis Energy Transduction by Protein Machinery*, https://doi.org/10.1007/978-981-10-8459-1_10

The hydrogen bonding acts as an attractive component of intermolecular interaction with the protein, and the hydrophobic effect can be operative as a repulsive component. Since these interaction components are connected to the underlying structure of the solution, all-atom treatment of protein solvation (hydration) is preferable to be conducted with explicit solvent (Karino and Matubayasi 2011; Weber and Asthagiri 2012; Kokubo et al. 2013).

The structure of a protein in solution is determined by the cooperation and/or competition of its intramolecular (structural) energy E_{intra} and the solvation free energy $\Delta\mu$. When $P(\mathbf{X})$ is the probability distribution function of the protein solute being at a certain structure $\mathbf{X}$, it is expressed as

$$-k_B T \log P(\mathbf{X}) = E_{\text{intra}}(\mathbf{X}) + \Delta\mu(\mathbf{X}) + \text{constant} \tag{10.1}$$

where k_B is the Boltzmann constant and T is the temperature. In Eq. 10.1, E_{intra} and $\Delta\mu$ are treated as functions of $\mathbf{X}$ and are to be evaluated at their computations with fixed $\mathbf{X}$. Within the context of statistical thermodynamics, $P(\mathbf{X})$ is the function that captures the structural feature of the protein. When a globular protein keeps its folded structure at the native condition, for example, this means that $P(\mathbf{X})$ is significant only in the set of folded structures. In the unfolding of the protein with heat and/or addition of denaturant, the dominant portion of $P(\mathbf{X})$ shifts to an unfolded set. Equation 10.1 thus shows that the structure of a protein is related to the energetics represented by E_{intra} and $\Delta\mu$; the former is the intramolecular energy of the protein, and the latter is determined by the intermolecular interaction between the protein and solvent. The solvation effect is then quantified by the solvation free energy $\Delta\mu$, and its atomistic treatment is desired for elucidating the protein structure within the framework of statistical thermodynamics.

This chapter presents all-atom treatments of the solvation effect on protein through combination of molecular simulation and the theory of solutions in the energy representation. Molecular simulation is suitable for treating the energetics of protein in solution since it is performed under a given set of intra- and intermolecular potential functions, and in principle, the energetics can be examined at any details. E_{intra} can be readily computed, indeed, whereas the numerically exact computation of $\Delta\mu$ is challenging in practice for such large molecules as protein. Still, a statistical thermodynamic discussion of protein structure can be conducted on the basis of Eq. 10.1 once $\Delta\mu$ is determined. An approximate but accurate method of $\Delta\mu$ evaluation is thus desirably established in the all-atom analysis that faithfully takes into account such interactions as hydrogen bonding and hydrophobic effect.

In Sect. 10.2, we introduce a fast scheme for the $\Delta\mu$ calculation (Matubayasi and Nakahara 2000, 2002, 2003; Sakuraba and Matubayasi 2014). We describe the concept and theoretical background briefly and provide benchmark results with amino-acid analogs. The free-energy scheme is then employed to analyze the hydration effect on the structural fluctuation of the protein at equilibrium (Karino and Matubayasi 2011; Kamo et al. 2016; Yamamori et al. 2016) and the interaction component responsible for protein denaturation by urea (Yamamori et al. 2016;

Yamamori and Matubayasi 2017). In Sect. 10.3, we focus on the equilibrium fluctuation of protein structure in water. The compensation between E_{intra} and $\Delta\mu$ is demonstrated, and the intermolecular interaction governing the structural fluctuation is identified. In Sect. 10.4, the topic is the urea effect on protein structure. The free energy of transfer of a protein from pure-water solvent to a urea–water mixed solvent is treated, and the interaction component governing the transfer energetics is specified. The present chapter is concluded in Sect. 10.5.

10.2 Energy-Representation Method as an Endpoint Calculation of the Solvation Free Energy

The solvation free energy is the free-energy change for turning on the solute–solvent interaction potential. Its calculation requires much demand with explicit solvent for such large molecules as protein, though, when the standard method of free-energy perturbation or thermodynamic integration is to be employed (Shirts and Pande 2005). In the standard method of free-energy calculation, in fact, a number of intermediate states need to be introduced to connect the solvent system without solute and the solution system of interest that involves full coupling of the solute–solvent interaction. The computational load for the solvation free energy $\Delta\mu$ increases in proportion to the number of the intermediate states, which are often not physically meaningful and are employed only for numerical convenience. To avoid the use of intermediate states, we developed the method of energy representation (Matubayasi and Nakahara 2002, 2003; Sakuraba and Matubayasi 2014). It is a theory of distribution functions in solution and is formulated by adopting the solute–solvent pair interaction energy as the coordinate for the distribution functions constituting the free-energy functional. Among a variety of approximate free-energy methods (Levy et al. 1991; Luzhkov and Warshel 1992; Åqvist et al. 1994; Carlson and Jorgensen 1995; Kast 2001; Vener et al. 2002; Fdez Galván et al. 2003; Freedman and Truong 2004; Imai et al. 2006; Chuev et al. 2007; Yamamoto 2008; Frolov et al. 2011), the energy-representation method is unique in terms of the accuracy, efficiency, and range of applicability (Takahashi et al. 2004; Matubayasi et al. 2006, 2008; Karino et al. 2010; Takahashi et al. 2008; Kawakami et al. 2012; Takemura et al. 2012; Mizukami et al. 2012; Saito et al. 2013; Karino and Matubayasi 2013; Frolov 2015; Ishizuka et al. 2015; Date et al. 2016). Figure 10.1 illustrates the scheme of $\Delta\mu$ computation in the method of energy representation. The simulation is performed only at the initial and final states (endpoints) for turning on the solute–solvent interaction, and $\Delta\mu$ is obtained through an approximate functional from a set of energy distribution functions computed in the endpoint simulations. With this scheme, $\Delta\mu$ can be feasibly calculated in explicit solvent for a protein with a few hundred residues (Karino and Matubayasi 2011; Yamamori et al. 2016; Yamamori and Matubayasi 2017; Takemura et al. 2012; Mizukami et al. 2012; Saito et al. 2013), while it was observed for small molecules

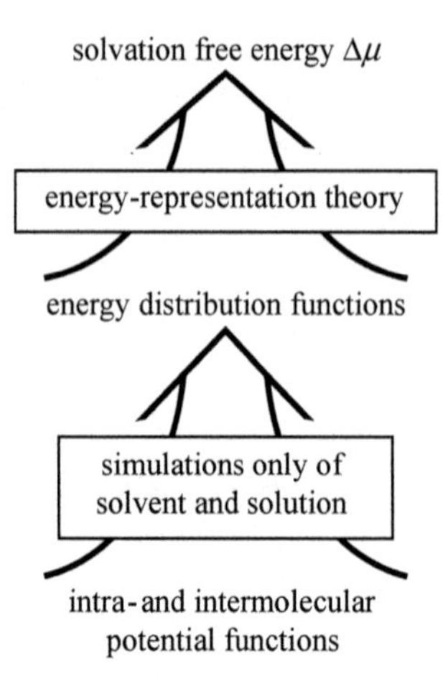

Fig. 10.1 Computation scheme of the solvation free energy in the method of energy representation

that the error from the approximation in the functional is not larger than the error due to the use of force field (Karino et al. 2010; Karino and Matubayasi 2013; Frolov 2015).

In the energy-representation method, the value of the solute–solvent pair interaction v of interest is employed as the coordinate ε for the solute–solvent distribution, and the instantaneous distribution $\hat{\rho}$ at a snapshot configuration of solute and solvent is defined as (Matubayasi and Nakahara 2000)

$$\hat{\rho}(\varepsilon) = \sum_i \delta(v(\psi, \mathbf{x}_i) - \varepsilon) \tag{10.2}$$

where ψ denotes the configuration of the solute molecule, $\mathbf{x}_i$ refers to the configuration of the ith solvent molecule, and the sum is taken over all the solvent molecules. Let $\rho(\varepsilon)$ and $\rho_0(\varepsilon)$ be the ensemble averages of Eq. 10.2 in the solution system of interest (final state of solute insertion) and in the solvent system with the

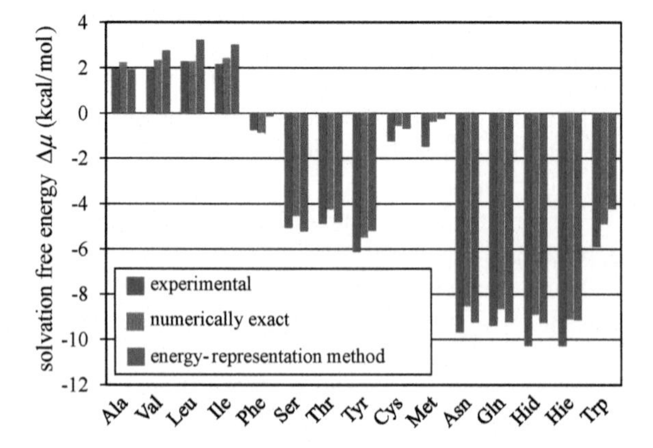

Fig. 10.2 Solvation free energies $\Delta\mu$ from experiments and computations adopting the OPLS-AA force field for the solutes and TIP3P for water. Hid and Hie refer to the neutral, protonated states of His, in which the proton is attached to the δ and ε nitrogens, respectively

solute uncoupled (initial state), respectively; in the latter, the solute is placed in the system as a test (virtual) particle and does not disturb the solvent configuration. $\rho(\varepsilon)$ corresponds to the averaged histogram of the solute–solvent pair energy in the solution, and when the solvent is homogeneous without solute, $\rho_0(\varepsilon)$ is equal to the product of the (number) density of bulk solvent and the density of states for the solute–solvent pair potential. The solvation free energy $\Delta\mu$ is then expressed exactly as

$$\Delta\mu = \int d\varepsilon \varepsilon \rho(\varepsilon) - k_B T \int d\varepsilon \left[(\rho(\varepsilon) - \rho_0(\varepsilon)) - \rho(\varepsilon) \log\left(\frac{\rho(\varepsilon)}{\rho_0(\varepsilon)}\right) - (\rho(\varepsilon) - \rho_0(\varepsilon))\,\Omega(\varepsilon) \right] \tag{10.3}$$

where $\Omega(\varepsilon)$ represents the contribution due to the change in the solvent–solvent correlation caused by the insertion of the solute. In the current implementation of the energy-representation method, the $\Omega(\varepsilon)$ term is treated by combined hypernetted-chain (HNC)-type and Percus–Yevick (PY)-type approximations, with its explicit form and methodological details given in references (Matubayasi and Nakahara 2002; Sakuraba and Matubayasi 2014). In Eq. 10.3, the first term is the average sum of the solute–solvent interaction energy in the solution system of interest, and the second term takes into account the effect of solvent reorganization including the excluded-volume effect.

The performance of an approximate method may be quantified by examining the mean absolute deviation (MAD) defined as

$$\frac{1}{n}\sum_i \left|\Delta\mu_{i,1} - \Delta\mu_{i,2}\right| \tag{10.4}$$

where $\Delta\mu_{i,1}$ and $\Delta\mu_{i,2}$ denote the solvation free energies of the ith solute in the first and second sets of the data, respectively, and n is the number of solutes examined. The computed values from the energy-representation method are shown in Fig. 10.2 for the neutral amino-acid analog solutes listed in Table 10.1 (Shirts and Pande 2005; Wolfenden et al. 1981). When the TIP3P model is employed for water,

Table 10.1 Correspondence between the amino-acid side chains and their analog solutes

Amino acid	Analog solute	Amino acid	Analog solute
Ala	Methane	Cys	Methanethiol
Val	Propane	Met	Methyl ethyl sulfide
Leu	*iso*-butane	Asn	Acetamide
Ile	*n*-Butane	Gln	Propionamide
Phe	Toluene	Hid	4-Methylimidazole
Ser	Methanol	Hie	4-Methylimidazole
Thr	Ethanol	Trp	3-Methylindole
Tyr	*p*-Cresol		

MAD between the energy-representation method and the numerically exact method is 0.5 and 0.4 kcal/mol, respectively, for the OPLS-AA and Amber99sb force fields (Jorgensen et al. 1983, 1996; Hornak et al. 2006). MAD for the experimental values is actually 0.7 and 1.3 kcal/mol against the numerically exact values with OPLS-AA and Amber99sb, respectively, while it is 0.7 and 1.6 kcal/mol against the approximate values obtained with the energy-representation method (Karino et al. 2010; Karino and Matubayasi 2013). MAD with the experimental value is similar between the exact and approximate methods, and thus, the method of energy representation can be as useful as the exact $\Delta\mu$ calculation for practical purpose. Benchmark computations are more demanding for larger solutes. Alanine dipeptide and β-cyclodextrin were treated by an exact method in water, and it was seen that while the deviation between the exact and approximate values of $\Delta\mu$ enhances with the solute size, the accuracy is kept for the $\Delta\mu$ difference between conformational states (Harris et al. 2017).

10.3 Equilibrium Fluctuation in Pure-Water Solvent: The Compensation Between the Intramolecular and Hydration Effects and the Governing Interaction Identified by Correlation Analysis

The structure of protein fluctuates through the interaction with solvent. The structural distribution during the equilibrium fluctuation is connected to the intramolecular and intermolecular interactions of protein through Eq. 10.1. To examine the relationship between the protein structure and the solvation (hydration) effect in terms of energetics (Karino and Matubayasi 2011; Kamo et al. 2016; Yamamori et al. 2016), Fig. 10.3 shows the correlation between the intramolecular (structural) energy E_{intra} and the solvation free energy $\Delta\mu$ for Trp-cage and cytochrome *c* (indexed by PDB codes of 1L2Y and 1HRC, respectively Neidigh et al. 2002; Bushnell et al. 1990) in pure-water solvent. In Fig. 10.3, each point corresponds to a single (snapshot) structure of protein that was sampled during the equilibrium fluctuation in water. It is then seen that in the variation of the protein structure, E_{intra} and $\Delta\mu$ are anti-correlated to each other with a slope of -1 from the least-square fit. The variation ranges of E_{intra} and $\Delta\mu$ are on the order of 100 kcal/mol, and the range of $(E_{\mathrm{intra}} + \Delta\mu)$ variation is much smaller. It is thus evidenced in Fig. 10.3 that the solvent water induces and compensates a large fluctuation of the protein structural energy.

It is then of interest to identify the component of protein–water interaction that governs the variation of $\Delta\mu$. To do so, we will correlate $\Delta\mu$ against three components in the free-energy functional expressed as Eq. 10.3. The first two are the electrostatic and van der Waals components of the average sum of the solute–solvent interaction energy in the solution system of interest (protein–water system). The first term of Eq. 10.3 corresponds to the average sum and is further given as a

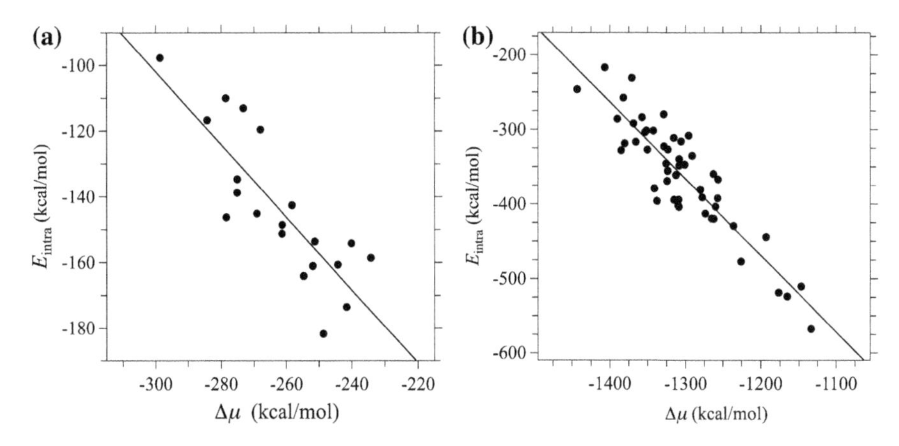

Fig. 10.3 Correlation plot between the protein intramolecular (structural) energy E_{intra} and the solvation free energy $\Delta\mu$ for (**a**) Trp-cage and (**b**) cytochrome *c*, where a single point in the figure corresponds to a single (snapshot) structure of protein. The solid line represents the linear regression from the least-square fit of E_{intra} against $\Delta\mu$. In (**a**), the slope is -1.1 with the correlation coefficient of -0.83, and in (**b**), the slope and the correlation coefficient are -1.0 and -0.90, respectively

sum of the electrostatic and van der Waals components denoted as $\langle v \rangle_{\mathrm{elec}}$ and $\langle v \rangle_{\mathrm{vdW}}$, respectively (Karino and Matubayasi 2011; Yamamori et al. 2016). They are intuitively appealing parts of $\Delta\mu$ and are computed at each (fixed) structure $\mathbf{X}$ of the protein. The third component examined for the correlation to $\Delta\mu$ is the excluded-volume effect $\Delta\mu_{\mathrm{excl}}$, which makes a major contribution to the second term of Eq. 10.3. This effect refers to the free-energy penalty corresponding to the displacement of solvent molecules from the region to be occupied by the solute at solvation. The excluded-volume effect is a repulsive component of solute–solvent interaction and is pointed out by Kinoshita as a decisive factor for structure formation of large molecule (Kinoshita 2013). Just as $\langle v \rangle_{\mathrm{elec}}$ and $\langle v \rangle_{\mathrm{vdW}}$, $\Delta\mu_{\mathrm{excl}}$ is intuitively appealing but is not an observable. It can be defined only by adopting a model of solvation and can be naturally introduced within the framework of the energy-representation method by restricting the domain of integration in Eq. 10.3 to a high-energy region above ε^c, where ε^c is the threshold for specifying the excluded-volume domain (Karino and Matubayasi 2011; Yamamori et al. 2016). It is set with a requirement that the domain of $\varepsilon > \varepsilon^c$ corresponds to the overlap of the solute with a solvent molecule and is (numerically) inaccessible in the solution system of interest ($\rho(\varepsilon) = 0$ for $\varepsilon > \varepsilon^c$). Here, ε^c is taken to be 20 and 25 kcal/mol for Trp-cage and cytochrome *c*, respectively, while the following discussion holds irrespective of the (reasonable) setting of the ε^c value.

In Fig. 10.4, the (total) solvation free energy $\Delta\mu$ is correlated against $\langle v \rangle_{\mathrm{elec}}$, $\langle v \rangle_{\mathrm{vdW}}$, and $\Delta\mu_{\mathrm{excl}}$. The correlation is clearly present for the electrostatic component $\langle v \rangle_{\mathrm{elec}}$ of the protein–water interaction energy. A protein structure with more negative $\Delta\mu$ is more favored by solvent water and appears with stronger $\langle v \rangle_{\mathrm{elec}}$. The slope is 2.0 for the least-square fit of $\langle v \rangle_{\mathrm{elec}}$ against $\Delta\mu$, and this strong correlation

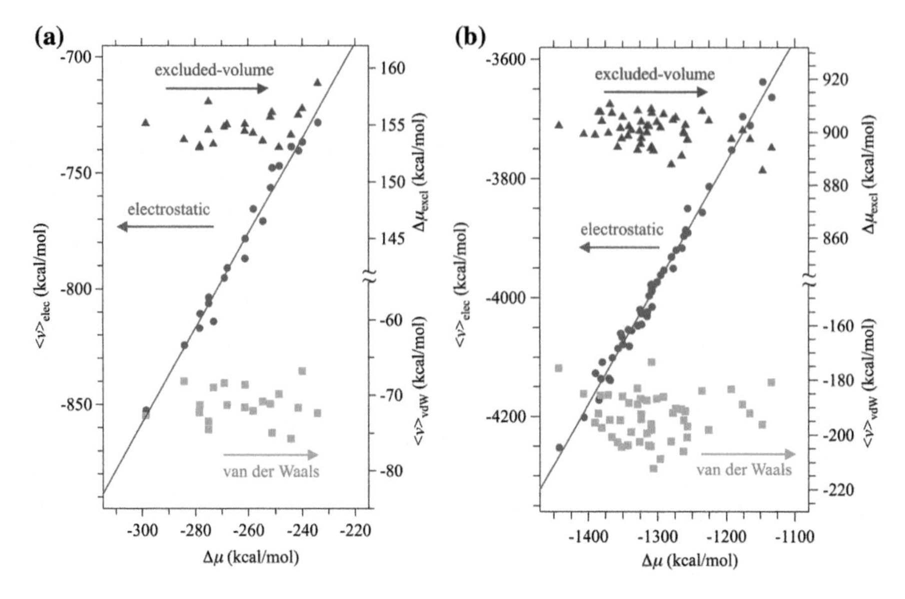

Fig. 10.4 Correlation plots of $\Delta\mu$ against the electrostatic component $\langle v \rangle_{elec}$ and the van der Waals component $\langle v \rangle_{vdW}$ of the average sum of the solute–solvent interaction energy in the solution system and the excluded-volume component $\Delta\mu_{excl}$ for (**a**) Trp-cage and (**b**) cytochrome *c*, where a single point in the figure corresponds to a single (snapshot) structure of protein. The solid line represents the linear regression from the least-square fit of $\langle v \rangle_{elec}$ against $\Delta\mu$. The slope is 2.0 with the correlation coefficient of 0.99 in both (**a**) and (**b**)

can be interpreted in terms of the exchange of hydrogen bonds between the protein solute and the solvent water. When one hydrogen bond within the protein and one hydrogen bond in water solvent are broken and transformed to two hydrogen bonds between protein and water, the change in E_{intra} corresponds to half the (sign-reversed) change in $\langle v \rangle_{elec}$. The correlation between $\Delta\mu$ and $\langle v \rangle_{elec}$ in Fig. 10.4 then follows since the variations of E_{intra} and $\Delta\mu$ compensate against each other as shown in Fig. 10.3. On the other hand, the van der Waals component $\langle v \rangle_{vdW}$ and the excluded-volume component $\Delta\mu_{excl}$ do not correlate to $\Delta\mu$ in Fig. 10.4. Their ranges of variation are smaller by order of magnitude than the variation range of $\langle v \rangle_{elec}$, and according to Fig. 10.4, therefore, the equilibrium fluctuation of protein structure in pure-water solvent is governed by the electrostatic interaction between protein and water with minor contributions from the van der Waals interaction and the excluded-volume effect.

In the above, the interaction component governing the equilibrium fluctuation of protein in pure-water solvent was analyzed in terms of correlations. It should be noted, in fact, that the effect of each component or parameter within intramolecular and intermolecular interaction potentials cannot be singled out in general when the molecular structure is varied in realistic manner. This is because an interaction component or parameter cannot be independently tuned for real system, and in such situations, a correlation analysis can be useful for identifying the factors governing the phenomena of interest. Indeed, a correlation is often enough for interpretation and prediction, though it does not necessarily imply a causal relationship. $\Delta\mu$ carries all the information of solvation effect, and our analysis reveals which interaction component in $\Delta\mu$ varies with the (total) $\Delta\mu$.

10.4 Energetic Mechanism of Unfolding by Urea

The structure of a flexible solute molecule such as protein changes in response to the surrounding environment. The solvent environment can then be varied by addition of a cosolvent, and the urea-induced unfolding of protein is a drastic example of modifying the structure of a protein solute by (mixed) solvent. The unfolding mechanism by urea has been treated with a variety of standpoints (Stumpe and Grubmüller 2007; Lindgren and Westlund 2010; Horinek and Netz 2011; Kokubo et al. 2011; Paul and Paul 2015; Ikeguchi et al. 2001; van der Vegt et al. 2004; Hua et al. 2008; Canchi et al. 2010; Canchi and García 2011, 2013; Moeser and Horinek 2014). In this section, the urea effect is addressed in the context of solvation energetics. According to Eq. 10.1, addition of the urea cosolvent leads to the change in the probability distribution function of the structure $\mathbf{X}$ of the protein solute through

$$P^{\mathrm{mix}}(\mathbf{X}) \propto P^{\mathrm{wat}}(\mathbf{X}) \exp\left(-\left(\Delta\mu^{\mathrm{mix}}(\mathbf{X}) - \Delta\mu^{\mathrm{wat}}(\mathbf{X})\right)/k_B T\right) \tag{10.5}$$

where P^{wat} and $\Delta\mu^{\mathrm{wat}}$ are the distribution function and the solvation free energy in pure-water solvent, respectively, and P^{mix} and $\Delta\mu^{\mathrm{mix}}$ are those in the mixed solvent of water and urea. Equation 10.5 holds for each (snapshot) structure $\mathbf{X}$ and shows that upon addition of the cosolvent, the population increases more for a structure $\mathbf{X}$ with more negative $(\Delta\mu^{\mathrm{mix}} - \Delta\mu^{\mathrm{wat}})$. $(\Delta\mu^{\mathrm{mix}} - \Delta\mu^{\mathrm{wat}})$ thus plays a key role in investigating the cosolvent effect on the structure of flexible solute. This difference is called transfer free energy and is the free-energy change upon transfer of the solute from pure-water solvent to the mixed solvent of water and cosolvent. The focus of the present section is the free energy of transfer from pure-water solvent to the mixture of urea with water (Yamamori et al. 2016; Yamamori and Matubayasi 2017). We conduct correlation analyses for $(\Delta\mu^{\mathrm{mix}} - \Delta\mu^{\mathrm{wat}})$ of cytochrome *c* and T4-lysozyme (indexed by PDB codes of 1HRC and 1LYD, respectively Bushnell et al. 1990; Rose et al. 1988).

The correlation between the transfer free energy and a structural index is shown in Fig. 10.5 for the two proteins. A negative correlation is seen, and a more negative $(\Delta\mu^{\mathrm{mix}} - \Delta\mu^{\mathrm{wat}})$ corresponds to a protein structure exposed more to solvent. The transfer free energy is a thermodynamic measure to quantify the difference in the solvent environments with and without the cosolvent, and Fig. 10.5 sets a basis for addressing the urea effect on the protein structure in terms of the solvation energetics.

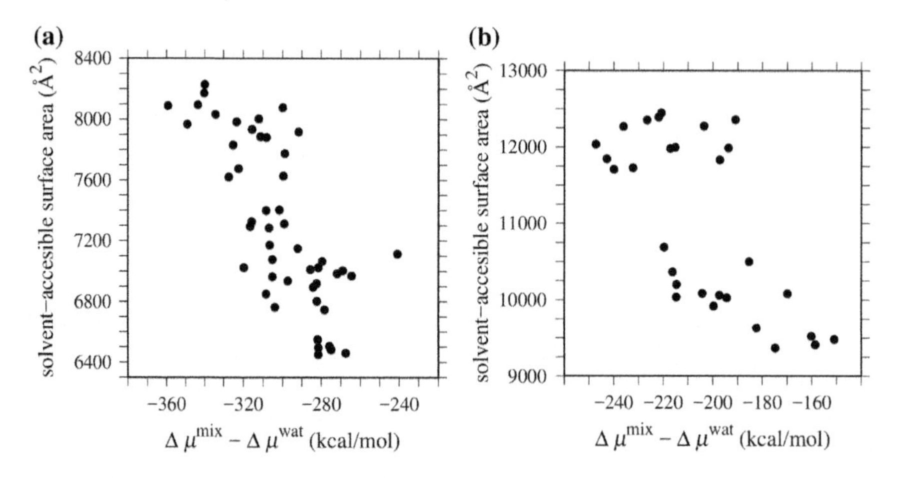

Fig. 10.5 Correlation plot of the solvent-accessible surface area with the free energy of transfer $(\Delta\mu^{mix} - \Delta\mu^{wat})$ of (**a**) cytochrome *c* from pure-water solvent to 8 M mixture of urea with water and of (**b**) T4-lysozyme to 3 M mixture, where a single point in the figure corresponds to a single (snapshot) structure of protein. The solvent-accessible surface area is calculated by defining the protein surface as the one at contact with the spherical probe that mimics the water molecule and has a radius of 1.4 Å. The correlation coefficient is −0.75 and −0.66 in (**a**) and (**b**), respectively

To specify the intermolecular interaction that governs the effect of added urea, we examine the correlations of the transfer free energy against the changes upon transfer of the electrostatic and van der Waals components of the solute–solvent interaction energy and of the excluded-volume effect. The correlations were analyzed in the previous section for the solvation free energy and the electrostatic, van der Waals, and excluded-volume components in pure-water solvent, while the focus in this section is on the differences between the mixed solvent of urea and water and the pure solvent of water. In Fig. 10.6, the transfer free energy $(\Delta\mu^{mix} - \Delta\mu^{wat})$ is correlated against the changes upon transfer of the solute–solvent energetics written as $\left(\langle v\rangle_{elec}^{mix} - \langle v\rangle_{elec}^{wat}\right)$, $\left(\langle v\rangle_{vdW}^{mix} - \langle v\rangle_{vdW}^{wat}\right)$, and $\left(\Delta\mu_{excl}^{mix} - \Delta\mu_{excl}^{wat}\right)$, where $\langle v\rangle_{elec}^{mix}$ and $\langle v\rangle_{vdW}^{mix}$ are the electrostatic and van der Waals components of the average sum of the solute–solvent interaction energy in the mixed solvent, respectively, and $\langle v\rangle_{elec}^{wat}$ and $\langle v\rangle_{vdW}^{wat}$ are the corresponding quantities in pure-water solvent. $\Delta\mu_{excl}^{mix}$ and $\Delta\mu_{excl}^{wat}$ denote the excluded-volume components in the mixed solvent and pure water, respectively, and are introduced through Eq. 10.3 by employing a high-energy cutoff as done in the analysis for Fig. 10.4 (Yamamori et al. 2016; Yamamori and Matubayasi 2017). According to Fig. 10.6, $(\Delta\mu^{mix} - \Delta\mu^{wat})$ correlates well with the van der Waals component $\left(\langle v\rangle_{vdW}^{mix} - \langle v\rangle_{vdW}^{wat}\right)$ of the solute–solvent energy of transfer. On the other hand, the correlation is weaker against the electrostatic component $\left(\langle v\rangle_{elec}^{mix} - \langle v\rangle_{elec}^{wat}\right)$ and the excluded-volume component $\left(\Delta\mu_{excl}^{mix} - \Delta\mu_{excl}^{wat}\right)$, and the range of variation is smaller for these components than for the van der Waals component. The transfer energetics is thus governed by the van der Waals component in the sense that among the electrostatic, van der Waals, and excluded-volume components, the free energy of transfer reflects the van der Waals most sensitively upon variation of the protein structure, which implies in combination with Fig. 10.5 that an extended structure of protein is favored with addition of urea due to the strengthened van der Waals interaction. Actually, the anti-correlation between the urea and water contributions is observed for each of $\left(\langle v\rangle_{\mathrm{elec}}^{\mathrm{mix}} - \langle v\rangle_{\mathrm{elec}}^{\mathrm{wat}}\right)$, $\left(\langle v\rangle_{\mathrm{vdW}}^{\mathrm{mix}} - \langle v\rangle_{\mathrm{vdW}}^{\mathrm{wat}}\right)$, and $\left(\Delta\mu_{\mathrm{excl}}^{\mathrm{mix}} - \Delta\mu_{\mathrm{excl}}^{\mathrm{wat}}\right)$ (Yamamori et al. 2016; Yamamori and Matubayasi 2017). Furthermore, the separated contributions from urea and water compensate each other for the electrostatic component; the electrostatic interaction of the protein with water is well replaced by that with urea (Stumpe and Grubmüller 2007; Lindgren and Westlund 2010; Horinek

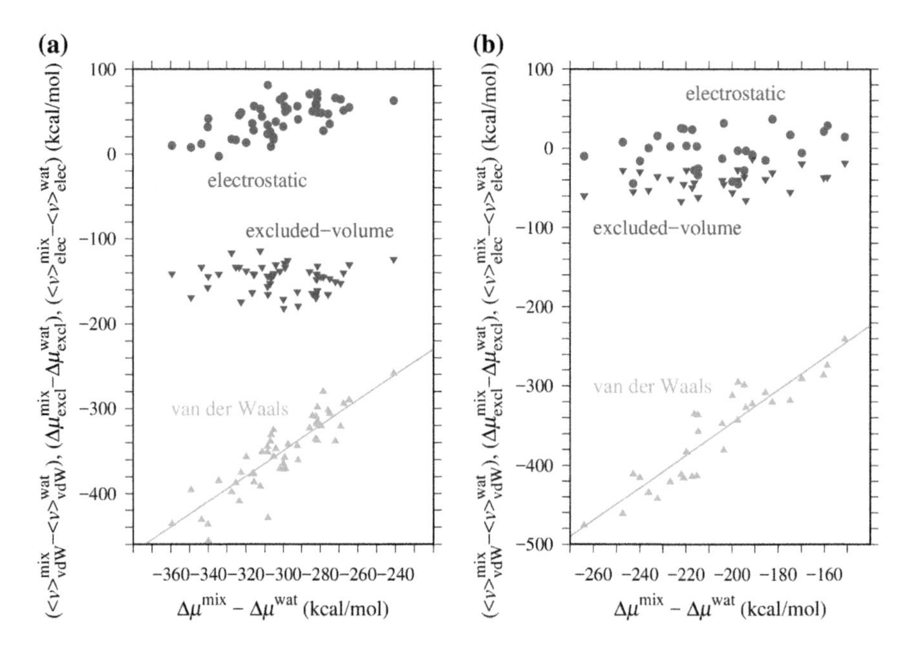

Fig. 10.6 Correlation plots of the transfer free energies $(\Delta\mu^{\mathrm{mix}} - \Delta\mu^{\mathrm{wat}})$ of (**a**) cytochrome *c* and (**b**) T4-lysozyme against the transfer quantities of the electrostatic and van der Waals components of the solute–solvent energy and the excluded-volume component, where a single point in the figure corresponds to a single (snapshot) structure of protein and the protein is transferred from pure-water solvent to 8 M and 3 M mixtures of urea and water for (**a**) and (**b**), respectively. $\langle v\rangle_{\mathrm{elec}}^{\mathrm{mix}}$ and $\langle v\rangle_{\mathrm{vdW}}^{\mathrm{mix}}$ denote the electrostatic and van der Waals components of the average sum of the solute–solvent interaction energy in the mixed solvent, respectively, and $\langle v\rangle_{\mathrm{elec}}^{\mathrm{wat}}$ and $\langle v\rangle_{\mathrm{vdW}}^{\mathrm{wat}}$ are the corresponding quantities in pure-water solvent. $\Delta\mu_{\mathrm{excl}}^{\mathrm{mix}}$ and $\Delta\mu_{\mathrm{excl}}^{\mathrm{wat}}$ are the excluded-volume components in the mixed solvent and pure water, respectively. The solid line represents the linear regression from the least-square fit of $\left(\langle v\rangle_{\mathrm{vdW}}^{\mathrm{mix}} - \langle v\rangle_{\mathrm{vdW}}^{\mathrm{wat}}\right)$ against $(\Delta\mu^{\mathrm{mix}} - \Delta\mu^{\mathrm{wat}})$. In (**a**), the slope is 1.6 with the correlation coefficient of 0.89, and in (**b**), the slope and the correlation coefficient are 2.1 and 0.92, respectively

and Netz 2011; Kokubo et al. 2011; Paul and Paul 2015). The excluded-volume component is similarly compensated upon transfer, while the compensation is only partial for the van der Waals component and the transfer value of the (total) van der Waals interaction correlates well with the urea contribution. Figures 10.5 and 10.6 therefore show that with addition of urea, a protein unfolds to extended structures through direct, van der Waals interaction with the urea cosolvent (Stumpe and Grubmüller 2007; Lindgren and Westlund 2010; Horinek and Netz 2011; Kokubo et al. 2011; Ikeguchi et al. 2001; van der Vegt et al. 2004; Hua et al. 2008; Canchi et al. 2010; Canchi and García 2011, 2013).

10.5 Conclusions

A fast and accurate scheme for all-atom computation of the free energy of protein solvation was developed by combining molecular simulation and the energy-representation theory of solutions. It is an endpoint method of free-energy calculation, and the free-energy analysis of a protein with a few hundred residues has become feasible with explicit solvent. The applications were then described for the equilibrium fluctuation in pure-water solvent and for the transfer from pure water to the mixed solvent of urea and water. It was seen in the former that the electrostatic interaction between protein and water plays a key role in the equilibrium fluctuation of the protein structure. It was observed in the latter, on the other hand, that the transfer energetics is governed by the van der Waals component and that a protein unfolds through its direct, van der Waals interaction with urea.

In statistical thermodynamics, the solvation free energy quantifies the solvent effect on the protein structure and is desirably evaluated with all-atom model. The endpoint method in the present chapter is a useful route to the evaluation, and an improvement of free-energy functional on a sound theoretical basis will further extend the utility of protein simulation with explicit solvent. Furthermore, the structure of a protein or its complex can be controlled by the variation of solvent environment. An example is the solubilizing reagent for dissolving protein aggregates such as inclusion body and amyloid. Its choice corresponds to the engineering of intermolecular interaction, and the correlation analysis as described in the present chapter can be helpful for systematic exploration of a solubilizing cosolvent.

References

Åqvist J, Medina C, Samuelsson JE (1994) A new method for predicting binding affinity in computer-aided drug design. Prot Eng 7:385–391

Bushnell GW, Louie GV, Brayer GD (1990) High-resolution three-dimensional structure of horse heart cytochrome *c*. J Mol Biol 214:585–595

Canchi DR, García AE (2011) Backbone and side-chain contributions in protein denaturation by urea. Biophys J 100:1526–1533

Canchi DR, García AE (2013) Cosolvent effects on protein stability. Annu Rev Phys Chem 64:273–293
Canchi DR, Paschek D, García AE (2010) Equilibrium study of protein denaturation by urea. J Am Chem Soc 132:2338–2344
Carlson HA, Jorgensen WL (1995) An extended linear response method for determining free energies of hydration. J Phys Chem 99:10667–10673
Chuev GN, Fedorov MV, Crain J (2007) Improved estimates for hydration free energy obtained by the reference interaction site model. Chem Phys Lett 448:198–202
Date A, Ishizuka R, Matubayasi N (2016) Energetics of nonpolar and polar compounds in cationic, anionic, and nonionic micelles studied by all-atom molecular dynamics simulation combined with a theory of solutions. Phys Chem Chem Phys 18:13223–13231
Fdez Galván I, Sánchez ML, Martín ME, Olivares del Valle FJ, Aguilar MA (2003) Geometry optimization of molecules in solution: joint use of the mean field approximation and the free-energy gradient method. J Chem Phys 118:255–263
Freedman H, Truong TN (2004) Coupled reference interaction site model/simulation approach for thermochemistry of solvation: theory and prospects. J Chem Phys 121:2187–2198
Frolov AI (2015) Accurate calculation of solvation free energies in supercritical fluids by fully atomistic simulations: probing the theory of solutions in energy representation. J Chem Theory Comput 11:2245–2256
Frolov AI, Ratkova EL, Palmer DS, Fedorov MV (2011) Hydration thermodynamics using the reference interaction site model: speed or accuracy? J Phys Chem B 115:6011–6022
Harris RC, Deng N, Levy RM, Ishizuka R, Matubayasi N (2017) Computing conformational free energy differences in explicit solvent: an efficient thermodynamic cycle using an auxiliary potential and a free energy functional constructed from the end points. J Comp Chem 38:1198–1208
Horinek D, Netz RR (2011) Can simulations quantitatively predict peptide transfer free energies to urea solutions? Thermodynamic concepts and force field limitations. J Phys Chem A 115:6125–6136
Hornak V, Abel R, Okur A, Strockbine B, Roitberg A, Simmerling C (2006) Comparison of multiple amber force fields and development of improved protein backbone parameters. Proteins 65:712–725
Hua L, Zhou R, Thirumalai D, Berne BJ (2008) Urea denaturation by stronger dispersion interactions with proteins than water implies a 2-stage unfolding. Proc Natl Acad Sci USA 105:16928–16933
Ikeguchi M, Nakamura S, Shimizu K (2001) Molecular dynamics study on hydrophobic effects in aqueous urea solutions. J Am Chem Soc 123:677–682
Imai T, Harano Y, Kinoshita M, Kovalenko A, Hirata F (2006) A theoretical analysis on hydration thermodynamics of proteins. J Chem Phys 125:024911 (7 pp)
Ishizuka R, Matubayasi N, Tu KM, Umebayashi Y (2015) Energetic contributions from the cation and anion to the stability of carbon dioxide dissolved in imidazolium-based ionic liquids. J Phys Chem B 119:1579–1587
Jorgensen WL, Chandrasekhar J, Madura JD, Impey RW, Klein ML (1983) Comparison of simple potential functions for simulating liquid water. J Chem Phys 79:926–935
Jorgensen WL, Maxwell DS, Tirado-Rives J (1996) Development and testing of the OPLS all-atom force field on conformational energetics and properties of organic liquids. J Am Chem Soc 118:11225–11236
Kamo F, Ishizuka R, Matubayasi N (2016) Correlation analysis for heat denaturation of Trp-cage miniprotein with explicit solvent. Protein Sci 25:56–66
Karino Y, Matubayasi N (2011) Free-energy analysis of hydration effect on protein with explicit solvent: equilibrium fluctuation of cytochrome *c*. J Chem Phys 134:041105 (4 pp)
Karino Y, Matubayasi N (2013) Interaction-component analysis of the urea effect on amino acid analogs. Phys Chem Chem Phys 15:4377–4391
Karino Y, Fedorov MV, Matubayasi N (2010) End-point calculation of solvation free energy of amino-acid analogs by molecular theories of solution. Chem Phys Lett 496:351–355

Kast SM (2001) Combinations of simulation and integral equation theory. Phys Chem Chem Phys 3:5087–5092
Kawakami T, Shigemoto I, Matubayasi N (2012) Free-energy analysis of water affinity in polymer studied by atomistic molecular simulation combined with the theory of solutions in the energy representation. J Chem Phys 137:234903 (9 pp). Erratum (2014) J Chem Phys 140:169903 (2 pp)
Kinoshita M (2013) A new theoretical approach to biological self-assembly. Biophys Rev 5:283–293
Kokubo H, Hu CY, Pettitt BM (2011) Peptide conformational preferences in osmolyte solutions: transfer free energies of decaalanine. J Am Chem Soc 133:1849–1858
Kokubo H, Harris RC, Asthagiri D, Pettitt BM (2013) Solvation free energies of alanine peptides: the effect of flexibility. J. Phys. Chem. B 117:16428–16435
Levy RM, Belhadj M, Kitchen DB (1991) Gaussian fluctuation formula for electrostatic free-energy changes in solution. J Chem Phys 95:3627–3633
Lindgren M, Westlund PO (2010) On the stability of chymotrypsin inhibitor 2 in a 10 M urea solution: the role of interaction energies for urea-induced protein denaturation. Phys Chem Chem Phys 12:9358–9366
Luzhkov V, Warshel A (1992) Microscopic models for quantum mechanical calculations of chemical processes in solutions: LD/AMPAC and SCAAS/AMPAC calculations of solvation energies. J Comput Chem 13:199–213
Matubayasi N, Nakahara M (2000) Theory of solutions in the energetic representation. I. Formulation. J Chem Phys 113:6070–6081
Matubayasi N, Nakahara M (2002) Theory of solutions in the energy representation. II. Functional for the chemical potential. J Chem Phys 117:3605–3616. Erratum (2003) J Chem Phys 118:2446
Matubayasi N, Nakahara M (2003) Theory of solutions in the energy representation. III. Treatment of the molecular flexibility. J Chem Phys 119:9686–9702
Matubayasi N, Liang KK, Nakahara M (2006) Free-energy analysis of solubilization in micelle. J Chem Phys 124:154908 (13 pp)
Matubayasi N, Shinoda W, Nakahara M (2008) Free-energy analysis of the molecular binding into lipid membrane with the method of energy representation. J Chem Phys 128:195107 (13 pp)
Mizukami T, Saito H, Kawamoto S, Miyakawa T, Iwayama M, Takasu M, Nagao H (2012) Solvation effect on the structural change of a globular protein: a molecular dynamics study. Int J Quant Chem 112:344–350
Moeser B, Horinek D (2014) Unified description of urea denaturation: backbone and side chains contribute equally in the transfer model. J Phys Chem B 118:107–114
Neidigh JW, Fesinmeyer RM, Andersen NH (2002) Designing a 20-residue protein. Nat Struct Mol Biol 9:425–430
Paul S, Paul S (2015) Exploring the counteracting mechanism of trehalose on urea conferred protein denaturation: a molecular dynamics simulation study. J Phys Chem B 119:9820–9834
Rose DR, Phipps J, Michniewicz J, Birnbaum GI, Ahmed FR, Muir A, Anderson WF, Narang S (1988) Crystal structure of T4-lysozyme generated from synthetic coding DNA expressed in Escherichia coli. Protein Eng Des Sel 2:277–282
Saito H, Iwayama M, Mizukami T, Kang J, Tateno M, Nagao H (2013) Molecular dynamics study on binding free energy of Azurin-Cytochrome c551 complex. Chem Phys Lett 556:297–302
Sakuraba S, Matubayasi N (2014) ERmod: Fast and versatile computation software for solvation free energy with approximate theory of solutions. J Comput Chem 35:1592–1608
Shirts MR, Pande VS (2005) Solvation free energies of amino acid side chain analogs for common molecular mechanics water models. J Chem Phys 122:134508 (12 pp)
Stumpe MC, Grubmüller H (2007) Interaction of urea with amino acids: implications for urea-induced protein denaturation. J Am Chem Soc 129:16126–16131
Takahashi H, Matubayasi N, Nakahara M, Nitta T (2004) A quantum chemical approach to the free energy calculations in condensed systems: the QM/MM method combined with the theory of energy representation. J Chem Phys 121:3989–3999

Takahashi H, Ohno H, Kishi R, Nakano M, Matubayasi N (2008) Computation of the free energy change associated with one-electron reduction of coenzyme immersed in water: a novel approach within the framework of the quantum mechanical/molecular mechanical method combined with the theory of energy representation. J Chem Phys 129:205103 (14 pp)
Takemura K, Guo H, Sakuraba S, Matubayasi N, Kitao A (2012) Evaluation of protein-protein docking model structures using all-atom molecular dynamics simulations combined with the solution theory in the energy representation. J Chem Phys 137:215105 (10 pp)
van der Vegt NFA, Trzesniak D, Kasumaj B, van Gunsteren WF (2004) Energy-entropy compensation in the transfer of nonpolar solutes from water to cosolvent/water mixtures. ChemPhysChem 5:144–147
Vener MV, Leontyev IV, Dyakov YuA, Basilevsky MV, Newton MD (2002) Application of the linearized MD approach for computing equilibrium solvation free energies of charged and dipolar solutes in polar solvents. J Phys Chem B 106:13078–13088
Weber V, Asthagiri D (2012) Regularizing binding energy distributions and the hydration free energy of protein cytochrome *c* from all-atom simulations. J Chem Theory Comput 8:3409–3415
Wolfenden R, Andersson L, Cullis PM, Southgate CCB (1981) Affinities of amino acid side chains for solvent water. Biochemistry 20:849–855
Yamamori Y, Matubayasi N (2017) Interaction-component analysis of the effects of urea and its alkylated derivatives on the structure of T4-lysozyme. J Chem Phys 146:225103 (13 pp)
Yamamori Y, Ishizuka R, Karino Y, Sakuraba S, Matubayasi N (2016) Interaction-component analysis of the hydration and urea effects on cytochrome *c*. J Chem Phys 144:085102 (14 pp)
Yamamoto T (2008) Variational and perturbative formulations of quantum mechanical/molecular mechanical free energy with mean-field embedding and its analytical gradients. J Chem Phys 129:244104 (15 pp)

Chapter 11
Uni-directional Propagation of Structural Changes in Actin Filaments

Taro Q. P. Uyeda, Kien Xuan Ngo, Noriyuki Kodera and Kiyotaka Tokuraku

Abstract When a protein molecule is bound with another, its structure is likely to change in one way or the other. The structure of a protein molecule in a protein complex is also likely to change when binding partner in the complex undergoes a conformational change. It is therefore no surprise that binding of an actin-binding protein to a protomer in an actin filament changes the structure of that actin protomer, and that the resultant conformational change in the actin protomer affects the structure of the neighboring protomers in the same filament. Moreover, eukaryotic actin appears to have evolved to efficiently spread the conformational change in the actin protomer initially bound with actin-binding protein over a long distance along the filament (cooperative conformational change), as has been observed in the cases of cofilin- and myosin-induced cooperative conformational changes. We speculate that the high degree of cooperativity in conformational changes in actin filaments enables cooperative binding of actin-binding proteins, which is necessary for actin filaments to perform specific functions by selectively interacting with a subset of actin-binding proteins among the large number of actin-binding proteins present in the cell. Interestingly, cooperative conformational changes propagate to only one direction along the filament, at least in the cases of cofilin and myosin II-induced

T. Q. P. Uyeda (✉) · K. X. Ngo
Faculty of Advanced Science and Engineering, Department of Physics, Waseda University, 3-4-1 Okubo, Shinjuku, Tokyo 169-8555, Japan
e-mail: t-uyeda@waseda.jp

K. X. Ngo
Laboratory for Molecular Biophysics, Brain Science Institute, RIKEN, Wako, Saitama 351-0198, Japan

N. Kodera
Department of Physics and Bio-AFM Frontier Research Center, Kanazawa University, Kanazawa, Ishikawa 920-1192, Japan

K. Tokuraku
Department of Applied Sciences, Muroran Institute of Technology, Muroran, Hokkaido 050-8585, Japan

© Springer Nature Singapore Pte Ltd. 2018

M. Suzuki (ed.), *The Role of Water in ATP Hydrolysis Energy Transduction by Protein Machinery*, https://doi.org/10.1007/978-981-10-8459-1_11

conformational changes. Functional significance of those uni-directional conformational changes in actin filaments is not known, but we propose that they play roles in directional signal transmission along one-dimensional polymer in cells, or in force generation by myosin.

Keywords Actin · Cooperativity · Polarity · Cofilin · Myosin

11.1 Introduction

Eukaryotic actin is an essential protein that is distinguished from other proteins due to its multiplicity of functions and extraordinary level of sequence conservation. An actin molecule is composed of four subdomains, with subdomains 1 and 2 forming the small domain and subdomains 3 and 4 forming the large domain (Fig. 11.1, left). The small and large domains are connected through subdomains 1 and 3, allowing a certain degree of rigid body movements between the two domains. One molecule of ATP or ADP is bound at the base of the cleft, and the nucleotide state as well as binding of various actin-binding proteins is known to induce conformational changes in actin, involving angle changes between the two domains.

Lower eukaryotes such as yeast and the cellular slime mold express a single species of actin, while multiple actin isoforms are expressed in higher eukaryotes. In mammalian non-muscle cells, β and γ actins are expressed but only β actin is essential (Belyantseva et al. 2009). The model plant *Arabidopsis thaliana* has 8 actin genes, but ACT8 is sufficient to support the growth (Kandasamy et al. 2009). Nonetheless, actin plays a large number of essential functions in eukaryotic cells. It is a major challenge to understand how a single species of protein is able to perform multiple different functions simultaneously in a common cytoplasm.

Monomeric actin reversibly polymerizes to form actin filaments (Fig. 11.1, right), and in most cases, filaments are the functional form of actin. Actin filaments are made of helically intertwined two protofilaments. Within each actin protofilament, hundreds or thousands of actin molecules associate in a head-to-tail fashion, and therefore, an actin protofilament is polar. Furthermore, the two actin protofilaments in an actin filament are aligned with the same polarity, such that an actin filament is also a polar structure with all component actin molecules (protomers) aligned in the same orientation in a filament. Consequently, the two ends of an actin filament are unequal. The end with subdomains 2 and 4 exposed is called the minus or pointed end, while the other end with subdomains 1 and 3 exposed is called the plus or barbed end. In vitro experiments demonstrated that the plus end is easier to polymerize (faster growth rate as well as lower critical concentration for polymerization) than the minus end (Kondo and Ishiwata 1976).

In general, polymerization and function of actin filaments are dictated by interaction with specific actin-binding proteins (ABPs) in vivo (Pollard and Cooper 2009). A cell expresses a large number of ABPs, but each actin filament does not interact with all the ABPs present in the cell. Rather, each actin filament interacts

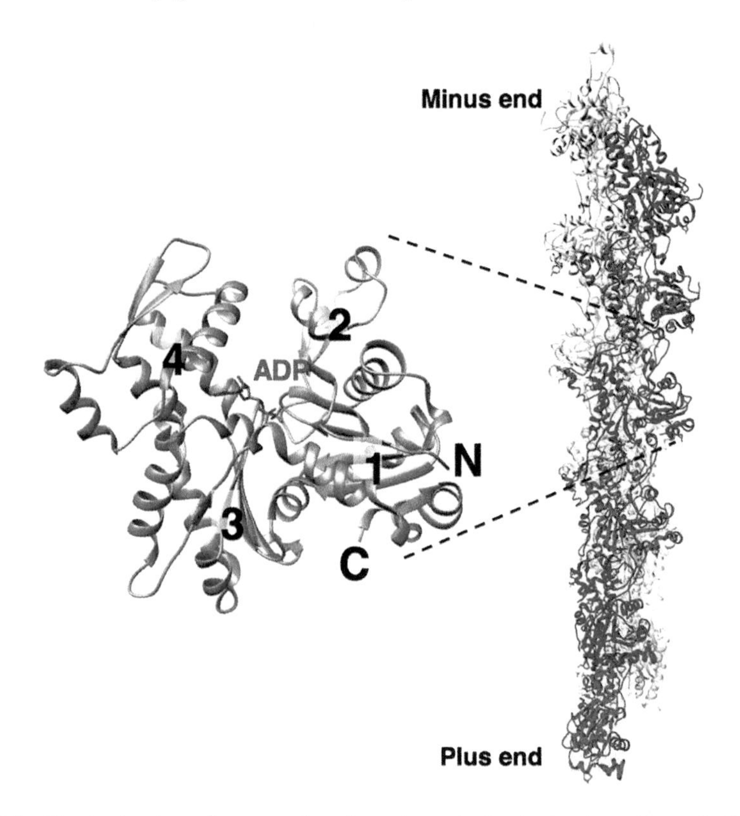

Fig. 11.1 Atomic structure of monomeric actin and pseudoatomic structure of an actin filament. Left: the crystal structure of monomeric actin, which carried an ADP molecule and polymerization of which was blocked by attaching tetramethylrhodamine to Cys374 (not shown) (Otterbein et al. 2001) (PDB code: 1J6Z). The numbers indicate subdomains, and N and C show the two termini. Right: structure of an actin filament composed of 12 actin protomers, revealed by cryoelectron microscopy (Murakami et al. 2010) (PDB code: 3G37). Protomers in one protofilament are colored in red and blue alternatingly, and those in the other protofilament in pink and light blue, respectively. The actin monomer and one of the blue protomers connected by two dashed lines are viewed from a similar direction, but the details are different primarily due to the conformational change associated with polymerization. These images were generated using the UCSF Chimera software

with only a small subset of the ABPs present to perform a specific task. It is therefore critically important to understand how interactions of actin filaments with different ABPs are regulated in a cell.

Because actin filaments are made of the same actin protein, it is generally believed that selective binding of ABPs to specific actin filaments is achieved by local biochemical regulation of the ABPs. However, it is recently suggested that an actin filament itself can choose which ABP to bind, by changing its structure (Galkin et al. 2012; Michelot and Drubin 2011; Romet-Lemonne and Jegou 2013; Schoenenberger et al. 2011; Tokuraku et al. 2009; Uyeda et al. 2011). Moreover, in certain cases, structural changes in actin filaments are shown to be cooperative, in a sense that a conformational change that occurred in one actin protomer is propagated to neighboring actin protomer in the same filament. However, the relationship between the propagation direction of the structural changes and the filament polarity is only poorly addressed. In this chapter, we would like to review this emerging new question and discuss its potential physiological significance, including its possible contribution to ATP-dependent directional movement of myosin along actin filaments and intracellular signaling.

11.2 Polymorphism of Pure Actin Filaments

Monomeric actin has been crystallized in complexes with ABPs that inhibit polymerization, but actin filaments have not been crystallized, and the atomic structure of the actin filaments has not been solved. It is well established that monomeric actin, as well as actin molecules within filaments, has different conformations depending on whether it has ATP, ADP and Pi, or ADP at its ATP-binding site (Graceffa and Dominguez 2003; Murakami et al. 2010; Oda et al. 2009). Polymerization of actin and its relationship with ATP hydrolysis in vitro is also well established (reviewed by Carlier 1990) and can be summarized as follows (Carlier 1990; Korn et al. 1987).

Interactions of actin protomers in filaments and monomers in solution are very different at the two ends; the association/dissociation reactions are much faster at the plus ends than at the minus ends. Thus, in the presence of ATP-actin monomer at concentrations close to the critical concentration for polymerization, ATP-actin monomer is incorporated into a filament at its plus end. ATP is then hydrolyzed, followed by slow Pi release to result in ADP-bound protomers. At the minus end, ATP-actin monomer is incorporated only at a slow rate when compared with the rates of ATP hydrolysis and Pi release, due to the slow dynamics of this end. Hence, ADP-actin protomer is often exposed at the minus end. Because ADP-actin protomer is faster to dissociate from the filament ends than the ATP-actin or ADP-Pi-actin protomer, ATP-actin monomer keeps binding to the plus end and ADP-actin protomer keeps dissociating from the minus end, resulting in steady, uni-directional flow of actin protomers (treadmill). When one watches one point near the plus end of a filament, protomers in the area initially carry ATP but it would appear as if a segment of protomers carrying ADP-Pi comes and passes from the minus-end direction, followed by a segment rich with molecules carrying ADP only. This is a well-recognized case of uni-directional propagation of conformational changes in actin filaments, which is dependent on ATP hydrolysis and the difference in the dynamics between the two filament ends. The nucleotide state of actin protomers can function as a timer after polymerization, and indeed, aged sections of actin filaments are preferentially severed by cofilin, because cofilin specifically binds and severs ADP-actin protomers (Blanchoin and Pollard 1999; Carlier et al. 1997).

Recent advances of cryoelectron microscopy combined with rapid freezing have yielded near-atomic picture of filaments prepared under standard conditions (Fujii et al. 2010; Galkin et al. 2010b, 2015; Murakami et al. 2010). One of those studies (Galkin et al. 2010b) reported structural polymorphism even in filaments made of skeletal muscle actin molecules most of which homogenously carry ADP. Galkin et al. identified six different conformations that can be discriminated by the resolution of electron microscopy. This study is even more intriguing, since Galkin et al. treated 17 consecutive actin protomers as a single particle in the single particle analysis, implying that at least 17 consecutive actin protomers in one filament usually take the same one of the six conformations (cooperative polymorphism). It should be noted, however, that another study, which was performed under a similar condition, reported a single conformation for ADP-carrying skeletal muscle actin filaments (Fujii et al. 2010), casting doubt on the cooperative polymorphism reported by Galkin et al. Further studies are needed to reconcile the discrepancy.

11.3 Actin Filaments Interacting with ABPs Are Cooperatively Polymorphic

11.3.1 Cofilin

Although polymorphism of pure ADP-actin filaments is still controversial, it is well established that ADP-actin filaments interacting with certain ABPs take different conformations. Among the ABP-induced conformational changes, those induced by cofilin has been well characterized.

Binding of cofilin to actin filaments is cooperative (De La Cruz 2005; Hawkins et al. 1993; Hayakawa et al. 2014; Hayden et al. 1993; McGough et al. 1997), so that cofilin added at a non-saturating concentration tends to form clusters along actin filaments, while leaving other segments of the filament mostly bare. McGough and Weeds observed actin filaments fully decorated by cofilin using electron microscopy and found that the helical pitch is shorter by ~25% (McGough et al. 1997). Galkin et al. extended this research using electron microscopy with higher resolution and reported that the super-twisting is accompanied by major structural change of each actin protomer, i.e., tilting between the two domains, and that the conformational change is propagated to bare zones neighboring the cofilin clusters (cooperative conformational change) (Galkin et al. 2011).

Electron microscopy does not yield information on dynamics of interactions, and furthermore, the standard high-resolution structural analysis of actin filaments using electron microscopy involves averaging over a long filament length and cannot yield information about the structure at the boundary between cofilin clusters and the neighboring bare zones. To complement the limitations of electron microscopy, therefore, a number of fluorescence microscopic studies were also performed in order to observe the dynamics of actin–cofilin interactions (e.g., Hayakawa et al.

2014; Suarez et al. 2011; Wioland et al. 2017). These studies suggested that severing of actin filaments by cofilin is promoted by structural discontinuities at or near the boundaries between cofilin-bound and bare sections of the filament.

11.3.2 Myosin

Although less well characterized than cofilin–actin interactions, ABP-induced cooperative conformational change in actin filaments was first demonstrated in actin–myosin interactions by a pioneering work by Oosawa et al. (1973). Oosawa and his colleagues labeled actin fluorescently and found that the fluorescence intensity increased when skeletal muscle HMM was added to the labeled actin filaments in the absence of ATP. Furthermore, the fluorescence increase saturated when only 1/10 (mol/mol) of HMM was added, demonstrating that the fluorescence increase is not due to direct interaction of the fluorophore with HMM, and is dependent on propagation of conformational changes from the actin protomer bound with HMM to the neighboring free protomers (cooperative polymorphism). Other biophysical measurements also reported myosin-induced cooperative conformational changes in actin (Fujime and Ishiwata 1971; Loscalzo et al. 1975; Miki et al. 1982; Siddique et al. 2005; Tawada 1969; Thomas et al. 1979). However, recent high-resolution cryoelectron microscopic study concluded that the structure of actin filaments fully decorated with skeletal muscle S1 (the motor domain of myosin) in the absence of ATP is hardly different from that of bare actin filaments (Fujii and Namba 2017). It may be that subtle structural changes that were undetected by cryoelectron microscopy, such as difference in side chain orientation, caused changes in fluorescence intensity detected by the studies mentioned above. Alternatively, only sparse myosin binding, but not full decoration, may be able to elicit structural changes in actin, as suggested by Miki et al. (1982).

Unfortunately, the standard electron microscopic method of obtaining high-resolution images of actin filaments involves averaging of large number of protomers in filaments. This is not suitable to reveal structure of individual protomers in filaments sparsely bound with S1, as actin protomers presumably take different conformations depending on the relationship with the bound S1 molecule. Thus, despite the critical importance, the nature of structural changes in actin filaments induced by myosin in the presence of ATP is still highly elusive, since S1 interacts with actin filaments only transiently in the presence of ATP.

To complement the limitations of electron microscopy, as was the case with cofilin binding, fluorescence microscopic studies were also performed in order to observe the dynamics of actin–myosin interactions. For example, Tokuraku et al., extending the pioneering electron microscopic work by Orlova and Egelman (1997), analyzed the process of cooperative binding of HMM to actin filaments in the presence of trace concentrations of ATP (Tokuraku et al. 2009). They found that substoichiometric amounts of HMM in the presence of 0.1 μM ATP binds cooperatively to a subset of actin filaments while leaving other filaments apparently bare, and proposed that actin filaments have at least two distinct conformations with significant difference in the affinity for myosin motor domain in the presence of ATP. However, the standard fluorescence microscopy only reports the presence of molecules but, unlike electron microscopy, yields no information on the structure of the molecules. Furthermore, the spatial resolution of fluorescence microscopy is way too low to gain insight into what is happening in each actin protomer during cooperative interaction with myosin. A third microscopic method, high-speed atomic force microscopy (HS-AFM), that has the time resolution of tens of ms and spatial resolution of several nm, was recently developed by Ando et al. (2013) and is expected to fill the gap between electron and fluorescence microscopies.

Strikingly, binding of cofilin to actin filaments is inhibited by 1 μM S1 in the presence of 0.1 mM ATP in vitro (Ngo et al. 2016), and myosin II and cofilin are segregated in cells. This suggests that cooperative conformational changes in actin filaments induced by S1 in the presence of ATP are physiologically relevant, even though the nature of cooperative conformational changes is not well understood. In the presence of saturating concentrations (e.g., 0.1 mM) of ATP, binding of 1 μM S1 to actin filaments is transient and most of the actin protomers are not bound with S1 at a given moment. Therefore, the fact that cofilin binding was strongly inhibited by 1 μM S1 in the presence of 0.1 mM ATP implies that the impact of structural change that occurred in an actin protomer that transiently interacted with S1 propagates over a long distance along the filament (long-range cooperativity), and/or the impact of the structural change remains long after the S1 had dissociated by ATP binding (memory effect). It may appear paradoxical that the structural impact of S1 binding to the actin structure is much larger in the presence of ATP than in its absence despite the much weaker overall affinity. It may be natural, however, that S1 energized with bound ADP and Pi, a predominant intermediate in the presence of ATP, impacts on the structure of actin much more profoundly even though the binding itself is very short, when compared with that in the absence of ATP.

11.3.3 Other ABPs

When two proteins interact with each other, the structure of each protein should change in one way or the other, when compared with that isolated free in solution. It is thus expected that binding of any ABP to an actin filament would change the structure of the bound actin protomer in one way or the other. Now, consider that the ABP-bound actin protomer is in direct contact with four actin protomers in the filament, and structural change in the ABP-bound actin protomer is expected to influence the structure of the four neighboring actin protomers. The resultant structural change in the neighboring actin protomer may influence the structure of further neighbors, in a domino-toppling manner. At the same time, the structural changes in those bare actin protomers would likely modulate the affinities for the first ABP and other ABPs. It would be therefore a surprise if ABP binding does not induce any structural change in the bound actin protomer, or if the structural change that occurred in the ABP-bound actin protomer is not at all propagated to the neighbors. The question is whether the induced structural changes are large enough to be detected or whether they surpass the thermal motion of the actin protomers.

Another important issue is whether the ABP-bound actin protomer can induce the same structural change in the neighboring protomers, a requirement for the propagation of the cooperative conformational changes. Galkin et al. (2012) speculated that actin filaments, particularly eukaryotic actin filaments, are evolved to be that way. Indeed, the facts that structural changes induced by cofilin and S1 + ATP induce cooperative binding along the filament suggest that cooperative conformational changes of actin filaments have intrinsic tendency to induce multiple neighboring protomers to take the same conformation. A large number of ABPs have been shown to induce cooperative conformational changes in actin filaments and/or to bind cooperatively to actin filaments. These ABPs include, in addition to S1 + ATP and cofilin, side-binding ABPs such as drebrin (Sharma et al. 2011, 2012), tropomyosin (Butters et al. 1993), caldesmon (Collins et al. 2011; Huang et al. 2010), fimbrin (Galkin et al. 2008; Hanein et al. 1997; Skau and Kovar 2010), α catenin (Hansen et al. 2013), and α actinin (Craig-Schmidt et al. 1981; Galkin et al. 2010a) and end-binding ABPs such as formin (Papp et al. 2006) and gelsolin (Orlova et al. 1995; Prochniewicz et al. 1996). Some of the ABP-induced structural changes are relatively subtle and detectable by high-resolution electron microscopy, but others, such as those induced by drebrin, as well as those induced by cofilin or S1 + ATP, involve major changes in helical pitch of the filament.

11.4 Uni-directional Propagation of Structural Changes of Actin Filaments Induced by Actin-Binding Proteins

Cooperative conformational changes induced by end-binding ABPs, formin and gelsolin, are uni-directional since those ABPs bind specifically to the plus ends of filaments so that the cooperative conformational changes can propagate only to the minus-end direction. Regarding cooperative conformational changes induced by side-binding ABPs, if succession of the same actin conformers in a filament is structurally favored as discussed above, the cooperative structural change in actin filaments should be propagated to both directions with respect to the filament polarity. However, structural asymmetry of the actin–actin interface may kinetically hinder the propagation to one direction, resulting in uni-directional propagation of the structural change.

We have addressed this question using HS-AFM and fluorescence microscopy. In one experiment, we followed the growth of HMM-GFP clusters along actin filaments in the presence of a trace concentration of ATP (Hirakawa et al. 2017). Intriguingly, the majority of HMM-GFP clusters grew in one direction (Fig. 11.2). In other cases, HMM-GFP clusters appeared to grow in two directions, but it was not possible to determine if these were truly bidirectional growth of the HMM-GFP

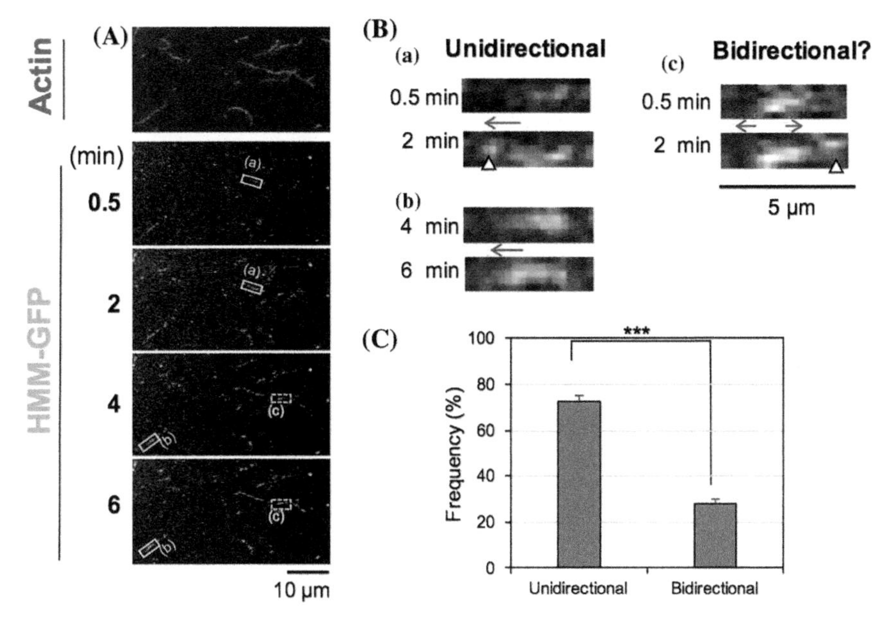

Fig. 11.2 HMM-GFP clusters grow uni-directionally along actin filaments. Fluorescence micrographs **A** show actin filaments labeled with rhodamine phalloidin (top) and HMM-GFP clusters in the same field taken at several time points after the addition of 10 nM HMM-GFP in the presence of 0.1 µM ATP. Enlarged images of HMM-GFP clusters, marked by (a), (b), and (c), are shown in (**B**). HMM-GFP clusters shown in (a) and (b) grew only to the left. The cluster shown in (c) appeared to grow in both directions. However, a closer examination suggested the possibility that the small fluorescent spot on the right (indicated by a triangle) might have newly formed independent of the larger cluster on the left, in which case the larger cluster on the left grew uni-directionally to the left. **C** Statistics of 46 clusters, with cases such as (c) classified as "bidirectional." Modified from (Hirakawa et al. 2017) with the permission of the publisher

clusters, or an artifact due to the limited spatial limitation of fluorescence microscopy. It is also yet to be determined if the growth directions of HMM-GFP clusters have a fixed relationship with the polarity of the filaments.

Growth of cofilin clusters has been investigated more thoroughly using HS-AFM (Ngo et al. 2015). Individual cofilin molecules are too small to visualize by HS-AFM, but cofilin clusters can be easily identified by two criteria: the thicker appearance of the filament and the 25% shorter helical pitch. Ngo et al. (2015) found that supertwisted conformation of the filament in cofilin clusters is propagated to the bare zone on the minus-end side of the cofilin cluster but not to the bare zone on the plus-end side of the cluster (Fig. 11.3). Ngo et al. (2015) took the advantage of the real-time observation capability of HS-AFM and further found that the cofilin clusters grow only to the minus-end direction. Galkin et al. (2001) previously proposed that the helical pitch of actin filaments naturally fluctuates locally and temporally, and that cofilin prefers to bind to regions of actin filaments with shorter helical pitch. Based on this idea, Ngo et al. (2015) suggested that cofilin molecules in solution preferentially bind to the supertwisted bare zone on the minus-end side of

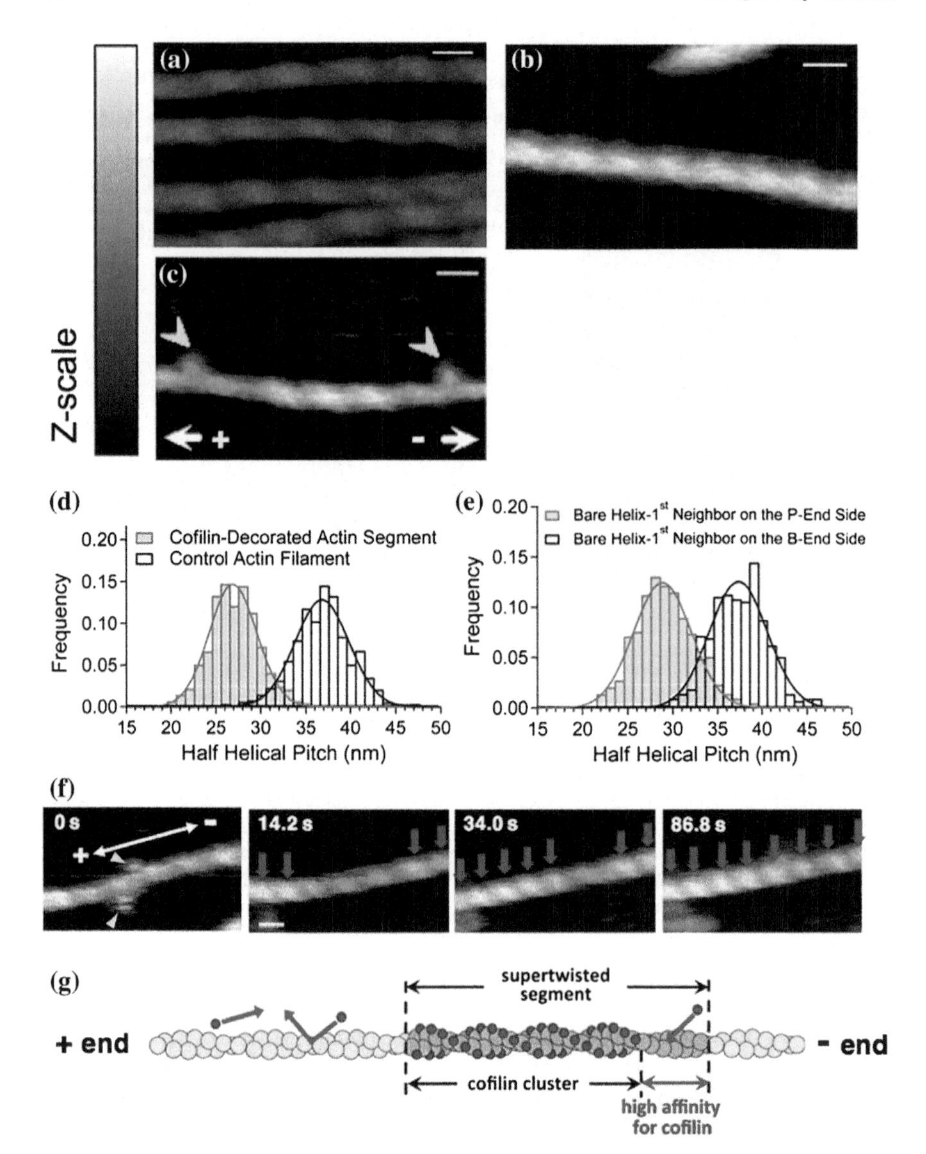

Z-scale
(a)
(b)
(c)
+
-
(d)
Cofilin-Decorated Actin Segment
Control Actin Filament
Frequency
Half Helical Pitch (nm)
(e)
Bare Helix-1st Neighbor on the P-End Side
Bare Helix-1st Neighbor on the B-End Side
Frequency
Half Helical Pitch (nm)
(f)
0 s
14.2 s
34.0 s
86.8 s
(g)
supertwisted segment
+ end
- end
cofilin cluster
high affinity for cofilin

◀**Fig. 11.3** High-speed atomic force microscopy (HS-AFM) observation of actin filaments interacting with cofilin. AFM visualizes height of objects on a flat surface, and lighter color in the AFM images represents taller areas of the object. Therefore, the crossover points, where the filament is taller because the two protofilaments are aligned vertically, appear lighter in the HS-AFM image of control actin filaments (**a**). The double-helical structure and individual actin protomers are also visible in this image. **b** Shows actin filaments fully bound with cofilin. Those filaments are thicker and have shorter half helical pitches (**d**), which are the distances between the adjacent crossover points, confirming previous electron microscopic studies. The filament shown in (**c**) has a short cofilin cluster in the middle. The polarity of this filament was identified on the basis of the tilted appearance of transiently binding S1 (the motor domain of skeletal muscle myosin II) molecules (yellow arrowheads). Pitches of half helices immediately neighboring cofilin clusters in this image and those similar to this were measured to construct the histograms shown in (**e**). Measurements were made separately for the plus-end neighbors and the minus-end neighbors. The half helical pitch on the minus-end side was as short as that in cofilin clusters (~27 nm), while that on the plus-end side was similar to that of the control filaments (~37 nm). Growth of the cofilin cluster was also uni-directional toward the minus-end direction, as shown in the time-lapse sequence shown in (**f**). The cluster on the left grew to the right (toward the minus end, as identified by the tilted transient binding of S1 indicated by yellow arrowheads), while that on the right did not grow to the left. Red arrows show positions of the crossover points within cofilin clusters, identified on the basis of greater height. These observations are summarized schematically in (**g**), based on the proposal by Galkin et al. (2001) that cofilin in solution prefers to bind to actin filaments with a shorter helical pitch. Modified from (Ngo et al. 2015). Notably, Ngo et al. (2015) demonstrated that the uni-directional spreading of the supertwisted structure in a cofilin cluster to the minus-end direction does not depend on the graded abundance of ADP-actin in the minus-end side

the preexisting cofilin cluster, resulting in uni-directional growth of the cofilin cluster to the minus-end direction. The uni-directional growth of the cofilin cluster further converts bare zone ahead with previously normal helical pitch to the supertwisted structure, driving continuous, uni-directional growth of cofilin cluster as long as the concentration of free cofilin in solution is above a certain threshold.

Different from the conclusion of Ngo et al. (2015), a recent fluorescence microscopic study reported that cofilin clusters grow bidirectionally (Wioland et al. 2017). This discrepancy may stem from the ~100-fold difference in spatial resolution between HS-AFM and fluorescence microscopy. Alternatively, interaction of negatively charged actin filaments with positively charged lipid surface in AFM experiments might have created large rotational friction and inhibited propagation of the structural change to the plus-end direction because structural coupling with the neighboring protomer is assumed weaker on the plus-end side. Regarding this possibility, it may be worth pointing out that actin filaments in cells are unlikely to rotate freely, owing to interaction with other ABPs and structures. In the fluorescence microscopic study by Wioland et al. (2017), actin filaments were anchored to the substrate either by bound gelsolin or spectrin, and those ABPs might have altered the property of actin filaments, as demonstrated experimentally for the case of gelsolin (Orlova et al. 1995). Both experimental systems have additional source of artifacts (mechanical tapping in the AFM experiment and stretching due to flow of solution in the fluorescence microscopy study). Identification of the factor that caused the discrepancy would shed light on the mechanism of uni-directional propagation of cofilin-induced conformational changes.

11.5 Possible Functional Implications of Uni-directional Propagation of Structural Changes of Actin Filaments

It is not obvious if the uni-directional conformational changes induced by ABP are functionally relevant, or they simply reflect structural polarity of the filaments with no specific functions. Answering this question experimentally in one way or the other is difficult when we do not have means to specifically perturb the propagation directionality of conformational changes in actin filaments. Nonetheless, it may be useful to envisage plausible models regarding the functional significance of the uni-directional propagation of conformational changes in vivo, so that we can design experiments to test them experimentally in vivo and in vitro. We would like to propose the following two models for future experimental examinations.

11.5.1 Role as Signaling Wires

If a specific state can propagate uni-directionally along a one-dimensional structure, this can potentially serve as a signaling wire. Thus, one possible function of the uni-directional propagation of conformational changes in actin filaments is to serve as a uni-directional signaling wire in the cells. This is plausible because actin filaments are aligned with the same polarity in certain regions of the cells, such as lamellipodia, filopodia, and microvilli (Fig. 11.4a). These polar filament arrays, not only actin filaments but also microtubules, are believed to serve as tracks for motor proteins to transport specific materials to a defined direction. The idea is that the actin filaments and microtubules in general could work as signaling wires without actually moving materials.

A more specific and realistic role of uni-directional propagation of a specific state along actin filaments can be pictured in interaction with cofilin. It is generally accepted that the graded abundance of ADP-actin, ADP-Pi-actin, and ATP-actin, i.e., more ADP-Pi-actin and ATP-actin near the cell periphery and more aged ADP-actin in the interior, drives selective disruption of aged actin filaments by cofilin for recycling because cofilin prefers to bind to ADP-actin (Blanchoin and Pollard 1999; Carlier et al. 1997). Qualitatively, uni-directional growth of cofilin clusters toward the minus-end direction, which is to the interior of the cell, would enhance the selectivity of disrupting older sections of the filaments. It is not known if the speed of the uni-directional growth of cofilin clusters is fast enough to be physiologically relevant, since the growth rate was too fast to measure by HS-AFM under physiological conditions (Ngo et al. 2015).

Focal adhesion in animal cells is another possible site where actin filaments may be able to work as signaling wires. Focal adhesion is a structural and signaling hub between actin filaments in stress fibers, and the extracellular matrix (Critchley 2000). Vinculin (Le Clainche et al. 2010) and formin (Zigmond 2004) in focal

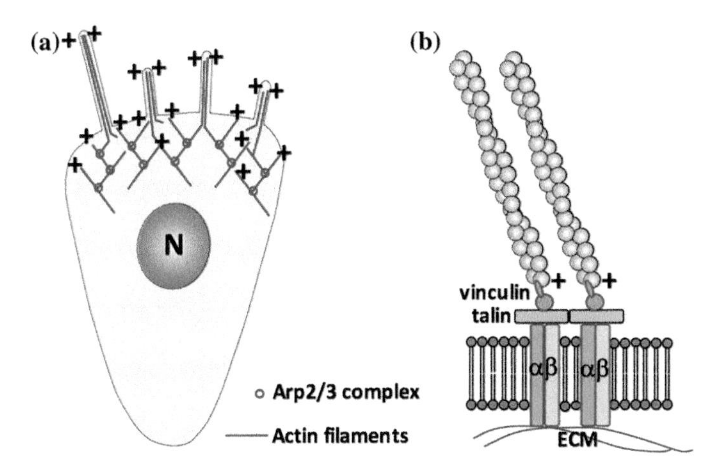

Fig. 11.4 Polar arrangements of actin filaments in cells. **a** In the lamellipodium along the leading edge of an amoeboid cell, there is a dendritic array of actin filaments, generated by the activity of the Arp2/3 complex. Thin projections called filopodia are mechanically supported by bundles of actin filaments. All of these actin filaments are aligned such that the plus ends are facing outward. Other thin projections on cell surfaces, such as microvilli in small intestine and stereocilia in ear, have similar polarized bundles of actin filaments. **b** At focal adhesions, integrin heterodimers (α and β) connect the extracellular matrix (ECM) to a large, focal adhesion complex of multiple proteins, of which only talin and vinculin are shown here. Vinculin binds to the plus ends of actin filaments (Le Clainche et al. 2010) in stress fibers

adhesions bind to the plus end of an actin filament, suggesting the possibility that actin filaments are bound to the focal adhesions at the plus ends. Therefore, uni-directional signaling along actin filaments to and from focal adhesions is structurally plausible.

11.5.2 Role in Force Generation by Myosin

Another possible role of uni-directional propagation of a specific state along actin filaments is to drive sliding motion of myosin motors in an ATP-dependent manner. In the prevailing swinging lever-arm theory, myosin is assumed to generate displacement by the swinging motion of the distal, rod-like lever-arm domain while strongly bound to an actin filament (Cooke et al. 1984; Holmes 1997) (Fig. 11.5). This model is supported by a number of structural and functional studies (Fisher et al. 1995; Ruff et al. 2001; Tsiavaliaris et al. 2004; Uyeda et al. 1996), but critically speaking, compelling evidence is limited. However, the situation is different with class V myosin, which moves slowly but processively along actin filaments using its very long lever-arm domain. Forkey et al. (2003) and Kodera et al. (Kodera et al. 2010) unequivocally demonstrated that the lever-arm swings as myosin V molecules move along actin filaments, although they do not necessarily

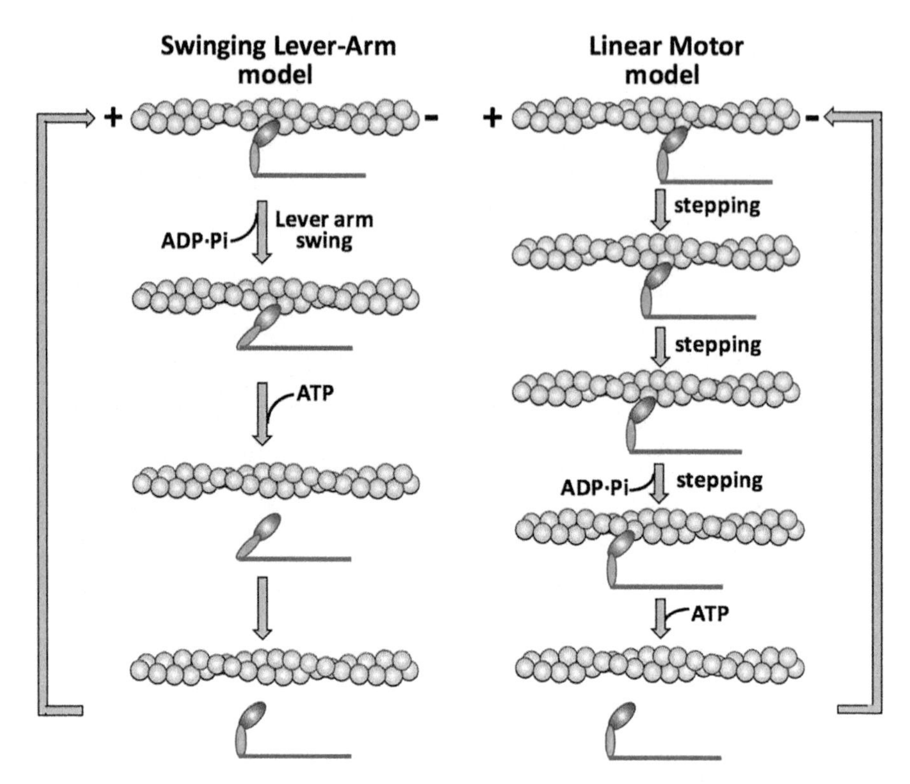

Fig. 11.5 Two models for the actin–myosin motility. In the prevailing "swinging lever-arm model," displacement is produced by an intramolecular conformational change within the motor domain of myosin II while it is strongly bound to an actin filament. After the stroke made by swinging of the lever arm (thin blue ellipse), ATP binds to the myosin motor to dissociate the complex and starts a new cycle. It is thus implied that each mechanical cycle is coupled with an enzymatic cycle in a one-to-one fashion (tight coupling). In contrast, the linear motor model assumes that the movement is generated by successive stepping of the myosin motor domain to the neighboring actin protomer. Each step is not necessarily coupled with hydrolysis of ATP (loose coupling). Red and blue large ellipses represent energized (i.e., carrying ADP and Pi) and basal states of the catalytic domain within the myosin motor, respectively

prove that the lever-arm swing produces active force or they swing as a passive consequence of movement. Furthermore, the sequential conformational change in the motor domain of myosin to swing the lever arm is generally assumed to be coupled with the enzymatic cycle to hydrolyze ATP on the motor domain (i.e., tight coupling). The swinging lever-arm mechanism and the tight coupling model are related to different aspects of the myosin's function, but those two are usually considered as two aspects of one coherent theory. In this model, actin is assumed to play two purely structural roles. They are to provide foothold for the myosin motor while it undergoes the swinging motion, and to stimulate the ATP hydrolysis cycle of myosin by accelerating Pi release from myosin motor carrying ADP and Pi (Lymn and Taylor 1971).

Different from this naïve assumption, certain modified actin filaments are able to bind strongly to myosin motors and to stimulate the ATPase activity, but are unable to move on class II myosin, to which muscle myosin belongs to. Such modifications include intramolecular crosslinking with glutaraldehyde (Prochniewicz and Yanagida 1990) and nicking of the so-called DNase loop with an endopeptidase (Schwyter et al. 1990). These results are difficult to reconcile in the framework of the simple lever-arm theory and suggest that some sort of dynamic property of actin filaments is necessary for the movement of myosin II. Strikingly, copolymerization of glutaraldehyde-treated actin with native actin-inhibited movement of the native protomers in the cofilament on myosin II-coated surfaces (Prochniewicz et al. 1993), implying that not only dynamics of each actin protomer but also long-range cooperative dynamics of the filament, is necessary for the movement of myosin II.

Myosin II is a non-processive motor but can produce fast movement when working in an ensemble. It is generally assumed that the force generation mechanism by class II myosin is basically the same as that of class V myosin, because of the sequence and structural similarities of the motor domains of the two myosins (Cheney et al. 1993). However, there are a number of pieces of evidence that argue oppositely, either directly or indirectly. For example, Kitamura et al. (1999) reported that, using a sophisticated single-molecule force measurement system, S1 of myosin II is able to produce multiple unitary displacements, each of which is the size of the actin protomer. This observation contradicts the tight coupling part of the coherent theory. Takano et al. (2010) presented results of a simulation study, which was consistent with the observation by Kitamura et al. (1999).

Phenotypes of certain mutant actins also suggest that the force-generating mechanism may be different between class II and V myosins. Gly146 is situated at the pivot between the two domains in an actin molecule, and mutation of this Gly to Val with a bulky side chain is expected to affect the interdomain motion within actin molecules. Strikingly, this G146 V mutation strongly inhibited movements on skeletal muscle myosin II, in terms of both speed and force generation, but had no inhibitory effect on movements on myosin V (Noguchi et al. 2012). DNase loop in subdomain 2 is well known to undergo major conformational changes depending on various conditions and is also deeply involved in communication with the adjacent actin protomer in a filament. Kubota et al. reported that a mutation in DNase loop (M47A) inhibits movements on myosin II, but not on myosin V (Kubota et al. 2009). The fact that two different mutations that are supposed to affect conformational changes of actin inhibited movements on myosin II but not on myosin V suggests that the motility mechanisms are qualitatively different between the two myosins, and motility of myosin II depends on conformational dynamics of actin filaments while that of myosin V does not.

All these data point to the possibility that dynamic conformational changes in actin filaments are required for motility of myosin II, which may move a long distance by making multiples of 5.5 nm steps in one ATP hydrolysis cycle (Figs. 11.5 and 11.6). If a myosin motor makes multiple 5.5 nm steps in a short period of time, this must involve dissociation from one actin protomer and rebinding to the next actin protomer on the plus-end side, rather than by making

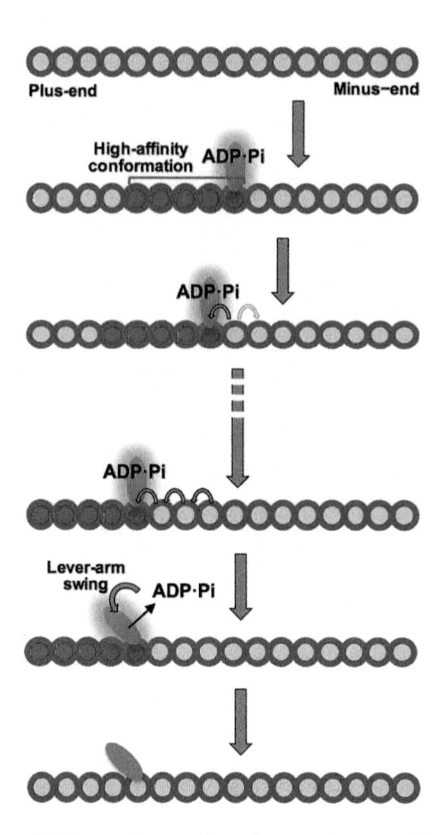

Fig. 11.6 A proposed model for stepping motion of a myosin motor. Binding of a myosin motor (ellipse) carrying ADP and Pi to an actin filament induces asymmetric cooperative conformational changes along the actin filament to create a zone of higher affinity on the plus-end side of the actin protomer to which the myosin motor is bound. Binding affinity between a myosin motor carrying ADP and Pi and an actin filament is intermediate, so that the bound myosin motor is able to diffuse along the filament, but the direction of diffusion is biased toward the plus end because the affinity is higher on that side. This is repeated multiple times if the load is small. At the end, ADP and Pi are released from the myosin motor, triggering the swing of the lever arm and establishment of high affinity rigor-like binding to bear tension. Binding of ATP to this bound myosin motor dissociates the complex to start a new chemomechanical cycle. Whether myosin-induced conformational changes propagate to the plus-end direction needs to be examined experimentally

multiple swings of the lever arm while bound to the same actin protomer. How then can this stepping be driven? A plausible scenario at this point is biased diffusion or biased Brownian motion, while the directional bias to random one-dimensional thermal motion is provided by structural asymmetry. More specifically, the bias may be provided purely by the structural asymmetry of the binding site, as has been suggested for a dynein-based engineered actin motor (Furuta et al. 2017). The simulation study by Takano et al. suggested that the bias of S1 movement along actin filaments is provided by the helical structure (Takano et al. 2010).

Alternatively, if binding of a myosin motor induces asymmetric structural changes around the bound actin protomer so that the affinity for a myosin motor is stronger on the plus-end side of the originally bound protomer than on the minus-end side, rebinding of the myosin head after transient detachment could be biased to the plus-end direction. Our recent observation that clusters of HMM-GFP tended to grow in one direction in the presence of low concentration of ATP (Hirakawa et al. 2017) supports the idea that binding of a myosin motor to an actin protomer in the presence of ATP can create an asymmetric affinity field around the bound actin protomer. This process may be repeated multiple times until Pi is released from the myosin motor to immobilize the myosin motor to the actin filament and induce the swinging of the lever arm, followed by ADP release and ATP binding to dissociate the complex. In this model, movement of myosin II is driven by two separate mechanisms, one that involves biased Brownian motion that is dependent on uni-directional conformational change in actin filaments, and the swing of the lever arm at the end of the ATPase cycle. Contribution of the former mechanism would be larger when the system is moving rapidly without producing large force, while the force produced by swinging of the lever arm would be predominant in the presence of high load. Myosin V might have evolved to move slowly but steadily using the latter mechanism only. Regarding the plausible physical principle of the biased Brownian motion, Suzuki and his colleagues discovered that the amount of hyper-mobile water around actin filaments increases when S1 interacts with the actin filaments (Suzuki et al. 2004), and recently proposed a model how a gradient of hyper-mobile water around the actin protomer to which S1 is bound can induce biased Brownian motion (Suzuki et al. 2017).

Experimental testing of biased Brownian motion is challenging, but is not totally out of the reach of our hands, with high-speed AFM and protein engineering in our hands.

References

Ando T, Uchihashi T, Kodera N (2013) High-speed AFM and applications to biomolecular systems. Annu Rev Biophys 42:393–414

Belyantseva IA, Perrin BJ, Sonnemann KJ, Zhu M, Stepanyan R, McGee J, Frolenkov GI, Walsh EJ, Friderici KH, Friedman TB, Ervasti JM (2009) Gamma-actin is required for cytoskeletal maintenance but not development. Proc Natl Acad Sci USA 106:9703–9708

Blanchoin L, Pollard TD (1999) Mechanism of interaction of Acanthamoeba actophorin (ADF/Cofilin) with actin filaments. J Biol Chem 274:15538–15546

Butters CA, Willadsen KA, Tobacman LS (1993) Cooperative interactions between adjacent troponin-tropomyosin complexes may be transmitted through the actin filament. J Biol Chem 268:15565–15570

Carlier MF (1990) Actin polymerization and ATP hydrolysis. Adv Biophys 26:51–73

Carlier MF, Laurent V, Santolini J, Melki R, Didry D, Xia GX, Hong Y, Chua NH, Pantaloni D (1997) Actin depolymerizing factor (ADF/cofilin) enhances the rate of filament turnover: implication in actin-based motility. J Cell Biol 136:1307–1322

Cheney RE, Riley MA, Mooseker MS (1993) Phylogenetic analysis of the myosin superfamily. Cell Motil Cytoskeleton 24:215–223

Collins A, Huang R, Jensen MH, Moore JR, Lehman W, Wang CL (2011) Structural studies on maturing actin filaments. Bioarchitecture 1:127–133

Cooke R, Crowder MS, Wendt CH, Bamett VA, Thomas DD (1984) Muscle cross-bridges: do they rotate? Adv Exp Med Biol 170:413–427

Craig-Schmidt MC, Robson RM, Goll DE, Stromer MH (1981) Effect of α-actinin on actin structure. Release of bound nucleotide. Biochim Biophys Acta 670:9–16

Critchley DR (2000) Focal adhesions—the cytoskeletal connection. Curr Opin Cell Biol 12: 133–139

De La Cruz EM (2005) Cofilin binding to muscle and non-muscle actin filaments: isoform-dependent cooperative interactions. J Mol Biol 346:557–564

Fisher AJ, Smith CA, Thoden J, Smith R, Sutoh K, Holden HM, Rayment I (1995) Structural studies of myosin: nucleotide complexes: a revised model for the molecular basis of muscle contraction. Biophys J 68:19S–26S; Discussion 27S–28S

Forkey JN, Quinlan ME, Shaw MA, Corrie JE, Goldman YE (2003) Three-dimensional structural dynamics of myosin V by single-molecule fluorescence polarization. Nature 422:399–404

Fujii T, Iwane AH, Yanagida T, Namba K (2010) Direct visualization of secondary structures of F-actin by electron cryomicroscopy. Nature 467:724–728

Fujii T, Namba K (2017). Structure of actomyosin rigour complex at 5.2 Å resolution and insights into the ATPase cycle mechanism. Nat Commun 8:13969

Fujime S, Ishiwata S (1971) Dynamic study of F-actin by quasielastic scattering of laser light. J Mol Biol 62:251–265

Furuta A, Amino M, Yoshio M, Oiwa K, Kojima H, Furuta K (2017) Creating biomolecular motors based on dynein and actin-binding proteins. Nat Nanotechnol 12:233–237

Galkin VE, Orlova A, Cherepanova O, Lebart MC, Egelman EH (2008) High-resolution cryo-EM structure of the F-actin-fimbrin/plastin ABD2 complex. Proc Natl Acad Sci USA 105:1494–1498

Galkin VE, Orlova A, Egelman EH (2012) Actin filaments as tension sensors. Curr Biol 22:R96–R101

Galkin VE, Orlova A, Kudryashov DS, Solodukhin A, Reisler E, Schröder GF, Egelman EH (2011) Remodeling of actin filaments by ADF/cofilin proteins. Proc Natl Acad Sci USA 108:20568–20572

Galkin VE, Orlova A, Lukoyanova N, Wriggers W, Egelman EH (2001) Actin depolymerizing factor stabilizes an existing state of F-actin and can change the tilt of F-actin subunits. J Cell Biol 153:75–86

Galkin VE, Orlova A, Salmazo A, Djinovic-Carugo K, Egelman EH (2010a) Opening of tandem calponin homology domains regulates their affinity for F-actin. Nat Struct Mol Biol 17:614–616

Galkin VE, Orlova A, Schröder GF, Egelman EH (2010b) Structural polymorphism in F-actin. Nat Struct Mol Biol 17:1318–1323

Galkin VE, Orlova A, Vos MR, Schröder GF, Egelman EH (2015) Near-atomic resolution for one state of F-actin. Structure. 23:173–182

Graceffa P, Dominguez R (2003) Crystal structure of monomeric actin in the ATP state. Structural basis of nucleotide-dependent actin dynamics. J Biol Chem 278:34172–34180

Hanein D, Matsudaira P, DeRosier DJ (1997) Evidence for a conformational change in actin induced by fimbrin (N375) binding. J Cell Biol 139:387–396

Hansen SD, Kwiatkowski AV, Ouyang CY, Liu H, Pokutta S, Watkins SC, Volkmann N, Hanein D, Weis WI, Mullins RD, Nelson WJ (2013) αE-catenin actin-binding domain alters actin filament conformation and regulates binding of nucleation and disassembly factors. Mol Biol Cell 24:3710–3720

Hawkins M, Pope B, Maciver SK, Weeds AG (1993) Human actin depolymerizing factor mediates a pH-sensitive destruction of actin filaments. Biochemistry 32:9985–9993

Hayakawa K, Sakakibara S, Sokabe M, Tatsumi H (2014) Single-molecule imaging and kinetic analysis of cooperative cofilin-actin filament interactions. Proc Natl Acad Sci USA 111:9810–9815

Hayden SM, Miller PS, Brauweiler A, Bamburg JR (1993) Analysis of the interactions of actin depolymerizing factor with G- and F-actin. Biochemistry 32:9994–10004

Hirakawa R, Nishikawa Y, Uyeda TQP, Tokuraku K (2017) Unidirectional growth of heavy meromyosin clusters along actin filaments revealed by real-time fluorescence microscopy. Cytoskeleton (Hoboken) 74:482–489

Holmes KC (1997) The swinging lever-arm hypothesis of muscle contraction. Curr Biol 7:112–118

Huang R, Grabarek Z, Wang CL (2010) Differential effects of caldesmon on the intermediate conformational states of polymerizing actin. J Biol Chem 285:71–79

Kandasamy MK, McKinney EC, Meagher RB (2009) A single vegetative actin isovariant overexpressed under the control of multiple regulatory sequences is sufficient for normal Arabidopsis development. Plant Cell 21:701–718

Kitamura K, Tokunaga M, Iwane AH, Yanagida T (1999) A single myosin head moves along an actin filament with regular steps of 5.3 nanometres. Nature 397:129–134

Kodera N, Yamamoto D, Ishikawa R, Ando T (2010) Video imaging of walking myosin V by high-speed atomic force microscopy. Nature 468:72–76

Kondo H, Ishiwata S (1976) Uni-directional growth of F-actin. J Biochem 79:159–171

Korn ED, Carlier MF, Pantaloni D (1987) Actin polymerization and ATP hydrolysis. Science 238:638–644

Kubota H, Mikhailenko SV, Okabe H, Taguchi H, Ishiwata S (2009) D-loop of actin differently regulates the motor function of myosins II and V. J Biol Chem 284:35251–35258

Le Clainche C, Dwivedi SP, Didry D, Carlier MF (2010) Vinculin is a dually regulated actin filament barbed end-capping and side-binding protein. J Biol Chem 285:23420–23432

Loscalzo J, Reed GH, Weber A (1975) Conformational change and cooperativity in actin filaments free of tropomyosin. Proc Natl Acad Sci USA 72:3412–3415

Lymn RW, Taylor EW (1971) Mechanism of adenosine triphosphate hydrolysis by actomyosin. Biochemistry 10:4617–4624

McGough A, Pope B, Chiu W, Weeds A (1997) Cofilin changes the twist of F-actin: implications for actin filament dynamics and cellular function. J Cell Biol 138:771–781

Michelot A, Drubin DG (2011) Building distinct actin filament networks in a common cytoplasm. Curr Biol 21:R560–R569

Miki M, Wahl P, Auchet JC (1982) Fluorescence anisotropy of labeled F-actin: influence of divalent cations on the interaction between F-actin and myosin heads. Biochemistry 21:3661–3665

Murakami K, Yasunaga T, Noguchi TQP, Gomibuchi Y, Ngo KX, Uyeda TQP, Wakabayashi T (2010) Structural basis for actin assembly, activation of ATP hydrolysis, and delayed phosphate release. Cell 143:275–287

Ngo KX, Kodera N, Katayama E, Ando T, Uyeda TQP (2015) Cofilin-induced unidirectional cooperative conformational changes in actin filaments revealed by high-speed atomic force microscopy. Elife 4:e04806

Ngo KX, Umeki N, Kijima ST, Kodera N, Ueno H, Furutani-Umezu N, Nakajima J, Noguchi TQP, Nagasaki A, Tokuraku K, Uyeda TQP (2016) Allosteric regulation by cooperative conformational changes of actin filaments drives mutually exclusive binding with cofilin and myosin. Sci Rep 6:35449

Noguchi TQP, Komori T, Umeki N, Demizu N, Ito K, Iwane AH, Tokuraku K, Yanagida T, Uyeda TQP (2012) G146V mutation at the hinge region of actin reveals a myosin class-specific requirement of actin conformations for motility. J Biol Chem 287:24339–24345

Oda T, Iwasa M, Aihara T, Maeda Y, Narita A (2009) The nature of the globular- to fibrous-actin transition. Nature 457:441–445

Oosawa F, Fujime S, Ishiwata S, Mihashi K (1973) Dynamic properties of F-actin and thin filament. Cold Spring Harbor Symp Quant Biol 37:277–285

Orlova A, Egelman EH (1997) Cooperative rigor binding of myosin to actin is a function of F-actin structure. J Mol Biol 265:469–474
Orlova A, Prochniewicz E, Egelman EH (1995) Structural dynamics of F-actin: II. Cooperativity in structural transitions. J Mol Biol 245:598–607
Otterbein LR, Graceffa P, Dominguez R (2001) The crystal structure of uncomplexed actin in the ADP state. Science 293:708–711
Papp G, Bugyi B, Ujfalusi Z, Barko S, Hild G, Somogyi B, Nyitrai M (2006) Conformational changes in actin filaments induced by formin binding to the barbed end. Biophys J 91:2564–2572
Pollard TD, Cooper JA (2009) Actin, a central player in cell shape and movement. Science 326:1208–1212
Prochniewicz E, Katayama E, Yanagida T, Thomas DD (1993) Cooperativity in F-actin: chemical modifications of actin monomers affect the functional interactions of myosin with unmodified monomers in the same actin filament. Biophys J 65:113–123
Prochniewicz E, Yanagida T (1990) Inhibition of sliding movement of F-actin by crosslinking emphasizes the role of actin structure in the mechanism of motility. J Mol Biol 216:761–772
Prochniewicz E, Zhang Q, Janmey PA, Thomas DD (1996) Cooperativity in F-actin: binding of gelsolin at the barbed end affects structure and dynamics of the whole filament. J Mol Biol 260:756–766
Romet-Lemonne G, Jegou A (2013) Mechanotransduction down to individual actin filaments. Eur J Cell Biol 92:333–338
Ruff C, Furch M, Brenner B, Manstein DJ, Meyhofer E (2001) Single-molecule tracking of myosins with genetically engineered amplifier domains. Nat Struct Biol 8:226–229
Schoenenberger CA, Mannherz HG, Jockusch BM (2011) Actin: from structural plasticity to functional diversity. Eur J Cell Biol 90:797–804
Schwyter DH, Kron SJ, Toyoshima YY, Spudich JA, Reisler E (1990) Subtilisin cleavage of actin inhibits in vitro sliding movement of actin filaments over myosin. J Cell Biol 111:465–470
Sharma S, Grintsevich EE, Hsueh C, Reisler E, Gimzewski JK (2012) Molecular cooperativity of drebrin1-300 binding and structural remodeling of F-actin. Biophys J 103:275–283
Sharma S, Grintsevich EE, Phillips ML, Reisler E, Gimzewski JK (2011) Atomic force microscopy reveals drebrin induced remodeling of f-actin with subnanometer resolution. Nano Lett 11:825–827
Siddique MS, Mogami G, Miyazaki T, Katayama E, Uyeda TQP, Suzuki M (2005) Cooperative structural change of actin filaments interacting with activated myosin motor domain, detected with copolymers of pyrene-labeled actin and acto-S1 chimera protein. Biochem Biophys Res Commun 337:1185–1191
Skau CT, Kovar DR (2010) Fimbrin and tropomyosin competition regulates endocytosis and cytokinesis kinetics in fission yeast. Curr Biol 20:1415–1422
Suarez C, Roland J, Boujemaa-Paterski R, Kang H, McCullough BR, Reymann AC, Guerin C, Martiel JL, De La Cruz EM, Blanchoin L (2011) Cofilin tunes the nucleotide state of actin filaments and severs at bare and decorated segment boundaries. Curr Biol 21:862–868
Suzuki M, Kabir SR, Siddique MS, Nazia US, Miyazaki T, Kodama T (2004) Myosin-induced volume increase of the hyper-mobile water surrounding actin filaments. Biochem Biophys Res Commun 322:340–346
Suzuki M, Mogami G, Ohsugi H, Watanabe T, Matubayasi N (2017) Physical driving force of actomyosin motility based on the hydration effect. Cytoskeleton (Hoboken) 74:512–527
Takano M, Terada TP, Sasai M (2010) Unidirectional Brownian motion observed in an in silico single molecule experiment of an actomyosin motor. Proc Natl Acad Sci USA 107:7769–7774
Tawada K (1969) Physicochemical studies of F-actin-heavy meromyosin solutions. Biochim Biophys Acta 172:311–318
Thomas DD, Seidel JC, Gergely J (1979) Rotational dynamics of spin-labeled F-actin in the sub-millisecond time range. J Mol Biol 132:257–273

Tokuraku K, Kurogi R, Toya R, Uyeda TQP (2009) Novel mode of cooperative binding between myosin and Mg^{2+}-actin filaments in the presence of low concentrations of ATP. J Mol Biol 386:149–162
Tsiavaliaris G, Fujita-Becker S, Manstein DJ (2004) Molecular engineering of a backwards-moving myosin motor. Nature 427:558–561
Uyeda TQP, Abramson PD, Spudich JA (1996) The neck region of the myosin motor domain acts as a lever arm to generate movement. Proc Natl Acad Sci USA 93:4459–4464
Uyeda TQP, Iwadate Y, Umeki N, Nagasaki A, Yumura S (2011) Stretching actin filaments within cells enhances their affinity for the myosin II motor domain. PLoS ONE 6:e26200
Wioland H, Guichard B, Senju Y, Myram S, Lappalainen P, Jegou A, Romet-Lemonne G (2017) ADF/cofilin accelerates actin dynamics by severing filaments and promoting their depolymerization at both ends. Curr Biol 27(1956–1967):e1957
Zigmond SH (2004) Formin-induced nucleation of actin filaments. Curr Opin Cell Biol 16:99–105

Chapter 12
Functional Mechanisms of ABC Transporters as Revealed by Molecular Simulations

Tadaomi Furuta and Minoru Sakurai

Abstract Active transport in cells is accomplished by a class of integral membrane proteins known as ATP-binding cassette (ABC) transporters. The energy source powering these molecular machines is the free energy generated by the binding of ATP molecules to nucleotide-binding domains (NBDs), as well as the free energy generated by ATP hydrolysis. The opening and closing motions of the NBDs are driven by these energies, which are propagated through transmembrane domains (TMDs) via mechanical transmission segments (coupling helices). As a result, the opening and closing motions of the TMDs are generated, which allow the uptake and release of substrates. In these processes, the chemical energy of ATP is converted into mechanical motion, a typical example of chemo-mechanical coupling. In this review, we describe the current understanding of this coupling mechanism, with a focus on the cooperative role of ATP and water.

Keywords ATP · Water · Chemo-mechanical coupling · ATPase

12.1 Introduction

ABC transporters belong to one of the largest transporter families; they use ATP to drive the transport of a wide range of substrates across membranes in both prokaryotic and eukaryotic cells. ABC transporters are divided into two subtypes, ABC importers and ABC exporters, on the basis of the direction of the transport reaction (Dassa and Bouige 2001). ABC importers, present only in prokaryotes, are involved in the uptake of nutrients and micronutrients from the extracellular milieu (Cui and Davidson 2011; Davidson et al. 2008) and are divided into three types: type I, type II, and energy-coupling factor (ECF). ABC exporters are present in both prokaryotes and eukaryotes. In humans, there are 48 distinct ABC transporters,

T. Furuta · M. Sakurai (✉)
Center for Biological Resources and Informatics,
Tokyo Institute of Technology, Tokyo, Japan
e-mail: msakurai@bio.titech.ac.jp

© Springer Nature Singapore Pte Ltd. 2018

M. Suzuki (ed.), *The Role of Water in ATP Hydrolysis Energy Transduction by Protein Machinery*, https://doi.org/10.1007/978-981-10-8459-1_12

which are expressed in many tissues, such as the intestines, liver, kidney, and brain (Moitra and Dean 2011). These play important roles in drug absorption, distribution, metabolism, and excretion (Szakács et al. 2008). For example, P-glycoprotein (P-gp or ABCB1) extrudes more than 200 chemically diverse substrate molecules, including therapeutic drugs, steroid hormones, and signaling molecules (Subramanian et al. 2016). Thus, it prevents therapeutic drugs from reaching their intracellular targets, which in turn causes resistance to their pharmacological actions, i.e., multidrug resistance (MDR). As a result, in certain cancer cells, chemotherapy is rendered ineffective (Fletcher et al. 2010). In several pathogenic bacteria, ABC exporters extrude diverse antibiotics into the extracellular space, thus conferring antibiotic resistance to these bacteria (Lomovskaya et al. 2007). MDR has recently become one of the most important challenges facing modern medicine. However, atomic-level understanding of the mechanism of substrate transport of ABC transporters remains unclear, and its elucidation is crucial for the development of novel drugs, including specific inhibitors for MDR.

ABC transporters share a common structural organization comprised of two nucleotide-binding domains (NBDs), which bind and hydrolyze ATP in the cytoplasm, and two transmembrane domains (TMDs), which form the translocation pathway for substrates (Hollenstein et al. 2007). In importers, the TMDs and NBDs are separate polypeptide chains. In bacterial exporters, a TMD is fused to an NBD, generating a “half transporter.” Many eukaryotic ABC exporters express all four domains in a single polypeptide chain. Importers have 10–20 TM helices, and exporters have a conserved core of twelve TM helices; each TMD has six TM helices. The TMDs make contact with the NBDs through coupling helices (CHs) that are located in two intracellular loops (ICL1 and ICL2). The primary amino acid sequences of the NBDs are highly conserved among all species, whereas the TMDs have significant diversity in both primary sequences and tertiary structures relevant to their polyspecificity for substrates.

To transport substrates through the membrane, the substrate-binding sites can alternate their access to the intracellular and extracellular sides of the membrane, which is called “alternating access model” (Jardetzky 1966); the two corresponding conformational states are referred to as inward-facing (IF) and outward-facing (OF) states, respectively. One of the most challenging issues in the study of ABC transporters is the elucidation of the mechanism by which such alternating motions are coupled with the binding of ATP molecules to the NBDs and their subsequent hydrolysis, i.e., “chemo-mechanical coupling.” Recently, the crystal structures of various ABC transporters have been determined for various reaction states, such as ATP-bound, ADP-bound, and ATP-free. On the basis of these findings, several types of reaction cycles have been proposed (Locher 2016). For example, in the ATP-free (apo) state of certain ABC exporters, the protein adopts an IF conformation where the two NBDs are separated and, consequently, the TMDs are open to the cytoplasmic side (Fig. 12.1a). Upon the binding of two ATP molecules, the NBDs dimerize, which may pull the coupling helices toward each other; consequently, the TMDs take on an OF conformation. Upon the hydrolysis of ATP, the reverse transition occurs to reset the transporter to the ground state for the next step;

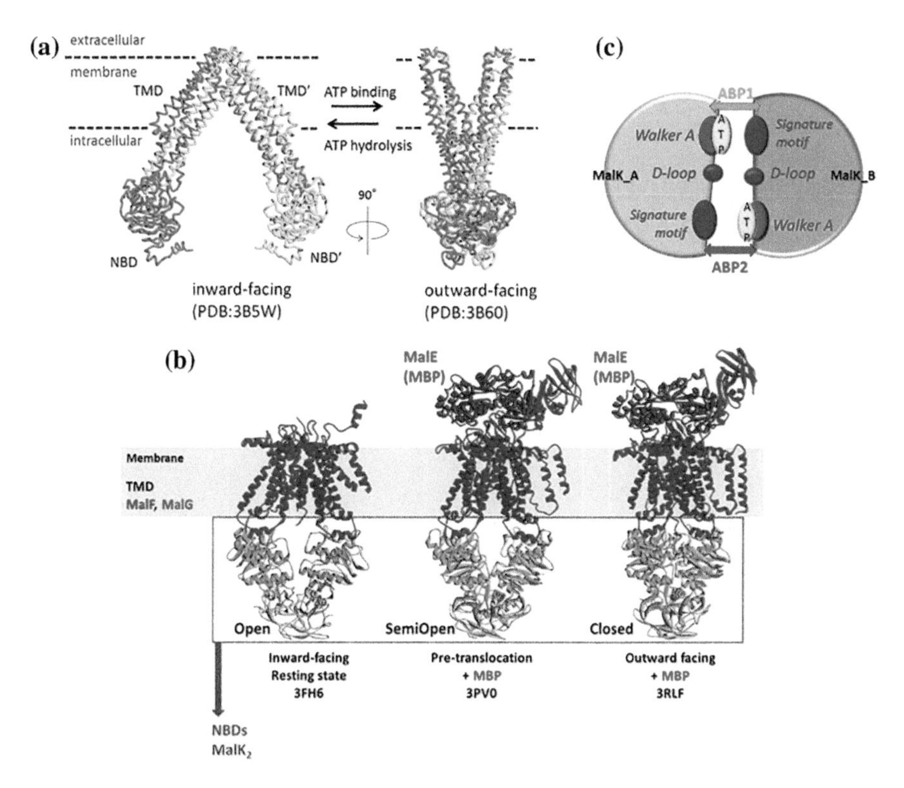

Fig. 12.1 Structures of ABC transporters. **a** The representative exporter MsbA. **b** The representative type I importer maltose transporter (MalFGK$_2$-E). Maltose-binding protein (MBP) corresponds to MalE. **c** Schematic representation of NBDs. ATP and several conservative motifs are represented. An ATP molecule is sandwiched between the Walker A motif and the signature motif to form an ATP binding pocket (ABP)

the NBD dimer opens, consequently causing the conversion to an IF conformation. The first half of the process is thought to be driven by the free energy of ATP binding to the NBDs and the second half of the process by the energy released by ATP hydrolysis. However, from the analysis of the available snapshot structures alone, it is difficult to understand the mechanism underlying the chemo-mechanical coupling.

Molecular simulations, including molecular dynamics (MD) simulations, free energy calculations, and QM/MM simulations, are powerful tools available to investigate the structure and mechanism of ABC transporters. Many excellent reviews concerning the structure and dynamics of ABC transporters have already been published. Thus, this mini review is not intended as a comprehensive review of the literature (Ferreira et al. 2015; George and Jones 2012; Liu et al. 2016); instead, we focus on the atomic-level understanding of the roles of ATP and water in the underlying mechanism of ABC transporters.

12.2 Role of ATP and Water in the Mechanism of the NBD Engine

The NBDs are the engine domains of the ABC transporter molecular machine. This engine works using the energy generated from ATP binding and hydrolysis. Upon the binding of two ATP molecules, the NBDs dimerize, which causes a conformational transition of the TMDs into a higher energy state, from the IF state to the OF state. Following a framework analogous to internal combustion engines, this process is often termed the "power-stroke" of transport (Dawson and Locher 2007; Higgins and Linton 2004; McDevitt et al. 2008; Smith et al. 2002). From the highly conserved nature of NBDs from different ABC transporters, it is generally believed that there is a universal mechanism for the NBD dimerization/dissociation that drives the power-stroke.

12.2.1 Thermodynamics of NBD Dimerization

Maltose transporter is one of the most extensively investigated type I importers; it is composed of heterodimeric TMDs (denoted MalF and MalG) and $MalK_2$ (Chen 2013) (Fig. 12.1b). $MalK_2$ consists of MalK_A and MalK_B, and each MalK domain has a regulatory domain (RD) and the NBD. According to the crystal structures, $MalK_2$ forms different conformations depending on the stage of its reaction cycle; open, semi-open, and closed states have been identified.

Hsu et al. (2015) performed conventional MD simulations and accelerated MD (aMD) simulations for *E. coli* $MalK_2$ with and without ATP binding for the above three states. aMD is an enhanced sampling method that enables the calculation of the free energy landscape in the molecular events of biomolecules, such as proteins (Hamelberg et al. 2004; Markwick and McCammon 2011). When MD simulations started from a semi-open state, the protein with ATP converged toward the closed state, where ATP molecules were correctly sandwiched between two conserved motifs, the Walker A and signature motifs (Fig. 12.1c). When MD simulations started from an open state, the protein with ATP converged to a semi-open state; however, further closing was not observed. In contrast, in the absence of ATP, $MalK_2$ was never closed, remaining open regardless of the starting structures. These results were rationalized from the free energy landscapes obtained from aMD simulations. One of the most interesting findings was that there was a downhill free energy gradient from the semi-open state to closed state when ATP molecules bound to the NBDs; the free energy difference between the two states was approximately 40 kcal/mol. In addition, from the free energy map, it was found that an asymmetric pathway was energetically favorable for the closure of the two ATP binding pockets (ABPs); one ABP was closed first, followed by the closing of the other ABP. These results provided direct evidence that ATP binding generates a free energy gradient available to drive the mechanical motion; i.e., the free energy from ATP binding was converted into mechanical force.

In general, due to their flexibility, proteins may remain in different substrates with or without their cognate substrates. To describe these phenomena, two models have been proposed, the "induced fit" and the "conformational selection" models (Changeux and Edelstein 2011; Hatzakis 2014). The former suggests that ligand binding drives the ligand-free (apo) conformation directly to the ligand-bound form, in which the ligand may remodel the protein energy landscape. The latter, in contrast, suggests that the dynamics of the protein are inherent and that the binding of the ligand to the protein merely shifts the equilibrium from the apo to the ligand-bound conformation. There is still controversy regarding whether the induced fit or the conformational selection model governs the protein–ligand binding mechanism. A study by Hsu et al. (2015) clearly indicated that the induced fit mechanism governs $MalK_2$ dimerization, because the conformational distribution was extended toward the closed state only when the ligand ATP was bound.

Hayashi et al. (2014) discussed the origin of the free energy gain obtained by $MalK_2$ dimerization by decomposing it into several thermodynamic quantities, including solvation contributions, where the solvation (hydration) contributions were calculated using a statistical thermodynamic theory of liquids called the three-dimensional reference interaction site model (3D-RISM); the internal energy was evaluated directly from the molecular mechanics (MM). In addition, the translational, rotational, and vibrational entropy changes upon dimerization were calculated using a conventional statistical mechanics method. According to their calculations, the free energy change upon dimerization of $MalK_2$ was -4.11 ± 0.41 kcal/mol and $+3.79 \pm 0.16$ kcal/mol for the ATP- and ADP-bound states, respectively. The former value is in a fairly good agreement with the experimental value (-7.01 kcal/mol) estimated from the dissociation constant for the $MalK_2$ dimer. Therefore, the $MalK_2$ dimer is stably formed when ATP molecules are bound, whereas such a dimer is energetically unfavorable in the ADP-bound state. One feature of this study was that the free energy change was decomposed smartly into the enthalpic and entropic contributions; the former included the conformational energy and hydration enthalpy, and the latter included the configurational entropy and hydration entropy. It was found that in the ATP-bound state, the energetic gain (361.58 kcal/mol) arising from direct MalK –MalK interactions (internal energy) was negated by a dehydration penalty (378.14 kcal/mol). However, a large gain of hydration entropy (-53.71 kcal/mol) led to the net free energy gain (-4.11 kcal/mol).

It is likely that such a large hydration entropy gain may arise from the construction of a closely packed structure of the $MalK_2$ dimer in the presence of ATP. The precise packing (i.e., shape complementarity) between the proteins at the atomic level resulted in a large decrease in the total excluded volume (-1725.63 Å^3) upon dimerization, and thus an increase in the total volume available to the translational displacement of water molecules in the system, which in turn provided a remarkably large contribution to the hydration entropy gain. This is known as the excluded volume effect (Yoshidome et al. 2008). In the case of the

ADP-bound $MalK_2$, on the other hand, such an improved shape complementarity between the interface regions of each MalK might be partially collapsed due to the release of two inorganic phosphates (Pi). Such a decrease of the precise packing could lead to a loss in the overlap of the excluded volume (−35.5 $Å^3$), resulting in the loss of the hydration entropy gain upon dimerization.

The computational method used in the above study by Hayashi et al. (2014) was a type of endpoint method that could calculate the free energy difference between the initial state (isolated proteins) and the final state (complex), but did not provide the free energy profile during the binding process. Furukawa-Hagiya et al. (2014) investigated the free energy profile during the dimerization process of two NBDs in the human cystic fibrosis transmembrane conductance regulator (CFTR), a chloride channel belonging to the ABC protein superfamily. They first constructed a homology model of the inward-facing CFTR and performed MD simulations to explore the dynamics of the CFTR in a membrane environment. As a result, two NBDs of the protein rapidly approached each other within 10 ns, forming a head-to-tail dimer in which the ATP molecules were sandwiched between the Walker A and signature motifs. This short timescale suggests that the strong long-range attractive force functions between the NBDs. To reveal the origin of this attractive force, they calculated the hydration free energy and its components (enthalpy and entropy) for the simulation trajectories using 3D-RISM. Then, the dimerization process was decomposed into three stages: (1) up to a first partial contact of the two NBDs from the separated state (0.0–1.4 ns), an approach stage; (2) the process leading to face-to-face contact of the NBDs (1.4–4.0 ns); and (3) a structural adjustment stage (4.0~ ns) to form more favorable interactions between the two NBDs.

In stage (1), the protein internal energy increased with time, whereas the solvation free energy decreased. It was noted that the decrease in the solvation free energy was dominated by its enthalpy component. Therefore, it was the solvation enthalpy that played a key role in initiating the approach of the largely separate two NBDs. In the initial structure, a significant number of water molecules existed in the interdomain space that were likely responsible for the generation of a long-range attractive force. In stage (2), the number of hydrogen bonds between the two NBDs increased, leading to a decrease in the internal energy. However, in general, a favorable residue–residue interaction was formed at the expense of a residue–water interaction, i.e., a dehydration penalty. This led to a loss (i.e., increase) of the solvation enthalpy. The internal energy gain was found to be almost completely negated by the increase in the solvation enthalpy. As a result, the entropic contribution became relatively more important in this stage. The solvation entropy increases upon dehydration because trapped water molecules around hydrophilic residues are "liberated" to the bulk, which is the excluded volume effect as described above. The resulting decrease in the entropic term of $-T\Delta S$ was −14.3 kcal/mol, which is the main driving force. In the final stage (3), the number of hydrogen bonds and hydrophobic contacts increases, implying the occurrence of structural adjustments between the NBD–NBD interfacial region and, simultaneously, the further liberation of interfacial water molecules. Thus, both the enthalpy and entropy comparably contribute to the free energy, although the largest contributor to the driving force is still the solvation entropy. Taken together, a long-range hydration force of electrostatic and enthalpic origin acting on ATP and its interaction counterpart in the ABP is most likely the major thermodynamic force that drives the two NBDs to approach each other, from a large separation toward a distance allowing direct contact. It should be noted that the two NBDs never dimerized in the apo state. Therefore, the above long-range attractive force likely originates from the interaction between the bound ATP molecules and the interdomain water molecules.

12.2.2 Kinetics of NBD Dimerization

Protein–protein interactions are generally categorized into two types, hydrophobic and hydrophilic, according to the dominant characteristics of the interaction. The dimerization of the NBDs is a typical hydrophilic protein–protein interaction, where an ATP binding site with ATP on one NBD makes contact with polar conserved motifs on the other NBD, leading to the formation of ABP. In general, hydrophilic interactions may be accompanied by a relatively large energetic penalty due to dehydration of interfacial water molecules. Accordingly, the NBD dimerization process is predicted to be unlikely to occur. However, as described above, the NBD dimerization in the CFTR was found to occur very rapidly (~10 ns), which suggests that this process is less inhibited than previously predicted. To solve the riddle of NBD dimerization, Sakaizawa et al. (2016) applied two types of MD simulations to a prokaryotic ABC transporter NBD dimer, *M. jannaschii* MJ0796 (NBDs): (1) Thermal fluctuation of each NBD was frozen during the dimerization (static simulation), and (2) thermal fluctuations of NBDs were explicitly taken into account during the dimerization (dynamic simulation). In the former simulation, the center of mass (COM) distance alone was decreased from 79 to 29 Å while maintaining each protein structure as fixed. The COM distance of 29 Å corresponded to the NBD dimer state in the crystal structure. The latter simulations were performed using targeted MD (tMD) methods, where the root-mean-square deviation (RMSD) was changed toward the target structure over the relevant time course, and the thermal fluctuations of NBDs were explicitly taken into account. The free energy profile, including solvation effects, was obtained from the MM calculations combined with 3D-RISM calculations, similar to the study described above by Furukawa-Hagiya et al. (2014).

In the range of 79–40 Å, the free energy profile was almost flat, with large fluctuations of ±5 kcal/mol in both the static and dynamic simulations. In the range of 40–29 Å, the two NBDs encountered each other near a COM distance of approximately 34 Å and subsequently surmounted the barrier to form the native dimer complex (COM = 29 Å). Interestingly, the barrier height was ~100 kcal/mol in the static system, whereas it was dramatically decreased to ~15 kcal/mol in the dynamic system. In the latter case, the protein association could easily occur under physiological conditions, whereas it was an extremely rare event in the former case. To elucidate the origin of the reaction barrier, the free energy was decomposed into the internal energy, the hydration enthalpy, and the hydration entropy. The internal energy and the hydration entropy both decreased as the COM distance decreased, while the hydration enthalpy increased significantly. The increase in the hydration enthalpy overcompensated for the decrease in the internal energy in magnitude, consequently resulting in the generation of the barrier. This unfavorable enthalpic contribution was significantly smaller in the dynamic system than in the static system around the barrier. This observation may be due to the dehydration penalty. For the occurrence of NBD–NBD contact, water molecules must escape from the interfacial region (dehydration of the interface); this process occurred more easily in the dynamic system than in the static system, which was supported by counting the number of the interfacial water molecules as a function of the COM distance.

Many diverse essential molecular processes within the cell are triggered by protein–protein interactions. Recently, several studies have shown that hydrophilic interactions are relatively more important than hydrophobic interactions (Ahmad et al. 2011; Ben-Naim 2006; Ulucan et al. 2014). The above findings, albeit obtained for a case study of NBD dimerization, provide insights concerning diverse protein–protein interaction processes.

12.2.3 The Free Energy of ATP Hydrolysis and the Mechanism of Its Generation

12.2.3.1 ATP Hydrolysis in Solution

ATP is the “molecular unit of currency” of intracellular energy transfer (Knowles 1980). When the phosphate–phosphate bonds (phospho-anhydride bonds) in ADP and ATP are hydrolyzed, a free energy of 7.3 kcal/mol is released at a standard state of 1 M (Netlson and Cox 2005). This bond is therefore called the high-energy phosphate bond. Molecular machines, such as ABC transporters and motor proteins (e.g., F1–Fo motor, myosin, kinesin), use ATP as an energy source. According to standard biochemistry textbooks, the following three factors are responsible for the “high-energy” character of the phospho-anhydride bond: (1) the resonance stabilization effect, (2) electrostatic repulsions, and (3) the hydration effect. It has been hypothesized that the third factor favors ATP hydrolysis due to the greater degree of solvation (hydration) of the Pi and ADP products relative to that of ATP, which further stabilizes the products relative to the reactants. However, quantum chemical calculations combined with a theory of solution, such as the continuum model (Colvin et al. 1995; Hayes et al. 1978), 3D-RISM (Yoshida 2014), and energy representation (Takahashi et al. 2017), have provided more accurate insight into this effect.

Table 12.1 shows the hydrolysis energy of pyrophosphate as a model of ATP in different ionized conditions. The second and third columns indicate the data from Colvin et al. (1995) and this work, respectively. In both studies, the solvent effect was evaluated using a polarizable continuum model (PCM). Our data accurately reproduce the experimental data as a whole, although the hydrolysis energy of reaction 1 was overestimated. Table 12.2 shows the decomposition of the hydrolysis energy into the intra- and intermolecular energies. The total hydrolysis energy is governed by the electronic energy and the electrostatic solvation energy, while the contributions of thermal and non-electrostatic solvation energies are small. The electronic energy and electrostatic solvation energy tended to cancel each other out, resulting in a small exothermic (negative) hydrolysis energy of ~10 kcal/mol. Even in an intrinsically endothermic reaction (positive contribution of the electronic energy), such as reaction 2, the total reaction energy was ultimately exothermic due to the hydration effect. For the extremely exothermic reactions, such as reactions 3–5, the excessive reaction energies were reduced by hydration to ~10 kcal/mol. As a result, the moderately exergonic ATP hydrolysis reaction becomes capable of directly coupling with energetically unfavorable (endergonic) reactions in cells.

Table 12.1 Hydrolysis energy ΔG^0 (kcal/mol) of pyrophosphate

Reaction no.	Hydrolysis reaction	Colvin et al.[a]	This work[b]	Exp.[c]
1	$H_4P_2O_7 + H_2O \rightarrow 2H_3PO_4$	−9.3	−13.3	−9.5
2	$H_3P_2O_7^{1-} + H_2O \rightarrow H_3PO_4^{1-} + H_3PO_4$	0.1	−8.3	−7.5
3	$H_2P_2O_7^{2-} + H_2O \rightarrow 2H_2PO_4^{1-}$	2.7	−6.0	−7.7
4	$HP_2O_7^{3-} + H_2O \rightarrow H_2PO_4^{2-} + H_2PO_4^{1-}$	6.6	−6.2	−7.1
5	$P_2O_7^{4-} + H_2O \rightarrow 2HPO_4^{2-}$	−0.8	−10.6	−10.4

[a]Reprinted with permission from Colvin et al. (1995). Copyright (1995) American Chemical Society. The theoretical level used is MP2 6-311++G(d,p) PCM//HF 6-311++G(d,p)
[b]Determined via the Gaussian09 program at a theoretical level of MP2 6-311++G(d,p) PCM//HF 6-311++G(d,p) PCM
[c]Reprinted with permission from George et al. (1970)

Table 12.2 Decomposition of the hydrolysis energy into the intra- and intermolecular energies

Reaction no.	Intramolecular		Intermolecular (solvation)		Total	Exp.
	Electronic	Thermal[a]	Electrostatic	Non-electrostatic[b]		
1	−13.3	−2.1	3.65	−1.59	−13.3	−9.5
2	22.5	−2.4	−26.9	−1.53	−8.3	−7.5
3	−51.8	−1.3	48.7	−1.7	−6.0	−7.7
4	−126.6	0.03	122.2	−1.82	−6.2	−7.1
5	−281.2	−1.38	273.8	−1.81	−10.6	−10.4

[a]Translational, rotational, and vibrational free energies
[b]Cavitation + dispersion + repulsion

The reaction mechanism of ATP hydrolysis has also been widely investigated with both experimental and theoretical approaches. Generally, the mechanisms of phosphate ester hydrolysis can be classified into two extreme cases (Kamerlin et al. 2008). One is an associative mechanism, in which nucleophilic attack precedes phosphorus oxygen bond cleavage, i.e.,

$$\mathrm{ROPO_3^{n-}} + \mathrm{H_2O} \rightarrow [\mathrm{RO}\cdots\mathrm{PO_3}\cdots\mathrm{OH_2}]^{\ddagger} \rightarrow \mathrm{RO^{(n-1)-}} + \mathrm{H_2PO_4^-}$$

where $[\,]^{\ddagger}$ represents the transition state. The other is a dissociative mechanism, in which the leaving group departs prior to nucleophilic attack, and then the resulting PO_3^- is transformed to $H_2PO_4^-$, i.e.,

$$\mathrm{ROPO_3^{n-}} \rightarrow \mathrm{RO^{(n-1)-}} + \mathrm{PO_3^-}$$
$$\mathrm{PO_3^-} + \mathrm{H_2O} \rightarrow \mathrm{H_2PO_4^-}$$

These mechanisms represent only the two most limiting cases, and there are various levels of concerted reactions in which bond formation and breaking occur simultaneously with a single transition state.

Theoretical studies have also been performed to obtain more detailed information at the atomic level (Grigorenko et al. 2006; Harrison and Schulten 2012; Klähn et al. 2006; Wang et al. 2015; Yamamoto 2010). Klähn et al. (2006) investigated a series of hydrolysis reactions of phosphate monoester anions in aqueous solution at the DFT/COSMO level. The resultant two-dimensional free energy surface suggested that the potential surface of the associative and dissociative paths is very flat around the transition-state region, and multiple reaction pathways are possible. Grigorenko et al. (2006) investigated the hydrolysis of methyl triphosphate in aqueous solution using QM/MM calculations and suggested that hydrolysis may proceed via a stepwise dissociative mechanism. Yamamoto (2010) investigated free energy profiles for methyl monophosphate and triphosphate anions at the DFT/COSMO level and suggested that in low dielectric environments, the stepwise mechanism involving prior metaphosphate dissociation was preferred over a concerted one that involves a penta-coordinated transition state. Harrison and Schulten (2012) investigated the hydrolysis reaction mechanism of a full ATP molecule via QM/MM simulations using classical and DFT-based quantum dynamics and suggested that the dissociative mechanism has a lower activation barrier and earlier transition state crossing than the associative mechanism. Wang et al. (2015) applied a nudged elastic band method to identify the minimum energy path of ATP hydrolysis in aqueous solution and calculated the free energy profile along this path with a QM/MM method. The results suggested that the reaction proceeds through a concerted path before the system reaches the transition state and along an associative path after the transition state.

12.2.3.2 ATP Hydrolysis in the NBD

The highly conserved nature of the NBDs in sequence and structure implies that the hydrolysis of ATP shares a common mechanism in the ABC protein family. There are several key issues concerning the mechanism of ATP hydrolysis in a wide variety of ATP-driven proteins. The first of these is the debate regarding "associative" versus "dissociative," similar to the solution reaction described above. The second issue concerns the identification of the final acceptor of the lytic water proton. In one model, the ATP γ-phosphate group itself acts as a catalytic base to abstract the proton from lytic water; we refer to this mechanism as the substrate-assisted catalysis (SAC) model (Zaitseva et al. 2005). In another model, a catalytic base, such as the conserved glutamate residue located on the Walker B motif, abstracts a proton from the lytic water prior to or at the transition state (TS). This model is called the general base catalysis (GBC) model. The last key issue concerns how many water molecules are needed for the ATP hydrolysis reaction, and this question is usually referred to as the one-water (1 W) model versus two-water (2 W) model debate (Prasad et al. 2013).

For ABC transporters, theoretical study on the above issues is still in its infancy compared to the extensive study of motor proteins, such as myosin, kinesin, and F1-ATPase motors (Kiani and Fischer 2016). QM/MM calculations by Zhou et al. (2013) showed that the conserved H-loop residue His662 in *E. coli* HlyB, a bacterial ABC transporter, can act first as a general acid and then as a general base to facilitate proton transfer in ATP hydrolysis. Without the assistance of His662, direct proton transfer from the lytic water to ATP results in a substantially higher barrier height. Huang and Liao (2016) used the QM/MM nudged elastic band method to investigate the ATP hydrolysis reaction in $MalK_2$. In their study, the potential of mean force (PMF) along the reaction pathway was obtained, with an activation free energy value of 19.2 kcal/mol, which was in agreement with the experimental results. According to their results, the reaction can be summarized as a dissociative and 1 W model. Moreover, the glutamate residue (Glu159) in the Walker B motif acts as a general base to abstract the proton from lytic water; however, the rate-limiting step is the bond cleavage between the γ- and β-phosphates rather than the deprotonation of lytic water. As a result, their model followed an atypical GBC pathway.

Hsu et al. (2016) applied QM/MM metadynamics simulation to $MalK_2$ and explored the free energy profile along an assigned collective variable. As a result, it was determined that the activation free energy was approximately 10.5 kcal/mol, and the reaction released a free energy of approximately 3.8 kcal/mol (Fig. 12.2a). The dissociation of the ATP γ-phosphate appeared to be the rate-limiting step, which supported the dissociative model. Moreover, Glu159, located in the Walker B motif, acted as a base to abstract the proton from the lytic water, but was not the catalytic base (Fig. 12.2b). Therefore, the reaction corresponded to an atypical GBC model, consistent with the results of Huang and Liao (2016). Hsu et al. (2016) also observed two interesting proton transfers: transfer from the His192 ε-position nitrogen to the dissociated inorganic phosphate Pi and transfer from the

(a)

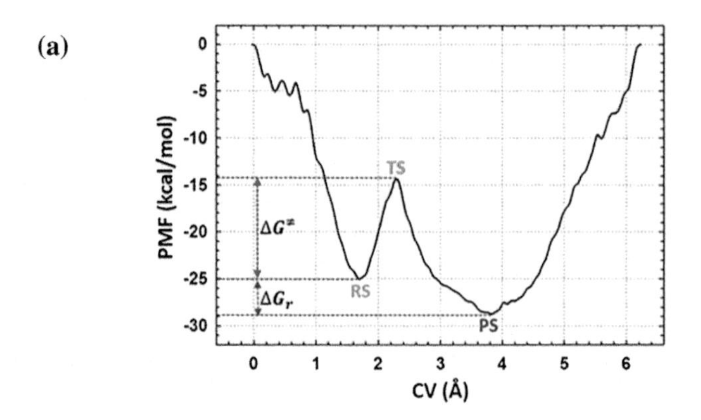

(b)

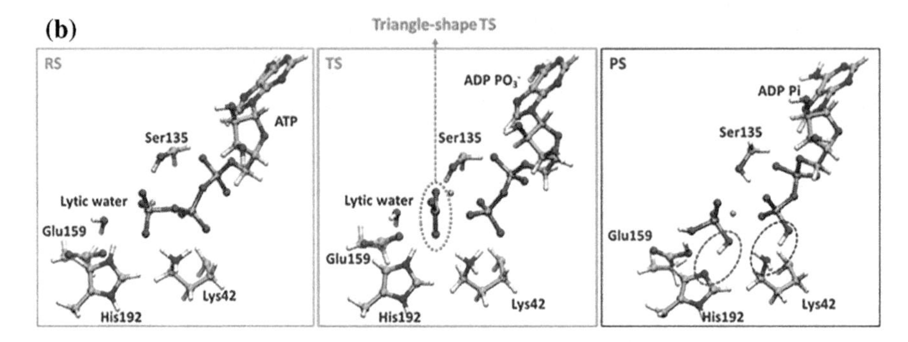

Fig. 12.2 **a** PMF along the reaction coordinate (CV), which represents the distance of $O3B_{ATP} - O_{water}$ minus $P_{\gamma} - O_{water}$. **b** Structures of the reactant (RS), transition (TS), and product (PS) states. Reprinted (adapted) with permission from Hsu et al. (2016). Copyright (2016) American Chemical Society

Lys42 side chain to the product ADP β-phosphate. These proton transfers could play a role in stabilizing the post-hydrolysis state.

12.3 Role of Coupling Helices (CHs) as a Mechanical Transmission

Binding and hydrolysis of ATP induce conformational changes in the NBDs that are transmitted to the TMDs via non-covalent interactions, leading to the conformational transition between IF and OF conformations. As described in the introduction, key architectural segments in this NBD–TMD transmission are CHs (Dawson and Locher 2006). Each TMD of an ABC exporter consists of the following structural elements: TM1-ECL1-TM2-ICL1-TM3-ECL2-TM4-ICL2-TM5-ECL3-TM6, where ECLs stand for extracellular loops. Coupling helix 1 (CH1) and coupling helix 2 (CH2) are located in ICL1 and ICL2, respectively; in the case of a full transporter, the CHs in the second half are called CH3 and CH4, respectively.

What are the structural and functional roles of CHs in ABC transporters? Mutation at the transmission interface can cause "uncoupling" or improper assembly of ABC transporters (Hollenstein et al. 2007). The mutations R659Q (in NBD1) or R378C (in CH2) of the human TAP1/2 lead to a loss of transport function (Chen et al. 1996; Oancea et al. 2009). The deletion of F508 (in NBD1) in the human CFTR, present in 90% of cystic fibrosis patients, causes multiple defects in the CFTR protein, leading to its impaired assembly during synthesis and reduced posttranslational stability (Molinski et al. 2012). The mutations in CHs or grooves of the NBDs of the human P-glycoprotein (P-gp) severely affect the maturation or function of the protein (Ferreira et al. 2017). Interestingly, on the transmission interface of the P-gp, it was found that the ATPase activity of the mutant F1086A (located in NBD2) could be restored when the nearby residue A266 (in CH2) was replaced with aromatic residues (Tyr, Phe, or Trp); i.e., CHs and grooves of the NBDs could act as "ball-and-socket" joints (Loo et al. 2013). From these results, the importance of CHs on transport function has been emphasized. However, the roles of each CH have not yet been revealed in detail.

Recently, Furuta et al. (2014) addressed the above issue through a combined study of biochemical experiments and MD simulations for the ABC transporter MsbA and its two mutants, in which all the amino acid residues of one CH (either CH1 or CH2) were mutated to Ala residues: (i) wild type (Wt), (ii) CH1 mutant (Mt1), and (iii) CH2 mutant (Mt2). The experiments showed that CH2 mutation caused a decrease in the ATPase activity of approximately 32% compared with that of the Wt; the k_{cat} values for Wt, Mt1, and Mt2 were 1.56, 1.43, and 0.504, respectively. In addition, a nearly equal decrease in the ATP binding affinity was observed for both Mt1 and Mt2; the K_m values for Wt, Mt1, and Mt2 were 0.189, 1.09, and 2.14, respectively. On the other hand, the findings from the MD simulations were summarized as follows: (1) The asymmetric opening of NBDs that is usually observed for the native NBDs was not observed for the apo state of Mt2, although it was observed for both Wt and Mt1, and (2) in Mt2, the OF conformation was slightly broken, even in the ATP-bound state. These dynamic and structural defects that would hinder the normal IF $\leftrightarrow$ OF conformational transition could account for the decrease in the ATPase activity (k_{cat}) associated with the CH2 mutation. In addition, it was found that the interaction between the adenosine part of ATP and the A-loop (Tyr351) was weakened in Mt1, and the interaction between the γ-phosphate of ATP and the H-loop (His357) was also weakened slightly in Mt2. These findings are consistent with the above experimental results indicating that the ATP binding affinity (K_m) is decreased in both mutants. Recently, owing to their functional importance, CHs have come to be considered "druggable" targets in the field of drug discovery for several ABC transporters (e.g., P-gp (Ferreira et al. 2017) and CFTR (Molinski et al. 2012)); the blocking of NBD–TMD communication or the impairment of efflux conformational changes may represent efficient therapeutic strategies.

In ABC importers, there is only one coupling helix (abbreviated CH_{im}) in each TMD, which is present in the ICL between TM3 and TM4 (in type I) and between TM6 and TM7 (in type II). Several single mutations in CM_{im} of the *E. coli* maltose transporter $MalFGK_2$-E also result in a significant decrease in transport function (Mourez et al. 1997).

12.4 Mechanisms of the IF↔OF Conformational Transition of TMDs

The NBDs in ABC transporters form dimers upon the binding of ATP molecules. Then, the ATP molecules act as a molecular glue. Thus, if the ATPs are hydrolyzed, the dimer would be separated. If the NBDs and the CHs are tightly joined, these NBD motions could cause the cytoplasmic ends of the TMDs to pull together and push apart, resulting in the IF↔OF conformational transitions of the TMDs. However, recent structural and various biophysical studies, including MD simulations, have suggested mechanistic diversity for the ATP-driven reaction mechanisms in ABC transporters (Locher 2016). Here, we describe a few representative models for each of the exporters and importers that have been proposed by MD simulations.

12.4.1 Concerted NBD–TMD Motion in Exporters

12.4.1.1 Alternating Access Mechanism

Furukawa-Hagiya et al. (2013) performed MD simulations for a homology model structure of CFTR in the IF state and compared the dynamics of the apo and ATP-bound states. In the ATP-bound state, the two NBDs made direct contact as a "canonical" NBD dimer, in which ATP molecules are sandwiched between the Walker A and signature motifs. However, in the apo state, only partial contact was observed, with a lateral movement (sliding) of the NBDs. The NBD dimerization induced by ATP binding was found to modify the collective motion of the whole protein to significantly activate the motion of the TMD regions. In particular, the opposite rotations in the NBDs and TMDs were generated, leading to a twisted structure of the TMDs and an increase in the fluctuations in the extracellular part of the TMDs. However, the opening of the Cl^- channel pore was not observed, likely due to an insufficient simulation time. Watanabe et al. (2013) performed MD simulations for the murine P-gp and observed an asymmetric closure of the NBDs in the presence of ATP over 100 ns of MD simulations, with or without the transport substrate verapamil in the drug-binding pocket (DBP). They observed that upon binding of the substrate in the DBP, the contact of the NBDs became tighter, but a limited simulation time did not allow for observation of the opening of the TMDs. Recently, Furuta et al. (2016) conducted multiple 100 ns MD simulations for the *T. maritima* heterodimeric exporter TM287/288 and revealed that the binding of substrate (Hoechst 33342) to the DBP enhanced the NBD dimerization through allosteric NBD–TMD communication, via comparisons between the substrate-free and substrate-bound simulations. In addition, they clarified that two ATPs were necessary to induce NBD dimerization, although the transition of the TMDs to the OF conformation was also not observed within the limited timescale.

As shown above, the IF ↔ OF conformational transition is beyond the timescales allowed by conventional MD simulations. Thus, to overcome this issue, the use of enhanced sampling MD methods, such as targeted MD (tMD) (Schlitter et al. 1993), is necessary. Weng et al. (2010) applied a tMD to the OF → IF transition pathway of the bacterial MsbA. The resultant MD trajectory revealed a clear spatiotemporal order of the conformational movements. Chang et al. (2013) applied tMD to the conformational transition of human P-gp and showed that the movement in the x-axis led to the closure of the NBDs, while movement in the y-axis adjusted the conformations of the NBDs to form the correct ABPs (the x- and y-axes define a plane parallel to the membrane). In addition, they suggested that six key segments located on the TMDs move along the y-axis as a result of NBD dimerization, which is in turn transmitted through the CHs to the rest of the TMDs, inducing the TMDs to adopt the OF conformation. In spite of these successful studies, tMD simulations may not necessarily be appropriate for addressing the IF ↔ OF transition problem, because a tMD study occasionally produces an unnatural conformation in which a pore is simultaneously open to both sides of the membrane. Moradi and Tajkhorshid (2013, 2014) developed a novel approach based on an extensive initial search for system-specific reaction coordinates and the use of non-equilibrium simulations and found that the twisting motions of the NBDs were coupled to the open–close motions of the TMDs; this suggested a novel hypothesis for the conformational dynamics of ABC exporters, termed the "doorknob" mechanism.

Another strategy to overcome the sampling problem and describe the large-scale conformational changes of proteins is to employ coarse-grained representations of the protein and/or environment models, such as elastic network models (ENMs), where the elastic forces acting on the particles obey Hooke's law (Tirion 1996). Recently, Xie et al. (2015) applied an adaptive anisotropic network model (aANM) to the allosteric transition of MsbA and obtained results consistent with the doorknob mechanism.

The above studies based on the tMD and aANM methods have provided detailed information concerning the pathway of the IF ↔ OF transition of MsbA. However, these methods are a type of interpolation method that is applicable only when both the initial and final structures are known. In other words, the conformational changes induced by ATP binding or ATP hydrolysis must be known in advance as input data for these models. Therefore, in principle, these studies provide no answer to the following questions: "Does the forward IF → OF transition actually occur with ATP binding alone at the NBDs, without ATP hydrolysis?" and "Does the backward OF → IF transition occur with ATP hydrolysis alone?" To answer these questions, it is necessary to examine how the protein responds to the application of an external perturbation from ATP. The solving of this problem is equivalent to elucidation of the mechanism of chemo-mechanical coupling in ABC transporters.

Recently, Arai et al. (2017) used a nonlinear ENM method to answer the first question. In this method, the equation of motion based on ENM with an additional term representing external forces was solved in the overdamped limit (Düttmann et al. 2012). They examined how the structure of MsbA changed from the initial IF conformation when two nodes were newly added to mimic the bound ATP molecules or when external forces, mimicking the effects of ATP binding, were applied to every single residue in the NBDs. In the former case, the NBDs dimerized tightly and the two TMDs subsequently opened toward the extracellular side, resulting in the transition to the OF conformation. Such a conformational transition was also reproduced by applying external forces which could cause rotational motion of the NBDs around the axis (x-axis) connecting the COM of the two NBDs. It should be noted that the application of an external force causing a translational motion along the x-axis alone led to the closure of the NBDs, but never caused the opening of the TMDs. The simulation also suggested that the NBD–TMD concerted motion occurs via the ICLs. More importantly, it was revealed that the ATP binding energy is converted into the distortion (elastic) energy of several transmembrane helices, such as TM3, TM4, and TM6. These results based on ENM are believed to originate from the inter-residue contact topology. Since most of the ABC exporters have the same structural topology as MsbA, the above results are helpful for understanding the transport mechanism of the other ABC exporters.

The use of a coarse-grained force field, such as MARTINI, is another powerful approach for analyzing the IF ↔ OF transition in ABC transporters. Recently, using the MARTINI force field, Wang and Liao (2015) performed coarse-grained MD simulations with umbrella sampling for the *C. elegans* P-gp, where the PMF for the IF ↔ OF transition was analyzed using the COM distance of the cytoplasmic portion (including NBDs) as a reaction coordinate. From the PMF result, it was found that there were three major minima (3.6–3.8, 3.0, and 2.8 Å) at the COM distance, which corresponded to an IF conformation resembling the crystal structure of the murine P-gp, a closed IF conformation whose NBDs are in loose contact, and an OF conformation resembling the crystal structure of the *S. aureus* Sav1866, respectively. The energy of these states was shown to increase; the OF conformation had a higher energy than the IF conformation, which is consistent with the results from the aforementioned nonlinear ENM study (Arai et al. 2017).

12.4.1.2 Outward-Only Mechanism

In an outward-only mechanism, an inward-facing cavity capable of interacting with a substrate is not formed during the substrate transport cycle. This mechanism has been proposed for the *C. jejuni* LLO flippase PglK (Perez et al. 2015). Recently, a similar mechanism that does not involve an IF conformation was proposed from MD simulations for the *S. aureus* Sav1866 (Xu et al. 2017). The simulations were performed for seven different nucleotide occupancy states, depending on whether the two APBs were occupied by ATP or ADP or were empty (apo). The different nucleotide occupancy states led to one of the two possible conformations of the TMD: one similar to the crystal structure with an outward-facing cavity (outward-open) and one in which the cavity had collapsed (outward-closed). The former remained stabilized by the hydrolysis of a single ATP. The latter allowed substrates to have partial access to the inner cavity, while the former allowed the substrates to transition between the inner and outer cavities with subsequent release to the extracellular medium. The conformational change was communicated to the TMD through a route involving the Q-loop and X-loop. These simulations may provide a basis for the unidirectionality of transport in ABC exporters.

12.4.2 Concerted NBD–TMD Motions in Importers

Distinct mechanisms have been proposed for type I and type II ABC importers (Locher 2016). Here, due to space constraints, we refer only to several recent studies concerning the mechanism of the *E. coli* maltose transporter because it provides a representative model of a generic mechanism of type I importers.

As described previously, the maltose transporter is composed of heterodimeric TMDs (called MalF and MalG) and homodimeric NBDs (called $MalK_2$). In addition to the TMDs and NBDs, the periplasmic maltose-binding protein MalE captures maltose and shuttles it to the transporter. Based on crystal structures and other experimental evidence, a hypothetical substrate translocation model has been proposed by Oldham and Chen (2011) and Chen (2013). In this model, the binding of a maltose-bound MalE to the periplasmic side of the TMDs was necessary to stabilize the pre-translocation state, and the binding of ATPs to the NBDs further triggered the conformational change from the inward-facing state to the outward-facing state, which in turn resulted in the translocation of maltose from MalE to the TMDs. Finally, ATP hydrolysis occurred, and the transporter returned to the inward-facing state, releasing the maltose from the intracellular side.

Li et al. (2014) performed aANM simulations to obtain a better understanding of the full dynamics of the maltose transporter and the conformational change from its resting state structure. They proposed a simplified model for its structural rearrangement from the inward-facing state (resting state) to the outward-facing state and additionally noted that there was a strong correlation in the motion between the CHs and the NBDs. This result strongly supports the "ball-and-socket joint" model (Oldham et al. 2007) and further strengthens the importance of the CHs, as they lead to the mechanical coupling between the TMDs and NBDs in the maltose transporter.

Recently, Weng et al. (2017) performed metadynamics simulations to clarify the role of MalE binding in the $MalFGK_2$ system. The resultant free energy surfaces showed that in the absence of MalE, lateral closing motions were energetically forbidden; however, upon MalE binding, more closed conformations, similar to the pre-translocation state, became increasingly stable. The significant effect of MalE binding on the free energy landscape appeared to partially support Oldham and Chen's model (described above). However, the simulation did not provide any information regarding the conformational change from the pre-translocation to the outward-facing state. This is likely because their protein model of $MalFGK_2$ did not include ATPs. On the other hand, aMD simulations for the $MalK_2$ system indicated that ATP binding to the NBDs is indispensable for their complete closure (Hsu et al. 2015). Upon ATP binding, the free energy profile relevant to the MalK_A–MalK_B interaction changed to a steep downhill shape, leading toward dimer formation; in contrast, in the absence of ATP, the open conformation was favored. Therefore, by combining the free energy landscapes identified by Weng et al. and those by Hsu et al., it can be hypothesized that both MalE and ATP binding are essential for the formation of the occluded (NBD-dimerized) state when starting from the inward-facing state. As expected, a recent MD study by Hsu et al. (2018) indicated that both the binding of MalE to the periplasmic side of the TMDs and the binding of ATP to MalK2 are necessary to facilitate the conformational change from the inward-facing state to the occluded state, in which $MalK_2$ is completely dimerized. MalE binding suppressed the fluctuation of the TMDs and MalF periplasmic region (MalF-P2) and thus prevented the incorrect arrangement of the MalF C-terminal (TM8) helix. Without MalE binding, the MalF TM8 helix showed a tendency to intrude into the substrate translocation pathway, hindering the closure of the $MalK_2$.

Oliveira et al. (2011) performed MD simulations on the outward-facing state of the maltose transporter and compared the dynamics of three different systems: ATP-bound, ADP Pi-bound, and ADP-bound systems. They found that an asymmetric opening of the ABPs occurred in the ADP-bound system and that the fluctuation of CHs increased during the opening process. Therefore, it was concluded that a significant conformational rearrangement of the ABP sites from the closed state to open state occurs after the release of the inorganic phosphate Pi in the post-ATP hydrolysis state and that the CHs may play important roles in mediating NBD–TMD communication.

12.5 Conclusion and Perspectives

In this review, we have described the current understanding of the mechanisms of the NBD engine, the CH transmission, and the power-stroke (IF ↔ OF conformational transition) caused by the NBD–TMD concerted motion. In particular, we focused on the IF → OF process, which is driven by the free energy gain associated with ATP binding. When NBDs are separated via water layers, a steep downhill potential is generated between them if ATP molecules bind to the ABPs. In other words, a long-range attractive force functions between the NBDs, and the origin of this force is the hydration enthalpy of the interdomain water molecules. The thermal fluctuations of the NBDs facilitate the escape of these water molecules from the interdomain region, which suppresses the increase in the reaction barrier accompanied by a dehydration penalty. Thus, the NBD dimerization is a relatively rapid process. The final dimer state is stabilized by the free energy gain from hydration entropy, corresponding to the so-called excluded volume effect. In exporters, the NBD–TMD concerted motion leads to the OF conformation of the TMDs. In particular, a twisting motion of the NBDs plays an important role in the generation of an opening motion of the TMDs. In the IF $\rightarrow$ OF transition, the ATP binding free energy is converted into the distortion (elastic) energies of several TM helices. In type I importers, such as the maltose transporter, the synergistic effect of ATP and the substrate-binding protein is necessary for the closure of NBDs.

As described above, the mechanisms underlying the chemo-mechanical coupling induced by ATP binding are currently being revealed by a wide array of diverse molecular simulation studies. However, it remains unclear how the ATP hydrolysis energy is converted into the mechanical motion. If the OF state energy is higher than the IF state energy, the OF state would spontaneously transform into the IF state when the ATP is hydrolyzed, because the molecular glue joining the two NBDs disappears with the ATP hydrolysis. The ATP hydrolysis would therefore be regarded as a switch, rather than an energy source. The mechanism of the post-hydrolysis process also remains unclear in the case of importers, such as the maltose transporter (Shilton 2015). The use of large-scale QM/MM MD simulations will be indispensable for addressing this question, which represents a significant challenge for researchers in the molecular simulation field.

Acknowledgements This work was supported in part by JSPS KAKENHI JP16H00825, JP16K12520, and JP15K00400. We cordially thank Mr. Sho Tanaka for his contribution to the calculated data shown in Tables 12.1 and 12.2.

References

Ahmad M, Gu W, Geyer T, Helms V (2011) Adhesive water networks facilitate binding of protein interfaces. Nat Commun 2:261

Arai N, Furuta T, Sakurai M (2017) Analysis of an ATP-induced conformational transition of ABC transporter MsbA using a coarse-grained model. Biophys Physicobiol 14:161–171

Ben-Naim A (2006) On the driving forces for protein-protein association. J Chem Phys 125:024901

Chang S-Y, Liu F-F, Dong X-Y, Sun Y (2013) Molecular insight into conformational transmission of human P-glycoprotein. J Chem Phys 139:225102

Changeux J-P, Edelstein S (2011) Conformational selection or induced fit? 50 years of debate resolved. F1000 Biol Rep 3:19

Chen HL, Gabrilovich D, Tampe R, Girgis KR, Nadaf S, Carbone DP (1996) A functionally defective allele of TAP1 results in loss of MHC class I antigen presentation in a human lung cancer. Nat Genet 13:210–213

Chen J (2013) Molecular mechanism of the *Escherichia coli* maltose transporter. Curr Opin Struct Biol 23:492–498

Colvin ME, Evleth E, Akacem Y (1995) Quantum chemical studies of pyrophosphate hydrolysis. J Am Chem Soc 117:4357–4362

Cui J, Davidson AL (2011) ABC solute importers in bacteria. Essays Biochem 50:85–99

Düttmann M, Togashi Y, Yanagida T, Mikhailov Alexander S (2012) Myosin-V as a mechanical sensor: an elastic network study. Biophys J 102:542–551
Dassa E, Bouige P (2001) The ABC of ABCs: a phylogenetic and functional classification of ABC systems in living organisms. Res Microbiol 152:211–229
Davidson AL, Dassa E, Orelle C, Chen J (2008) Structure, function, and evolution of bacterial atp-binding cassette systems. Microbiol Mol Biol Rev 72:317–364
Dawson RJP, Locher KP (2006) Structure of a bacterial multidrug ABC transporter. Nature 443:180–185
Dawson RJP, Locher KP (2007) Structure of the multidrug ABC transporter Sav 1866 from Staphylococcus aureus in complex with AMP-PNP. FEBS Lett 581:935–938
Ferreira RJ, Bonito CA, Ferreira MJU, dos Santos DJVA (2017) About P-glycoprotein: a new drugable domain is emerging from structural data. WIREs Comput Mol Sci 7:e1316
Ferreira RJ, Ferreira M-JU, dos Santos DJVA (2015) Reversing cancer multidrug resistance: insights into the efflux by ABC transports from in silico studies. WIREs Comput Mol Sci 5:27–55
Fletcher JI, Haber M, Henderson MJ, Norris MD (2010) ABC transporters in cancer: more than just drug efflux pumps. Nat Rev Cancer 10:147–156
Furukawa-Hagiya T, Furuta T, Chiba S, Sohma Y, Sakurai M (2013) The power stroke driven by ATP binding in CFTR as studied by molecular dynamics simulations. J Phys Chem B 117:83–93
Furukawa-Hagiya T, Yoshida N, Chiba S, Hayashi T, Furuta T, Sohma Y, Sakurai M (2014) Water-mediated forces between the nucleotide binding domains generate the power stroke in an ABC transporter. Chem Phys Lett 616–617:165–170
Furuta T, Sato Y, Sakurai M (2016) Structural dynamics of the heterodimeric ABC transporter TM287/288 induced by ATP and substrate binding. Biochemistry 55:6730–6738
Furuta T, Yamaguchi T, Kato H, Sakurai M (2014) Analysis of the structural and functional roles of coupling helices in the ATP-binding cassette transporter MsbA through enzyme assays and molecular dynamics simulations. Biochemistry 53:4261–4272
George AM, Jones PM (2012) Perspectives on the structure–function of ABC transporters: the switch and constant contact models. Prog Biophys Mol Biol 109:95–107
George P, Witonsky RJ, Trachtman M, Wu C, Dorwart W, Richman L, Richman W, Shurayh F, Lentz B (1970) "Squiggle-H2O". An enquiry into the importance of solvation effects in phosphate ester and anhydride reactions. Biochim Biophys Acta 223:1–15
Grigorenko BL, Rogov AV, Nemukhin AV (2006) Mechanism of triphosphate hydrolysis in aqueous solution: QM/MM simulations in water clusters. J Phys Chem B 110:4407–4412
Hamelberg D, Mongan J, McCammon JA (2004) Accelerated molecular dynamics: a promising and efficient simulation method for biomolecules. J Chem Phys 120:11919–11929
Harrison CB, Schulten K (2012) Quantum and classical dynamics simulations of ATP hydrolysis in solution. J Chem Theory Comput 8:2328–2335
Hatzakis NS (2014) Single molecule insights on conformational selection and induced fit mechanism. Biophys Chem 186:46–54
Hayashi T, Chiba S, Kaneta Y, Furuta T, Sakurai M (2014) ATP-induced conformational changes of nucleotide-binding domains in an ABC transporter. Importance of the water-mediated entropic force. J Phys Chem B 118:12612–12620
Hayes DM, Kenyon GL, Kollman PA (1978) Theoretical calculations of the hydrolysis energies of some "high-energy" molecules. 2. A survey of some biologically important hydrolytic reactions. J Am Chem Soc 100:4331–4340
Higgins CF, Linton KJ (2004) The ATP switch model for ABC transporters. Nat Struct Mol Biol 11:918–926
Hollenstein K, Dawson RJP, Locher KP (2007) Structure and mechanism of ABC transporter proteins. Curr Opin Struct Biol 17:412–418
Hsu W-L, Furuta T, Sakurai M (2018) The mechanism of nucleotide-binding domain dimerization in the intact maltose transporter as studied by all-atom molecular dynamics simulations. Proteins 86:237–247

Hsu W-L, Furuta T, Sakurai M (2015) Analysis of the free energy landscapes for the opening-closing dynamics of the maltose transporter ATPase MalK2 using enhanced-sampling molecular dynamics simulation. J Phys Chem B 119:9717–9725
Hsu W-L, Furuta T, Sakurai M (2016) ATP hydrolysis mechanism in a maltose transporter explored by QM/MM metadynamics simulation. J Phys Chem B 120:11102–11112
Huang W, Liao J-L (2016) Catalytic mechanism of the maltose transporter hydrolyzing ATP. Biochemistry 55:224–231
Jardetzky O (1966) Simple allosteric model for membrane pumps. Nature 211:969
Kamerlin SCL, Florián J, Warshel A (2008) Associative versus dissociative mechanisms of phosphate monoester hydrolysis: on the interpretation of activation entropies. ChemPhysChem 9:1767–1773
Kiani FA, Fischer S (2016) Comparing the catalytic strategy of ATP hydrolysis in biomolecular motors. Phys Chem Chem Phys 18:20219–20233
Klähn M, Rosta E, Warshel A (2006) On the mechanism of hydrolysis of phosphate monoesters dianions in solutions and proteins. J Am Chem Soc 128:15310–15323
Knowles JR (1980) Enzyme-catalyzed phosphoryl transfer reactions. Annu Rev Biochem 49: 877–919
Li CH, Yang YX, Su JG, Liu B, Tan JJ, Zhang XY, Wang CX (2014) Allosteric transitions of the maltose transporter studied by an elastic network model. Biopolymers 101:758–768
Liu H, Li D, Li Y, Hou T (2016) Atomistic molecular dynamics simulations of ATP-binding cassette transporters. WIREs Comput Mol Sci 6:255–265
Locher KP (2016) Mechanistic diversity in ATP-binding cassette (ABC) transporters. Nat Struct Mol Biol 23:487–493
Lomovskaya O, Zgurskaya HI, Totrov M, Watkins WJ (2007) Waltzing transporters and 'the dance macabre' between humans and bacteria. Nat Rev Drug Discov 6:56–65
Loo TW, Bartlett MC, Clarke DM (2013) Human P-glycoprotein Contains a greasy ball-and-socket joint at the second transmission interface. J Biol Chem 288:20326–20333
Markwick PRL, McCammon JA (2011) Studying functional dynamics in bio-molecules using accelerated molecular dynamics. Phys Chem Chem Phys 13:20053–20065
McDevitt CA, Crowley E, Hobbs G, Starr KJ, Kerr ID, Callaghan R (2008) Is ATP binding responsible for initiating drug translocation by the multidrug transporter ABCG2? FEBS J 275:4354–4362
Moitra K, Dean M (2011) Evolution of ABC transporters by gene duplication and their role in human disease. Biol Chem 392:29–37
Molinski S, Eckford P, Pasyk S, Ahmadi S, Chin S, Bear C (2012) Functional rescue of F508del-CFTR using small molecule correctors. Front Pharmacol 3
Moradi M, Tajkhorshid E (2013) Mechanistic picture for conformational transition of a membrane transporter at atomic resolution. Proc Natl Acad Sci USA 110:18916–18921
Moradi M, Tajkhorshid E (2014) Computational recipe for efficient description of large-scale conformational changes in biomolecular systems. J Chem Theory Comput 10:2866–2880
Mourez M, Hofnung M, Dassa E (1997) Subunit interactions in ABC transporters: a conserved sequence in hydrophobic membrane proteins of periplasmic permeases defines an important site of interaction with the ATPase subunits. EMBO J 16:3066–3077
Netlson DL, Cox MC (2005) Lehninger: principles of biochemistry, 4th edn. W. H. Freeman and Company, New York
Oancea G, O'Mara ML, Bennett WFD, Tieleman DP, Abele R, Tampé R (2009) Structural arrangement of the transmission interface in the antigen ABC transport complex TAP. Proc Natl Acad Sci USA 106:5551–5556
Oldham ML, Chen J (2011) Crystal structure of the maltose transporter in a pretranslocation intermediate state. Science 332:1202–1205
Oldham ML, Khare D, Quiocho FA, Davidson AL, Chen J (2007) Crystal structure of a catalytic intermediate of the maltose transporter. Nature 450:515

Oliveira ASF, Baptista AM, Soares CM (2011) Inter-domain communication mechanisms in an ABC importer: a molecular dynamics study of the MalFGK2E complex. PLoS Comput Biol 7: e1002128
Perez C, Gerber S, Boilevin J, Bucher M, Darbre T, Aebi M, Reymond J-L, Locher KP (2015) Structure and mechanism of an active lipid-linked oligosaccharide flippase. Nature 524:433
Prasad BR, Plotnikov NV, Warshel A (2013) Addressing open questions about phosphate hydrolysis pathways by careful free energy mapping. J Phys Chem B 117:153–163
Sakaizawa H, Watanabe HC, Furuta T, Sakurai M (2016) Thermal fluctuations enable rapid protein–protein associations in aqueous solution by lowering the reaction barrier. Chem Phys Lett 643:114–118
Schlitter J, Engels M, Krüger P, Jacoby E, Wollmer A (1993) Targeted molecular dynamics simulation of conformational change-application to the T$\leftrightarrow$R transition in insulin. Mol Simul 10:291–308
Shilton Brian H (2015) Active transporters as enzymes: an energetic framework applied to major facilitator superfamily and ABC importer systems. Biochem J 467:193–199
Smith PC, Karpowich N, Millen L, Moody JE, Rosen J, Thomas PJ, Hunt JF (2002) ATP binding to the motor domain from an ABC transporter drives formation of a nucleotide sandwich dimer. Mol Cell 10:139–149
Subramanian N, Condic-Jurkic K, O'Mara ML (2016) Structural and dynamic perspectives on the promiscuous transport activity of P-glycoprotein. Neurochem Int 98:146–152
Szakács G, Váradi A, Özvegy-Laczka C, Sarkadi B (2008) The role of ABC transporters in drug absorption, distribution, metabolism, excretion and toxicity (ADME–Tox). Drug Discov Today 13:379–393
Takahashi H, Umino S, Miki Y, Ishizuka R, Maeda S, Morita A, Suzuki M, Matubayasi N (2017) Drastic compensation of electronic and solvation effects on ATP hydrolysis revealed through large-scale QM/MM simulations combined with a theory of solutions. J Phys Chem B 121:2279–2287
Tirion MM (1996) Large amplitude elastic motions in proteins from a single-parameter, atomic analysis. Phys Rev Lett 77:1905–1908
Ulucan O, Jaitly T, Helms V (2014) Energetics of hydrophilic protein-protein association and the role of water. J Chem Theory Comput 10:3512–3524
Wang C, Huang W, Liao J-L (2015) QM/MM investigation of ATP hydrolysis in aqueous solution. J Phys Chem B 119:3720–3726
Wang Z, Liao J-L (2015) Probing structural determinants of ATP-binding cassette exporter conformational transition using coarse-grained molecular dynamics. J Phys Chem B 119:1295–1301
Watanabe Y, Hsu W-L, Chiba S, Hayashi T, Furuta T, Sakurai M (2013) Dynamics and structural changes induced by ATP and/or substrate binding in the inward-facing conformation state of P-glycoprotein. Chem Phys Lett 557:145–149
Weng J-W, Fan K-N, Wang W-N (2010) The conformational transition pathway of ATP binding cassette transporter MsbA revealed by atomistic simulations. J Biol Chem 285:3053–3063
Weng J, Gu S, Gao X, Huang X, Wang W (2017) Maltose-binding protein effectively stabilizes the partially closed conformation of the ATP-binding cassette transporter MalFGK2. Phys Chem Chem Phys 19:9366–9373
Xie XL, Li CH, Yang YX, Jin L, Tan JJ, Zhang XY, Su JG, Wang CX (2015) Allosteric transitions of ATP-binding cassette transporter MsbA studied by the adaptive anisotropic network model. Proteins 83:1643–1653
Xu Y, Seelig A, Bernèche S (2017) Unidirectional transport mechanism in an ATP dependent exporter. ACS Cent Sci 3:250–258
Yamamoto T (2010) Preferred dissociative mechanism of phosphate monoester hydrolysis in low dielectric environments. Chem Phys Lett 500:263–266
Yoshida N (2014) Efficient implementation of the three-dimensional reference interaction site model method in the fragment molecular orbital method. J Chem Phys 140:214118

Yoshidome T, Kinoshita M, Hirota S, Baden N, Terazima M (2008) Thermodynamics of apoplastocyanin folding: comparison between experimental and theoretical results. J Chem Phys 128:225104

Zaitseva J, Jenewein S, Jumpertz T, Holland IB, Schmitt L (2005) H662 is the linchpin of ATP hydrolysis in the nucleotide-binding domain of the ABC transporter HlyB. EMBO J 24:1901–1910

Zhou Y, Ojeda-May P, Pu J (2013) H-loop histidine catalyzes ATP hydrolysis in the *E. coli* ABC-transporter HlyB. Phys Chem Chem Phys 15:15811–15815

Chapter 13
Statistical Thermodynamics on the Binding of Biomolecules

Tomohiko Hayashi

Abstract The binding of biomolecules in water plays an essential role in the expression of life phenomena. In this chapter, we show that the underlying mechanism of this binding can be clarified by calculating the thermodynamic quantities based on statistical mechanics. The three types of biomolecule binding are analyzed within a theoretical framework: (I) the binding between a soft peptide (a portion of protein) and a rigid RNA, (II) the one-to-many molecular recognition by a soft peptide accompanying target-dependent structuring, and (III) the actin–myosin binding. Types (I) and (II) are related to pharmacological applications, and type (III) is an elementary process for muscle contraction. These apparently different binding processes share the same underlying mechanism, which can be characterized using a unified theoretical framework. The binding is driven by a large gain of water entropy in the entire system. This gain primarily originates from the reduction of "water crowding," which is attributed to a large overlap of the biomolecule excluded volumes (EV) upon binding, referred to as the entropic EV effect. Such a large EV overlap is achieved by the formation of sufficiently high shape complementarity on an atomic level within the binding interface. The electrostatic complementarity within the interface is ensured as much as possible to compensate for the energetic loss due to dehydration. Although the elimination of biomolecule fluctuations within the binding interface causes a large conformational entropy loss, it is surpassed by these complementarity formations when the binding is accomplished.

Keywords Binding free energy · Prion protein · RNA aptamer
p53 protein · Actomyosin

T. Hayashi (✉)
Institute of Advanced Energy, Kyoto University, Uji, Kyoto 611-0011, Japan
e-mail: thayashi@iae.kyoto-u.ac.jp

© Springer Nature Singapore Pte Ltd. 2018

M. Suzuki (ed.), *The Role of Water in ATP Hydrolysis Energy Transduction by Protein Machinery*, https://doi.org/10.1007/978-981-10-8459-1_13

13.1 Introduction

Life is composed of biomolecules and solvent molecules. The biomolecules are self-organized (as if spontaneously) in an aqueous solution. Since the self-organization process is coupled with a variety of life phenomena, elucidation of its mechanism is a crucial task to unveil the mechanism of life. The binding (or the noncovalent association) of biomolecules in water, which is the main topic in this chapter, is a fundamental self-organization process.

Let us consider the binding equilibrium between two biomolecules in an aqueous solution. Experimentally, this equilibrium is quantified by a dissociation constant K_D. The constant K_D is related to the binding free energy, denoted here as $\Delta G^{\text{Binding}}$, and $\Delta G^{\text{Binding}} = -RT\ln(K_D)$, where R is the gas constant and T is the absolute temperature (Gilson et al. 1997; Zhou and Gilson 2009). Hereinafter, T is equal to 298 K, unless otherwise noted. $\Delta G^{\text{Binding}}$ is the decrease in system free energy due to the binding of two biomolecules ($\Delta G^{\text{Binding}} < 0$).

The methods of predicting $\Delta G^{\text{Binding}}$ are of great practical value. A practical use of these methods is computer-aided drug discovery (Zhou and Gilson 2009; Huang and Jacobson 2007; Shirts et al. 2010; Chodera et al. 2011; Leach et al. 2006; Keskin et al. 2016). By targeting the biomolecules relevant to the expressions of diseases, artificial compounds that can bind strongly to them have been designed (Huang and Jacobson 2007; Shirts et al. 2010). Thus far, the high-accuracy methods for $\Delta G^{\text{Binding}}$ prediction have been proposed using molecular dynamics (MD) simulation (Chodera et al. 2011). In another class of methods (Leach et al. 2006; Keskin et al. 2016), $\Delta G^{\text{Binding}}$ is represented by a scoring function; it is composed of the physically meaningful (or meaningless) components in which the weighting coefficients are multiplied. The coefficients are determined in advance on the basis of a sufficiently large set of experimental $\Delta G^{\text{Binding}}$, so that the calculated $\Delta G^{\text{Binding}}$ can be best fitted to the experimental one.

Although the experimental $\Delta G^{\text{Binding}}$ for the biomolecule bindings has been successfully reproduced by various methods, the underlying mechanism is still unclear. In general, the binding is discussed primarily by considering the intermolecular (or intramolecular) direct interactions. The water roles are often discussed only by distinguishing hydrophobic and hydrophilic groups of biomolecules (the former tends to be buried within the binding interface, whereas the latter tends to be exposed to the water side (Keskin et al. 2016)). However, it is frequently mentioned in the literature that the direct attractive interactions between biomolecules are regarded as the driving force for the binding, without discussing the breaking of biomolecule–water interactions. Water is often modeled as a dielectric continuum, or only the water molecules in the vicinity of biomolecules are taken into account. On the other hand, it has been pointed out that the water molecules in the entire system are crucial for binding through both the energetic and entropic effects (Hayashi et al. 2014; Hayashi et al. 2015; Oshima et al. 2016). Certainly, such water effects have been fully incorporated in the MD simulation studies based on all-atom models with explicit water (Chodera et al. 2011). Nevertheless, it is difficult to decompose the water effect into energetic and entropic components within this framework.

The purpose of this chapter is to clarify the underlying mechanisms in a variety of biomolecule bindings. For this, we show that the calculation of thermodynamic quantities based on statistical mechanics offers definitive information. This chapter is organized as follows. Section 13.2 presents the importance of water entropy on the binding. In Sect. 13.3, the method of calculating $\Delta G^{\text{Binding}}$, which is adopted in the following sections, is briefly presented. In Sects. 13.4 and 13.5, several studies related to the biomolecule bindings are reviewed. In Sect. 13.6, we propose a unifying theoretical framework.

13.2 Entropic Excluded-Volume Effect: The Physical Origin of Water-Entropy Gain upon Binding of Biomolecules

Water entropy plays an important role in the self-organization processes of biomolecules (Hayashi et al. 2014, 2015, 2017; Oshima et al. 2011, 2016; Kinoshita 2008, 2013; Yoshidome et al. 2008; Harano et al. 2008; Kinoshita and Yoshidome 2009; Yoshidome and Kinoshita 2012; Chiba et al. 2012; Oshima and Kinoshita 2015). Using the statistical mechanical theories for molecular liquids (Hansen and McDonald 2006), the change in water entropy upon the binding of two biomolecules in water is decomposed into the biomolecule–solvent pair (PA) and many-body (MB) correlation components (Oshima and Kinoshita 2015). By applying the morphometric approach (MA) (König et al. 2004; Roth et al. 2006), each component is further decomposed into the excluded volume (EV) and solvent-accessible surface (SAS) terms (Oshima and Kinoshita 2015). The four constituents, namely (PA, EV), (PA, SAS), (MB, EV), and (MB, SAS), are thus obtained. The investigations based on these decompositions lead to the following results: (1) the binding of biomolecules accompanies the water-entropy gain, and (2) the gain originates mainly from (PA, EV) and (MB, EV).

The physical origin of water-entropy gain is interpreted as follows. The presence of a biomolecule generates a space in which the centers of water molecules cannot enter (see Fig. 13.1) (Harano and Kinoshita 2005). EV is the volume of this space. The total volume available to the translational displacement of water molecules in the system is reduced by the appearance of EV, causing a water-entropy loss. The binding of two biomolecules accompanies an overlap of the two excluded spaces, leading to a decrease in the total EV by the overlapped space (see the right side of Fig. 13.1). Thus, the total volume available to the translational displacement of water molecules increases, resulting in a water-entropy gain. This is the physical interpretation of (PA, EV). Moreover, the presence of a water molecule also generates the EV for the other water molecules, and the water molecules in the entire system are entropically correlated. This correlation is referred to as “water crowding

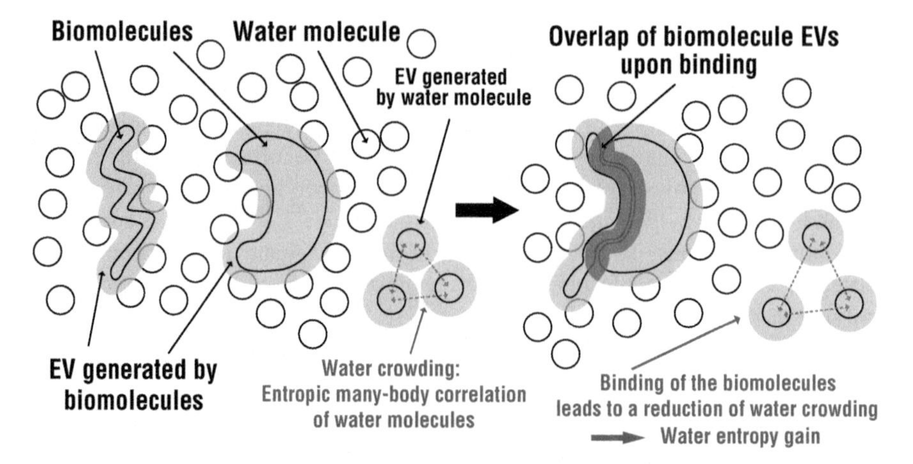

Fig. 13.1 Illustration of entropic excluded-volume effect. We emphasize that the excluded volume (EV) marked in green includes the volume of the molecule itself

(Oshima and Kinoshita 2015)." Due to the appearance of the biomolecule, the water crowding in the bulk becomes more significant. This enhancement in water crowding causes a water-entropy loss. The binding of two biomolecules (the overlap of EVs) leads to a reduction in water crowding, leading to a water-entropy gain. This is the physical interpretation of (MB, EV).

Interestingly, the contribution of (MB, EV) to the water entropy is substantially larger than that of (PA, EV) (Oshima and Kinoshita 2015). The many-body correlation among a large number of water molecules is very important. In the later sections, the combined effect of (PA, EV) and (MB, EV) is referred to as "entropic EV effect (Yoshidome and Kinoshita 2012; Kinoshita 2013; Oshima and Kinoshita 2015)." Note that the entropic EV effect is observed *in any solution system* (e.g., solutes in an organic solvent) and becomes stronger as the molecular size of the solvent decreases and/or the solvent number density increases (Kinoshita 2013). Due to the water–water hydrogen-bonding (H-bonding) interaction, water can exist in a dense fluid at ambient temperature and pressure despite its exceptionally small molecular size. Therefore, the entropic EV effect is strong enough for water to drive the binding of biomolecules.

13.3 Calculation of the Thermodynamic Quantities

Here, the all-atom model of biomolecules and the molecular model of water are adopted. $\Delta G^{\text{Binding}}$ is divided into a few physically insightful processes. A process is denoted by Δ^{δ}, and the change in free energy due to the process is denoted by $\Delta^{\delta}G$. We calculate $\Delta^{\delta}G$ and its constituents (A) to (F) below.

(A) Change in bonded energy, comprised of the bond-stretching, angle-bending, and torsional terms, denoted by $\Delta^{\delta}E_{\mathrm{B}}$,
(B) Energy change due to the gain (or loss) of the intra- and inter-biomolecular electrostatic and van der Waals (vdW) interactions, denoted by $\Delta^{\delta}E_{\mathrm{ES}}$ and $\Delta^{\delta}E_{\mathrm{vdW}}$, respectively,
(C) Energy change due to the gain (or loss) of the biomolecule–water electrostatic and vdW interactions,
(D) Change in water–water interaction energy (referred to as the water-reorganization energy (Hayashi et al. 2014, 2015; Oshima et al. 2016; Imai et al. 2007)),
(E) Change in water entropy, denoted by $\Delta^{\delta}S_{\mathrm{water}}$,
(F) Change in conformational entropy of biomolecules, denoted by $\Delta^{\delta}S_{\mathrm{C}}$.

Note that (C) + (D) is the change in hydration energy, denoted by $\Delta^{\delta}\varepsilon_{\mathrm{ES}}$ and $\Delta^{\delta}\varepsilon_{\mathrm{vdW}}$, respectively, for the electrostatic and vdW components. Therefore, $\Delta^{\delta}G$ is given by

$$\Delta^{\delta}G = \Delta^{\delta}E_{\mathrm{total}} - T\Delta^{\delta}S_{\mathrm{water}} - T\Delta^{\delta}S_{\mathrm{C}},$$
$$\Delta^{\delta}E_{\mathrm{total}} = \Delta^{\delta}E_{\mathrm{B}} + \Delta^{\delta}(E_{\mathrm{ES}} + \varepsilon_{\mathrm{ES}}) + \Delta^{\delta}(E_{\mathrm{vdW}} + \varepsilon_{\mathrm{vdW}}),$$

where $\Delta^{\delta}E_{\mathrm{total}}$ is the total energy change due to Δ^{δ}. The quantitative comparisons of the contributions of (A) to (F) upon Δ^{δ} are thus performed. (A) and (B) are calculated on the basis of a molecular mechanical potential, for which we employ the Amber99SB force field (Hornak et al. 2006). (C) and (D) are calculated using the three-dimensional reference interaction site model (3D-RISM) theory (Beglov and Roux 1995, 1996; Kovalenko and Hirata 1999; Ratkova et al. 2015; Hayashi et al. 2016) combined with the cSPC/E model (Luchko et al. 2010) for water and the Kovalenko–Hirata closure equation (Kovalenko and Hirata 1999). (E) is calculated using the angle-dependent integral equation (ADIE) theory (Kinoshita 2013; Hayashi et al. 2016; Kusalik and Patey 1988a, b; Kinoshita and Bérard 1996; Cann and Patey 1997) in combination with the MA (König et al. 2004; Roth et al. 2006). In the ADIE theory, water is modeled as a multipolar model (Kusalik and Patey 1988a, b), in which a point dipole and a point quadrupole of tetrahedral symmetry are embedded, and the influence of molecular polarizability of water is taken into account by employing the self-consistent mean field theory (Kusalik and Patey 1988a, b). The hypernetted-chain (HNC) approximation (Hansen and McDonald 2006) is adopted as the closure equation. (F) is estimated using a simple but reliable method based on statistical mechanics.

The 3D-RISM and ADIE theories are statistical mechanical theories related to molecular liquids. It has been confirmed that the version of 3D-RISM theory that we employed is advantageous for calculating (C) and (D) (Truchon et al. 2014; Oshima and Kinoshita 2016), whereas it underestimates the contribution of (E) (Hayashi et al. 2016). On the other hand, the ADIE theory gives sufficiently accurate results for the hydrophobic and hydrophilic hydration of solutes with simple shapes (Kinoshita 2008; Kinoshita and Yoshidome 2009); however, the direct application of the theory to a biomolecule with complex polyatomic structure is currently not desirable due to the technical difficulty in obtaining numerical solutions. Instead, (E) is calculated using a combination of the ADIE theory with MA (see Sect. 13.2). Thus, only the advantageous aspects of each of these theories are employed.

In general, $\Delta G^{\mathrm{Binding}}$ is experimentally measured under isobaric condition, whereas the theoretical calculations are made under isochoric condition (Cann and Patey 1997). However, as discussed in our original papers (Hayashi et al. 2014; Hayashi et al. 2015; Oshima et al. 2016), the changes in thermodynamic quantities under the isobaric condition are almost the same as those under the isochoric condition, for the cases described in the following sections.

13.4 Mechanism of Binding Between Intrinsically Disordered Peptides and Their Target Molecules

In an aqueous solution, proteins fold into their unique three-dimensional (3D) structures. However, there are peptides (or protein portions) that do not take a specific structure in isolated states. They are referred to as intrinsically disordered peptides (denoted by IDPs). In this section, we present the mechanism of binding between IDPs and the target biomolecules that accompany the target-dependent structural change of IDPs.

13.4.1 Case 1: Binding of an IDP Region of Prion Protein to a Partner RNA Aptamer

Within the N-terminal half of the prion protein ($\mathrm{PrP^C}$), the two portions, P1 and P16, are IDPs and flexible (Mashima et al. 2009, 2013). R12 is an artificially designed RNA (RNA aptamer) and binds specifically to P1 and P16 (Mashima et al. 2009). This binding stabilizes the normal cellular form of $\mathrm{PrP^C}$, which is expected to prevent prion diseases (Mashima et al. 2013). Therefore, the elucidation of this specific binding mechanism presents a challenge for medical applications.

The 3D structure of the 2 × P16:2 × R12 complex has been determined using the solution NMR technique (Mashima et al. 2013). The prevailing picture of the P16–R12 binding, based on this NMR structure, is summarized as follows (Mashima et al. 2013). The binding is driven by (1) the electrostatic interaction between the RNA phosphate groups (negatively charged) and the positively charged side chains of the amino acid residues, and (2) the so-called "π–π stacking interaction," which is the vdW attractive interaction between the two aromatic rings, a guanine (Gua) nucleotide of R12, and a tryptophan (Trp) indole ring of P16. Note that the π–π stacking interaction, which is modeled by the vdW attractive interaction using simplified the force field method, is a good approximation, although an electron many-body effect is not explicitly taken into account (Hayashi and Kinoshita 2016).

We analyzed the changes in the thermodynamic quantities due to the P16–R12 binding (Hayashi et al. 2014). The starting coordinates are the ensemble of five structures (PDB code: 2RU7 (Hayashi et al. 2014), see the right side of Fig. 13.2). Since the upper and lower halves of the $2 \times$P16:$2 \times$R12 complex share almost the same 3D structure, the upper half is taken from each complex structure and is used as the structure of the R12:P16 complex. The P16 in the isolated state (it is disordered) is modeled as a set of random coils.

First, we consider the process in which P16 and R12 are assumed to exhibit no structural changes upon binding, denoted by $\Delta^{\text{Binding-fixed}}$ (see Fig. 13.2). The changes in thermodynamic quantity upon this process are given in Table 13.1. The contribution from water-entropy gain ($-T\Delta^{\text{Binding-fixed}}S_{\text{water}}$) is a large negative value, whereas the system energy change ($\Delta^{\text{Binding-fixed}}E_{\text{total}}$) is a small negative or positive value. Thus, the driving force for $\Delta^{\text{Binding-fixed}}$ is water-entropy gain.

The water-entropy gain arises from an overlap of the EVs of P16 and R12 due to the entropic EV effect (see Sect. 13.2 and Fig. 13.1) and is largely influenced by the geometrical characteristics of P16 and R12. To maximize the water-entropy gain, a high shape complementarity at the atomic level is formed within the P16–R12 binding interface. This type of shape complementarity is characterized as follows: (1) P16 is stacked onto a flat portion of R12. Such face-to-face stacking leads to a large decrease in the total EV. As illustrated in Fig. 13.3a and b, the decrease in EV due to the contact of flat moieties in a face-to-face manner is substantially larger than that of other shapes such as spheres. (2) The spherical and disk-shaped solutes illustrated in Fig. 13.3a and b possess no polyatomic structures. On the other hand, the entropic EV effect is largely dependent on the details of the polyatomic structures of P16 and R12. Within the R12–P16 binding interface, the backbone and side-chain atoms are closely packed as illustrated Fig. 13.3c. The decrease in EV (water-entropy gain) due to such packing is much larger than that caused by the contact of flat surfaces.

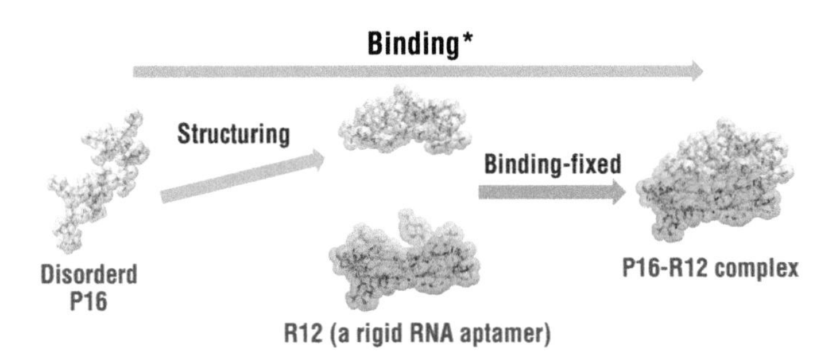

Fig. 13.2 Illustration of isolated P16, isolated R12, P16–R12 complex, and three processes (“Binding-fixed”, “Binding*”, and “Structuring”) considered in Sect. 13.4.1

Table 13.1 Free-energy change and its energetic and entropic components (in kcal/mol) upon P16–R12 binding in $\Delta^{\text{Binding-fixed}}$ (see Fig. 13.1). Five different models correspond to the upper halves of 2 × P16:2 × R12 complex structures in the PDB code: 2RU7. $\Delta^{\text{Binding-fixed}}G = \Delta^{\text{Binding-fixed}}E_{\text{total}} - T\Delta^{\text{Binding-fixed}}S_{\text{water}}$

Model	$\Delta^{\text{Binding-fixed}}G$	$\Delta^{\text{Binding-fixed}}E_{\text{total}}$	$-T\Delta^{\text{Binding-fixed}}S_{\text{water}}$
1	−36.16	5.47	−41.63
2	−39.80	−3.10	−36.71
3	−35.84	0.14	−35.98
4	−30.31	3.43	−33.73
5	−47.05	−6.58	−40.47

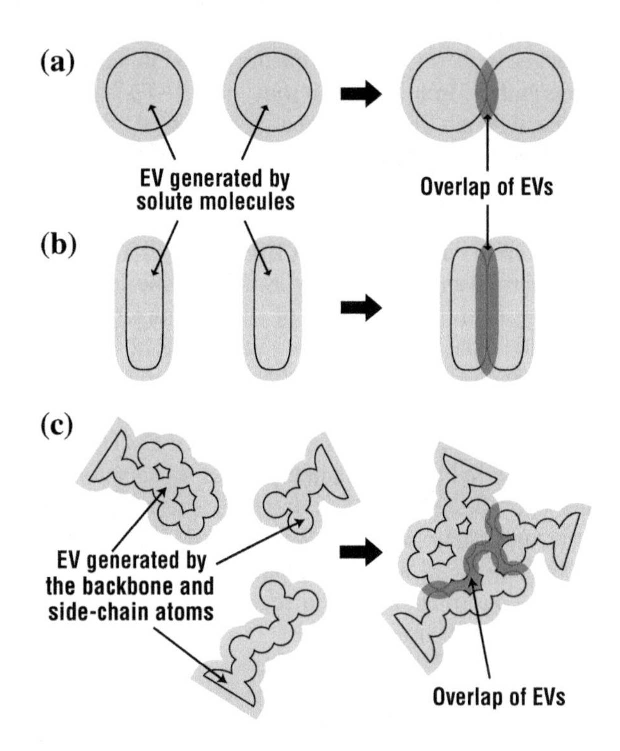

Fig. 13.3 Illustrations of **a** contact of spherical solutes in a point-to-point manner, **b** stacking of flat solutes in a face-to-face manner, and **c** close packing of the backbone and side-chain atoms

$\Delta^{\text{Binding-fixed}}E_{\text{total}}$ is further decomposed into the electrostatic and vdW components, as listed in Table 13.2. The former is positive ($\Delta^{\text{Binding-fixed}}(E_{\text{ES}} + \varepsilon_{\text{ES}}) > 0$) and opposes the binding, whereas the latter is negative ($\Delta^{\text{Binding-fixed}}(E_{\text{vdW}} + \varepsilon^{\text{vdW}}) < 0$). These components compensate each other, thus leading to a considerably small change in total energy (this change is either positive or negative).

The change in system energy is largely influenced by the dehydration of P16 and R12 upon binding (Fig. 13.4). The decrease in energy due to the gain of P16–R12 electrostatic interactions ($\Delta^{\text{Binding-fixed}}E_{\text{ES}} < 0$) is canceled by an increase in the electrostatic part of hydration energy ($\Delta^{\text{Binding-fixed}}\varepsilon_{\text{ES}} > 0$). Upon binding, there

Table 13.2 Further decomposition of energetic components (in kcal/mol) upon P16–R12 binding in $\Delta^{\text{Binding-fixed}}$. $\Delta^{\text{Binding-fixed}}E_{\text{total}} = \Delta^{\text{Binding-fixed}}E_{\text{B}} + \Delta^{\text{Binding-fixed}}(E_{\text{vdW}} + \varepsilon_{\text{vdW}}) + \Delta^{\text{Binding-fixed}}(E_{\text{ES}} + \varepsilon_{\text{ES}})$. $\Delta^{\text{Binding-fixed}}E_{\text{B}} = 0$ because P16 and R12 exhibit no structural change upon $\Delta^{\text{Binding-fixed}}$

Model	$\Delta^{\text{Binding-fixed}}E_{\text{total}}$	$\Delta^{\text{Binding-fixed}}(E_{\text{ES}} + \varepsilon_{\text{ES}})$	$\Delta^{\text{Binding-fixed}}(E_{\text{vdW}} + \varepsilon_{\text{vdW}})$	$\Delta^{\text{Binding-fixed}}E_{\text{ES}}$	$\Delta^{\text{Binding-fixed}}\varepsilon_{\text{ES}}$	$\Delta^{\text{Binding-fixed}}E_{\text{vdW}}$	$\Delta^{\text{Binding-fixed}}\varepsilon_{\text{vdW}}$
1	5.47	9.41	−3.94	−1048.45	1057.85	−54.92	50.98
2	−3.10	3.23	−6.33	−1017.00	1020.23	−52.84	46.51
3	0.14	2.19	−2.05	−1070.42	1072.61	−50.55	48.50
4	3.43	8.47	−5.04	−1001.51	1009.98	−50.81	45.77
5	−6.58	2.97	−9.55	−1021.89	1024.85	−53.50	43.95

occur losses in the P16–water and R12–water electrostatic interactions (contributing to the increase in energy) and a partial recovery of water–water interaction (contributing to the increase in energy). About half of the former is canceled by the latter; therefore, the net electrostatic-dehydration energy is positive. P16 possesses three Lys residues (positively charged), and they are in contact with the RNA phosphate groups (negatively charged) within the P16–R12 binding interface. These contacts are referred to as electrostatic complementarity (Fig. 13.4) and are essential for compensating the dehydration energy. However, the electrostatic complementarity is not perfect for the P16–R12 binding; therefore, the net change in the electrostatic energy of the system is positive, which hinders the binding.

On the other hand, $\Delta^{\text{Binding-fixed}}(E_{\text{vdW}} + \varepsilon_{\text{vdW}}) < 0$; however, its absolute value is considerably smaller than $\Delta^{\text{Binding-fixed}}E_{\text{vdW}}$. The reason for this is explained as follows. The π–π stacking interaction between the side chains of Gua and Trp results in a large gain in the vdW energy. However, it is found that the direct vdW interaction energy ($\Delta^{\text{Binding-fixed}}E_{\text{vdW}} < 0$) of P16–R12 is largely canceled by an increase in the vdW part of hydration energy ($\Delta^{\text{Binding-fixed}}\varepsilon_{\text{vdW}} > 0$). We argue that the term "π–π stacking interaction" is misleading, because it sounds as if the vdW attractive interaction is its only contributor, as in vacuum. For the stacking of two toluene molecules (an aromatic–aromatic interaction), it has been demonstrated (Hayashi and Kinoshita 2016) that more than half of the vdW attractive interaction due to the stacking is canceled out by the vdW-dehydration energy, and that there is a large gain in water entropy. In the P16–R12 binding, the Gua–Trp stacking also contributes to the binding through a large water-entropy gain and $\Delta^{\text{Binding-fixed}}(E_{\text{vdW}} + \varepsilon_{\text{vdW}})$.

Next, we consider the P16–R12 binding accompanied by the change in the global structure of P16, denoted by $\Delta^{\text{Binding*}}$ (see Fig. 13.2). The changes in thermodynamic quantity due to this process, $\Delta^{\text{Binding*}}G = \Delta^{\text{Binding*}}E_{\text{total}} - T\Delta^{\text{Binding*}}S_{\text{water}}$, are given in Table 13.3. Clearly, the contribution from water-entropy gain is the driving

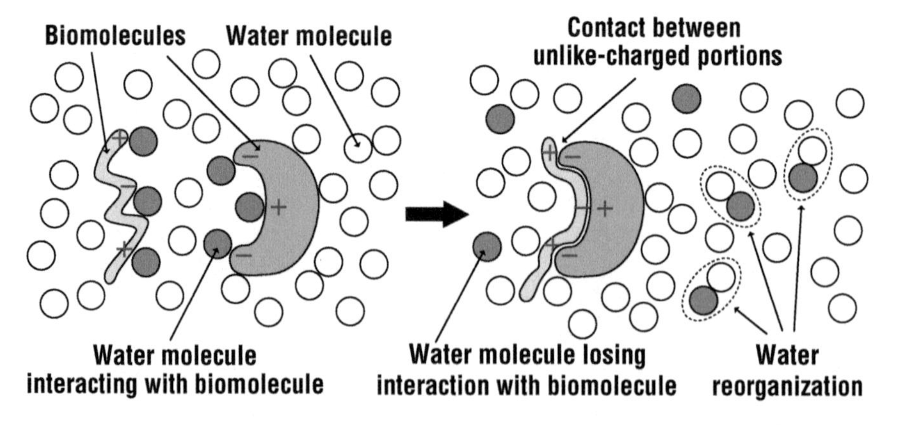

Fig. 13.4 Illustration of electrostatic complementarity, dehydration, and water reorganization

Table 13.3 Free-energy change and its energetic and entropic components (in kcal/mol) upon P16–R12 binding in $\Delta^{\mathrm{Binding}*}$. $\Delta^{\mathrm{Binding}*}G = \Delta^{\mathrm{Binding}*}E_{\mathrm{total}} - T\Delta^{\mathrm{Binding}*}S_{\mathrm{water}}$

Model	$\Delta^{\mathrm{Binding}*}G$	$\Delta^{\mathrm{Binding}*}E_{\mathrm{total}}$	$-T\Delta^{\mathrm{Binding}*}S_{\mathrm{water}}$
1	−21.67	36.26	−57.93
2	−14.04	40.32	−54.36
3	−23.75	27.43	−51.18
4	−4.59	45.10	−49.70
5	−11.53	48.84	−60.37

force ($-T\Delta^{\mathrm{Binding}*}S_{\mathrm{water}} < 0$), whereas the change in system energy is the hindering factor ($\Delta^{\mathrm{Binding}*}E_{\mathrm{total}} > 0$).

The difference between $\Delta^{\mathrm{Binding-fixed}}G$ and $\Delta^{\mathrm{Binding}*}G$ is due to the change in global structure (i.e., transition from the disordered state to the binding structure) of P16 upon binding, denoted by $\Delta^{\mathrm{Structuring-P16}}$. Note that $\Delta^{\mathrm{Binding}*}G = \Delta^{\mathrm{Binding-fixed}}G + \Delta^{\mathrm{Structuring}}G(\mathrm{P16})$. As shown in Table 13.4, the structuring of P16 is accompanied by the water-entropy gain ($\Delta^{\mathrm{Structuring}}S_{\mathrm{water}}(\mathrm{P16}) > 0$). The physical origin of the gain is also explained by the entropic EV effect. Upon structuring, P16 becomes more compact and the volume of the system EV decreases, leading to the water-entropy gain. On the other hand, the increase in energy due to this structuring ($\Delta^{\mathrm{Structuring}}E_{\mathrm{total}}(\mathrm{P16}) > 0$) is quite large, and the resultant $\Delta^{\mathrm{Structuring}}G(\mathrm{P16}) > 0$. Moreover, as discussed in the next paragraph, the structuring is accompanied by the conformational entropy loss, $\Delta^{\mathrm{Binding}*}S_{\mathrm{C}}$. This result is consistent with the experimental observation (Mashima et al. 2013); P16 takes no specific structure without its binding partner. In other words, the high shape complementarity within the P16–R12 binding interface, which is required for accomplishing the binding, is achieved by the energetic and entropic losses of P16 that accompany its structure formation.

The validity of the theoretical result is verified by comparing it with the experimental result (Mashima et al. 2013). Hereinafter, the theoretical and experimental values of the binding free energy are denoted by $\Delta G^{\mathrm{Binding-theory}}$ and $\Delta G^{\mathrm{Binding-experiment}}$, respectively. In a strict sense, $\Delta G^{\mathrm{Binding-experiment}}$ is not equivalent to $\Delta G^{\mathrm{Binding-theory}}$ (theoretically, K_{D} should be dimensionless). This is because the standard state for $\Delta G^{\mathrm{Binding-experiment}}$ is 1 mol/L and the activity coefficient is set at unity (experimentally, it is common to express K_{D} in molar unit), whereas $\Delta G^{\mathrm{Binding-theory}}$ is calculated for the biomolecules immersed in water at the

Table 13.4 Free-energy change and its energetic and entropic components (in kcal/mol) upon global structure change of P16. $\Delta^{\mathrm{Structuring}}G(\mathrm{P16}) = \Delta^{\mathrm{Structuring}}E_{\mathrm{total}}(\mathrm{P16}) - T\Delta^{\mathrm{Structuring}}S_{\mathrm{water}}(\mathrm{P16})$

Model	$\Delta^{\mathrm{Structuring}}G(\mathrm{P16})$	$\Delta^{\mathrm{Structuring}}E_{\mathrm{total}}(\mathrm{P16})$	$-T\Delta^{\mathrm{Structuring}}S_{\mathrm{water}}(\mathrm{P16})$
1	14.49	30.79	−16.30
2	25.76	43.42	−17.65
3	12.09	27.29	−15.19
4	25.71	41.68	−15.96
5	35.52	55.42	−19.90

infinite-dilution limit (Gilson et al. 1997). Nevertheless, the two values should be close to each other. Moreover, in the following section, we show that considering the conformational entropy loss upon binding, $\Delta^{\text{Binding*}}S_C$, significantly improves $\Delta G^{\text{Binding-theory}}$. Since R12 undergoes essentially no structural change upon binding, $\Delta^{\text{Binding*}}S_C$ is almost equal to the conformational entropy loss of P16, $\Delta^{\text{Structuring}}S_C$(P16), caused by the transition from the disordered state to the binding structure. It is roughly estimated in the following manner. Fitter (2003) studied the conformational entropy loss due to protein folding, and this loss is given by $-3k_BN_r\ln(r_u/r_f)$, where r_u and r_f are the radius parameters for the unfolded and folded states, respectively, N_r is the number of residues, and k_B is the Boltzmann constant. The radius parameters at 298 K were estimated by performing linear fitting in accordance with their temperature dependences. This method tends to significantly underestimate the loss. However, P16 possesses two proline residues, making its random coils rather compact, resulting in an exceptionally small loss. For this reason, it may be justified to adopt the following result as a rough estimation. The loss for P16 with $N_r = 12$ thus obtained was $-T\Delta^{\text{Binding*}}S_C \sim 8.83$ kcal/mol. Therefore, $\Delta G^{\text{Binding-theory}} = \Delta^{\text{Binding*}}G - T\Delta^{\text{Binding*}}S_C$ is in the range from −14.92 to 4.24 kcal/mol. The average value is −6.29 kcal/mol, which is comparable with $\Delta G^{\text{Binding-experiment}} = -6.45$ kcal/mol (Mashima et al. 2013). We can conclude that our theoretical results are reliable even in a quantitative sense.

The updated picture of the P16–R12 binding is summarized as follows. The binding is driven by the water-entropy gain with the achievement of high shape complementarity within the P16–R12 binding interface. Such an achievement is a consequence of the partner-dependent structuring of P16, which leads to the stacking of flat moieties and the precise packing within this binding interface at the atomic level. The electrostatic complementarity (i.e., contact of oppositely charged groups) in the P16–R12 complex is observed to overcome the dehydrations of R12 and P16 upon binding. However, it is not possible to achieve perfect electrostatic complementarity for the P16–R12 binding, which leads to an energetic loss. The shape complementarity due to the entropic EV effect is likely to be given priority. The loss of P16 conformational entropy, which opposes the binding, is surpassed by the water-entropy gain.

13.4.2 Case 2: One-to-Many Molecular Recognition by p53CTD Accompanying Target-Dependent Structure Formation

The extreme C-terminal peptide region of the tumor suppressor p53 protein, p53CTD, is an IDP (Hsu et al. 2013). This p53CTD recognizes a multitude of biomolecules and acts as a principal regulator for the tumor-suppressing activity of p53 (Vogelstein et al. 2000; Hoe et al. 2014). Therefore, research on molecular recognition by p53CTD may lead to the development of anticancer therapeutic methods.

Here, we consider the binding of p53CTD to four targets accompanying the formation of target-dependent structures, as illustrated in Fig. 13.5, namely the binding of p53CTD to S100B (ββ) (Rustandi et al. 1998), Sir-Af2 (Avalos et al. 2002), CBP bromodomain (Mujtaba et al. 2004), and Cyclin A (Lowe et al. 2002), by forming α-helix, β-sheet, and two different coils, respectively. Hereinafter, each binding is denoted as systems 1 to 4 (see Fig. 13.5). The structures of these complexes were determined by using NMR for systems 1 and 3 and by using X-ray crystallography for systems 2 and 4 (the PDB codes for the systems 1 to 4 are 1DT7, 1MA3, 1JSP, and 1H26, respectively).

The prevailing pictures of these bindings are target-dependent, and emphasis has been placed on p53CTD–target intermolecular interactions as follows: salt bridge and contact of hydrophobic portions (i.e., the so-called hydrophobic interaction) in system 1 (Rustandi et al. 1998), H-bonding and vdW interaction in system 2 (Avalos et al. 2002), contact of hydrophobic portions in system 3 (Mujtaba et al. 2004), and H-bonding and vdW interaction in system 4 (Lowe et al. 2002). The p53CTD–target electrostatic interaction has also been considered a driving force of the binding in system 1 (Rustandi et al. 1998). The salt bridge, H-bonding, and vdW interaction are p53CTD–target direct interactions, and the effect of water is considered only through the contact of hydrophobic portions.

We analyzed the changes in the thermodynamic quantities due to the four types of bindings (Hayashi et al. 2015). In the binding experiments, the portion of p53CTD used varies from system to system. The numbers of residues in systems 1 to 4 are 22, 18, 20, and 11, respectively. In systems 2 and 4, the structures of only 9 residues in the complexes could be detected using X-ray crystallography. This was

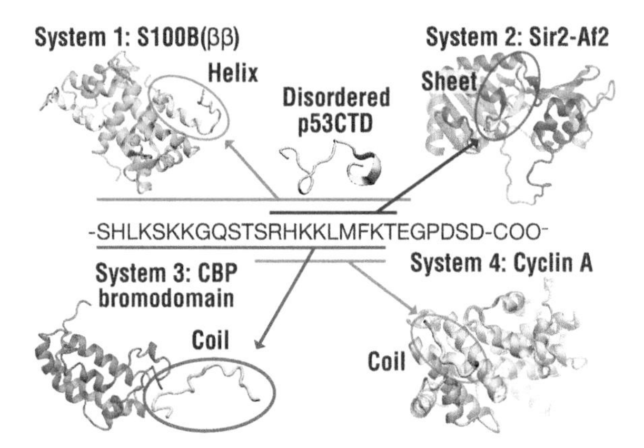

Fig. 13.5 Illustration of structure formations of p53CTD in binding to the four different target proteins. In systems 1, 2, 3, and 4, p53CTD binds to S100B (ββ), Sir-Af2, CBP bromodomain, and Cyclin A, respectively. In system 1, S100B (ββ) is present as a dimer, and two p53CTDs bind to it. In our theoretical calculations, the complex of the dimer and a p53CTD is regarded as the target, and the process where one more p53CTD binds to it is considered

probably due to the structural fluctuations of the remaining 9 and 2 residues in systems 2 and 4, respectively. Therefore, we assume that they do not participate in the binding, and the portions displayed in Fig. 13.5 are considered for p53CTD in our theoretical calculations. In the isolated state, p53CTD is modeled as a set of random coils for each system. We then assume that the numbers of residue structures due to binding, N_r, are 12, 9, 8, and 9 in systems 1 to 4, respectively (a more detailed discussion is provided in our original paper (Hayashi et al. 2015)).

The conformation entropy loss of p53CTD upon binding, $\Delta^{\mathrm{Structuring}}S_{\mathrm{C}}$(p53CTD), is estimated using N_r in the following manner:

(1) S_{C}(p53CTD) in the p53CTD–target complex is considered to be essentially zero.
(2) For the backbone of S_{C}(p53CTD) in the isolated state, we consider the following assumptions: There are two dihedral angles that can rotate, and each angle has three stable values. The number of possible combinations is $3^2 = 9$, to each of which the backbone conformations are uniformly distributed. Thus, the backbone contributions to S_{C}(p53CTD) is $N_r \times k_{\mathrm{B}} \cdot \ln(9)$.
(3) On the basis of a study by Doig and Sternberg (1995), we consider that the contribution from the side chain to S_{C}(p53CTD) is $N_r \times 1.7k_{\mathrm{B}}$ (the term, $1.7k_{\mathrm{B}}$, which is an average of the values for 20 residue species, implicitly includes all the factors affecting the side-chain conformational entropy of the protein).

Consequently, the relation $\Delta^{\mathrm{Structuring}}S_{\mathrm{C}}(\mathrm{p53CTD})/k_{\mathrm{B}} = -N_r \times \{\ln(9) + 1.7\}$ is obtained.

First, the process $\Delta^{\mathrm{Binding-fixed}}$, in which p53CTD and its targets are assumed to be fixed due to the binding (see Fig. 13.6), is considered. As shown in Table 13.5, we find that all the seemingly different binding processes are driven by the large gain of water entropy ($-T\Delta^{\mathrm{Binding-fixed}}S_{\mathrm{water}} < 0$), whereas the change in system energy ($\Delta^{\mathrm{Binding-fixed}}E_{\mathrm{total}}$) is a small negative or positive value. The water-entropy gain, which stems from the entropic EV effect (see Sect. 13.2 and Fig. 13.1), is caused by the sufficiently high shape complementarity on an atomic level within the p53CTD–target interface (see Fig. 13.3c).

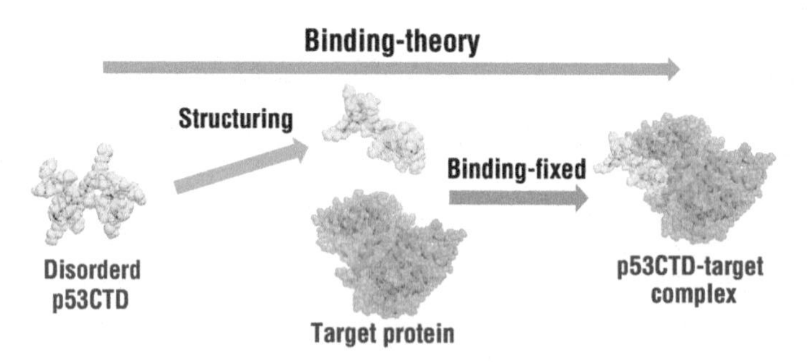

Fig. 13.6 Illustration of isolated p53CTD, isolated target protein, p53CTD–target complex, and three processes ("Binding-fixed", "Binding-theory", and "Structuring") considered in Sect. 13.4.2

Table 13.5 Free-energy change and its energetic and entropic components (in kcal/mol) upon p53CTD–target bindings in $\Delta^{\text{Binding-fixed}}$ (see Fig. 13.5). The p53CTD binds to S100B(ββ), Sir-Af2, CPB bromodomain, and Cyclin A, which are denoted by systems 1 to 4, respectively. PDB codes for the system 1 to 4 are 1DT7, 1MA3, 1JSP, and 1H26, respectively. $\Delta^{\text{Binding-fixed}}G = \Delta^{\text{Binding-fixed}}E_{\text{total}} - T\Delta^{\text{Binding-fixed}}S_{\text{water}}$. The 27, 10, and 8 structure models are considered for systems 1, 2, and 3, respectively, and the standard errors arise from the consideration of these multiple models (there is only a single structure model in system 4)

System	$\Delta^{\text{Binding-fixed}}G$	$\Delta^{\text{Binding-fixed}}E_{\text{total}}$	$-T\Delta^{\text{Binding-fixed}}S_{\text{water}}$
1	−54.53 ± 2.83	−0.57 ± 1.42	−53.96 ± 2.57
2	−66.27 ± 1.36	1.91 ± 1.36	−68.18 ± 0.01
3	−57.13 ± 4.95	11.96 ± 2.46	−69.09 ± 4.76
4	−59.32	13.29	−72.61

The term $\Delta^{\text{Binding-fixed}}E_{\text{total}}$ is further decomposed into physically insightful components, as listed in Table 13.6. The binding causes a large gain in the p53CTD–target electrostatic attractive interaction ($\Delta^{\text{Binding-fixed}}E_{\text{ES}} < 0$). However, it is accompanied by an even larger loss in the p53CTD–water and target–water electrostatic attractive interactions. This loss is partly compensated by a gain in the water–water electrostatic attractive interaction; however, the net change in the electrostatic component of system energy is positive ($\Delta^{\text{Binding-fixed}}\varepsilon_{\text{ES}} > 0$). The net change in the electrostatic component is positive. A similar discussion is applicable for the p53CTD–target, p53CTD–water, target–water, and water–water vdW attractive interactions; however, the net change in the vdW component of system energy is negative. The close packing within the p53CTD–target interface accompanies the water-entropy gain and the p53CTD–target vdW interaction energy gain. However, closer unlike-charged portions and more separated like-charged portions cannot always be assured by the closer packing, which leads to an incompleteness in the electrostatic complementarity.

When the target-dependent structure change of p53CTD is considered for calculating $\Delta G^{\text{Binding-theory}}$ (see Fig. 13.6), both the contributions from energetic and entropic components become higher (see Table 13.7). However, the resultant $\Delta G^{\text{Binding-theory}}$, in which $-T\Delta^{\text{Structuring}}S_{\text{C}}$ (p53CTD) is incorporated, has a much lower value than $\Delta G^{\text{Binding-experiment}}$ (Rustandi et al. 1998; Avalos et al. 2002; Mujtaba et al. 2004; Lowe et al. 2002). This is because the conformational entropy loss of the target due to the binding $\Delta^{\text{Structuring}}S_{\text{C}}$(target), which is difficult to be quantified, is not incorporated in $\Delta G^{\text{Binding-theory}}$. The term $\Delta^{\text{Structuring}}S_{\text{C}}$(target) is then roughly estimated in the following manner. Upon binding, the side chains of the target within the binding interface lose their flexibility. We assume that the number of such residues is in the range from 8 to 12 (same as the number of p53CTD residues that were structured). Using the relation $\Delta^{\text{Structuring}}S_{\text{C}}(\text{target})/k_{\text{B}} = -N_{\text{r}} \times -1.7$ (Doig and Sternberg 1995), $-T\Delta^{\text{Structuring}}S_{\text{C}}$(target) is found to be in the range from 8 to 12 kcal/mol. The agreement between $\Delta G^{\text{Binding-theory}}$ and

Table 13.6 Further decomposition of energetic components (in kcal/mol) upon p53CTD–target bindings in $\Delta^{\text{Binding–fixed}}$. $\Delta^{\text{Binding–fixed}}E_{\text{total}} = \Delta^{\text{Binding–fixed}}E_{\text{B}} + \Delta^{\text{Binding–fixed}}(E_{\text{vdW}} + \varepsilon_{\text{vdW}}) + \Delta^{\text{Binding–fixed}}(E_{\text{ES}} + \varepsilon_{\text{ES}})$. $\Delta^{\text{Binding–fixed}}E_{\text{B}} = 0$ because p53CTD and its targets exhibit no structural change upon $\Delta^{\text{Binding–fixed}}$

System	$\Delta^{\text{Binding–fixed}}E_{\text{total}}$	$\Delta^{\text{Binding–fixed}}(E_{\text{ES}} + \varepsilon_{\text{ES}})$	$\Delta^{\text{Binding–fixed}}(E_{\text{vdW}} + \varepsilon_{\text{vdW}})$	$\Delta^{\text{Binding–fixed}}E_{\text{ES}}$	$\Delta^{\text{Binding–fixed}}\varepsilon_{\text{ES}}$	$\Delta^{\text{Binding–fixed}}E_{\text{vdW}}$	$\Delta^{\text{Binding–fixed}}\varepsilon_{\text{vdW}}$
1	−0.57 ± 1.42	6.05 ± 1.70	−6.63 ± 0.99	−1349.09 ± 21.91	1355.14 ± 22.10	−55.27 ± 1.39	48.64 ± 1.00
2	1.91 ± 1.36	22.19 ± 1.55	−20.28 ± 0.39	−136.70 ± 1.87	158.88 ± 2.71	−74.46 ± 0.00	54.18 ± 0.40
3	11.96 ± 2.46	28.06 ± 4.13	−16.11 ± 2.39	−191.85 ± 21.41	219.92 ± 20.34	−61.92 ± 3.30	45.81 ± 1.69
4	13.29	26.40	−13.11	−657.49	683.88	−70.30	57.19

Table 13.7 Free-energy change and its energetic and entropic components (in kcal/mol) upon p53CTD and its target bindings in $\Delta^{\text{Binding-theory}}$. $\Delta^{\text{Binding-theory}}G = \Delta^{\text{Binding-theory}}E_{\text{total}} - T\Delta^{\text{Binding-theory}}S_{\text{water}} - T\Delta^{\text{Structuring}}S_{\text{C}}(\text{p53CTD})$. Experimental values of the binding free energy (Rustandi et al. 1998; Avalos et al. 2002; Mujtaba et al. 2004; Lowe et al. 2002), $\Delta^{\text{Binding-experiment}}G$, are also listed. The notation (−4.29, −4.06), for example, represents that the value is in the range from −4.29 to −4.06

System	$\Delta^{\text{Binding-theory}}G$	$\Delta^{\text{Binding-theory}}E_{\text{total}}$	$-T\Delta^{\text{Binding-theory}}S_{\text{water}}$	$-T\Delta^{\text{Structuring}}S_{\text{C}}(\text{p53CTD})$	$\Delta^{\text{Binding-experiment}}G$
1	−16.53 ± 2.67	14.06 ± 2.49	−58.04 ± 2.83	27.45	(−4.29, −4.06)
2	−31.85 ± 1.48	3.98 ± 1.47	−56.42 ± 0.59	20.59	(−3.78, −3.57)
3	−16.85 ± 4.06	29.18 ± 2.74	−64.33 ± 3.89	18.30	−3.89
4	−18.45 ± 0.53	37.82 ± 0.48	−76.86 ± 0.53	20.59	−4.97

Table 13.8 Free-energy change and its energetic and entropic components (in kcal/mol) upon target-dependent structure change of p53CTD. $\Delta^{\text{Structuring}}G$(p53CTD) = $\Delta^{\text{Structuring}}E_{\text{total}}$(p53CTD) $-T\Delta^{\text{Structuring}}S_{\text{water}}$(p53CTD)

System	$\Delta^{\text{Structuring}}G$(p53CTD)	$\Delta^{\text{Structuring}}E_{\text{total}}$(p53CTD)	$-T\Delta^{\text{Structuring}}S_{\text{water}}$(p53CTD)	$-T\Delta^{\text{Structuring}}S_{\text{C}}$(p53CTD)
1	38.00 ± 1.90	14.63 ± 1.95	−4.08 ± 1.87	27.45
2	34.42 ± 0.59	2.07 ± 0.56	11.76 ± 0.59	20.59
3	40.29 ± 2.95	17.22 ± 1.70	4.77 ± 2.42	18.30
4	40.87 ± 0.53	24.53 ± 0.48	−4.25 ± 0.53	20.59

$\Delta G^{\text{Binding-experiment}}$ is improved when $-T\Delta^{\text{Structuring}}S_C(\text{target})$ is incorporated in $\Delta G^{\text{Binding-theory}}$.

The updated picture of the p53CTD–target binding is summarized as follows. The binding is mainly driven by a large water-entropy gain and is characterized by sufficiently high shape complementarity as the first priority together with as much electrostatic complementarity as possible within the p53CTD–target interface. To achieve such complementarities, p53CTD forms a large diversity of structures in harmony with the target portions, where the p53CTD binds. Interestingly, as presented in Table 13.8, the structuring process of p53CTD is not always accompanied by the water-entropy gain, because the EV of p53CTD does not necessarily decrease upon binding. In any case, $\Delta^{\text{Structuring}}G(\text{p53CTD})$ is positive due to the energetic and conformational losses. This is the reason why p53CTD is disordered in the isolated state. The p53CTD is stabilized only by binding to its targets with the aid of hydration effects.

13.5 Mechanism of Actin–Myosin Binding

Actomyosin is an important biomolecular motor consisting of two kinds of proteins, namely actin and myosin (Sweeney and Houdusse 2010; Preller and Holmes 2013; Karagiannis et al. 2014). Actomyosin drives muscle contraction by binding (and separation) between actin fiber and myosin, which is coupled with the ATP hydrolysis cycle. In this section, we present the mechanism of actin–myosin binding in water.

In general, there are two modes in actin–myosin binding. One of them is strong binding, in which no nucleotide or ADP is bound to myosin, and the other is weak binding, in which ATP or ADP + Pi is bound to myosin. Through experiments involving the actin fiber and a myosin head (myosin subfragment S1) of rabbit skeletal muscle (Katoh and Morita 1996), it has been demonstrated that the enthalpic and entropic components of $\Delta G^{\text{Binding}}$ are both positive, irrespective of the presence of a nucleotide. This fact implies that the binding is entropically driven. In contrast, it has been concluded by several theoretical studies that the binding is driven by direct actin–myosin interactions, namely electrostatic and vdW attractive interactions, H-bonds, and salt bridges between actin and myosin (Lorenz and Holmes 2010; Okazaki et al. 2012; Várkuti et al. 2015).

We analyzed the changes in the thermodynamic quantities due to actin–myosin binding in a rigor state (state in which no nucleotide is bound to myosin) (Oshima et al. 2016). The binding of an actin trimer and myosin S1 is considered, and the trimer, S1, and their complex are regarded as actin, myosin, and actomyosin, respectively (see Fig. 13.7). Starting from the structure of actomyosin of *Dictyostelium discoideum*, which is the all-atom model constructed by Várkuti et al. (2015), we performed an MD simulation with explicit water to sample its fluctuating structures in water. Two MD simulations were also independently performed for the isolated actin and isolated myosin. Many (100) snapshot structures of actomyosin, actin, and myosin are thus generated.

We consider the two processes, as illustrated in Fig. 13.7:

(1) In $\Delta^{\text{Binding-fixed}}$, only the MD snapshot structures of actomyosin are employed, and the structures of actomyosin, actin, and myosin are obtained by simply separating them without any structural changes.
(2) In $\Delta^{\text{Binding-theory*}}$, the MD snapshot structures of actomyosin, isolated actin, and isolated myosin are employed. Thus, the fluctuation structures of actin and myosin and the changes in their structures due to binding are taken into account. Note that $\Delta^{\text{Binding-theory*}} = \Delta^{\text{Binding-fixed}} + \Delta^{\text{Structuring}}$.

As presented in Table 13.9, $\Delta^{\text{Binding-fixed}}$ is driven by a large gain of water entropy ($-T\Delta^{\text{Binding-fixed}}S_{\text{water}} < 0$). The water-entropy gain is caused by the entropic EV effect. Moreover, we find that the gain is mainly attributed to a reduction in "water crowding" (see Sect. 13.2 and Fig. 13.1; a more detailed discussion is provided in our original paper (Oshima et al. 2016)). The actin–myosin interface is closely packed due to the achievement of high shape complementarity on an atomic level, which is accompanied by a large reduction in water crowding. Interestingly, in $\Delta^{\text{Binding-theory*}}$, the water-entropy gain becomes higher than that in

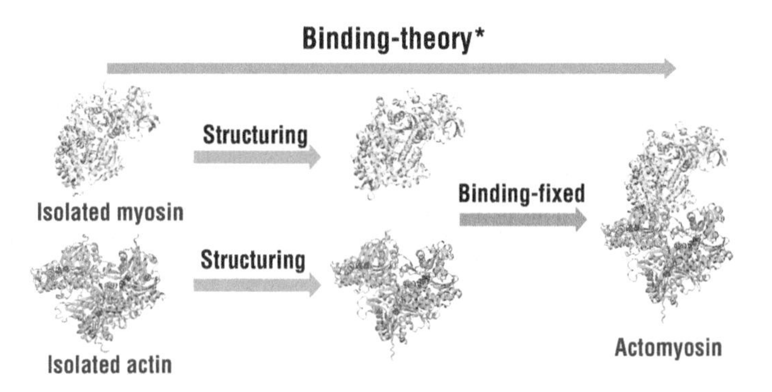

Fig. 13.7 Illustration of isolated myosin, isolated protein, actomyosin (action–myosin complex), and three processes ("Binding-fixed", "Binding-theory*", and "Structuring") considered in Sect. 13.5

Table 13.9 Free-energy change and its energetic and entropic components (in kcal/mol) upon actin–myosin binding in Δ^{δ}, where δ are "Binding-fixed" and "Binding-theory*" (see Fig. 13.7). $\Delta^{\delta}G = \Delta^{\delta}E_{\text{total}} - T\Delta^{\delta}S_{\text{water}}$. The 100 snapshot structures from a MD simulation for actomyosin are considered in $\Delta^{\text{Binding-fixed}}$, and those from MD simulations for actomyosin, actin, and myosin are considered in $\Delta^{\text{Binding-theory*}}$. The standard errors arise from the consideration of these snapshot structures

	$\Delta^{\delta}G$	$\Delta^{\delta}E_{\text{total}}$	$-T\Delta^{\delta}S_{\text{water}}$
δ = Binding–fixed	−72.60 ± 1.46	31.40 ± 0.85	−103.99 ± 1.21
δ = Binding-theory*	−64.12 ± 20.81	70.52 ± 17.38	−134.64 ± 10.45

$\Delta^{\text{Binding-fixed}}$ (see Table 13.9). Upon binding, actin and myosin change their structures in harmony with the actin–myosin binding site, leading to a closer packing within the binding interface (see Fig. 13.3c).

On the other hand, in $\Delta^{\text{Binding-fixed}}$, the total energy increases ($\Delta^{\text{Binding-fixed}}E_{\text{total}} > 0$), and the increase becomes higher in $\Delta^{\text{Binding-theory*}}$ (see Table 13.10). It is true that $\Delta^{\text{Binding-fixed}}$ results in a large gain of actin–myosin electrostatic attractive interaction ($\Delta^{\text{Binding-fixed}}E_{\text{ES}} < 0$), which is consistent with the previously reported MD simulation results. However, it is accompanied by an even larger loss of actin–water and myosin–water electrostatic attractive interactions. Although about half of the increase in energy due to this loss is recovered by the decrease in energy due to the structural reorganization of the water that is released upon binding, the remaining increase in energy is positive ($\Delta^{\text{Binding-fixed}}\varepsilon_{\text{ES}} > 0$). The resultant $\Delta^{\text{Binding-fixed}}(E_{\text{ES}} + \varepsilon_{\text{ES}})$ is positive and opposes the binding. This is because electrostatic complementarity within the actin–myosin interface cannot always be assured by the close interface packing (i.e., the formation of salt bridges or H-bonds between residues of actin and myosin is not perfect). The changes in the structures of actin and myosin upon binding reduce the electrostatic energy; however, this process is accompanied by an increase in the electrostatic-dehydration energy. Similarly, $\Delta^{\text{Binding-fixed}}E_{\text{vdW}} < 0$ and $\Delta^{\text{Binding-fixed}}\varepsilon_{\text{vdW}} > 0$. However, in both $\Delta^{\text{Binding-fixed}}$ and $\Delta^{\text{Binding-theory*}}$, the net changes in the vdW component of the system energy are lower than that in the electrostatic component. This suggests that the closely packed binding interface of actomyosin leads to a lowering of energy due to the actin–myosin vdW attractive interaction. Nevertheless, the actomyosin is destabilized in terms of the net change in system energy.

The experimental binding free energy, $\Delta G^{\text{Binding-experiment}}$, for actin–myosin binding is in the range from −8 to −9 kcal/mol (Takács et al. 2011; Várkuti et al. 2012). This value cannot readily be the $\Delta^{\text{Binding-theory*}}G$ (−64.12 kcal/mol), because the conformational entropy loss, $\Delta^{\text{Binding-theory*}}S_{\text{C}}$, is not incorporated in this theoretical value. For large proteins such as actomyosin, the calculation of $\Delta^{\text{Binding-theory*}}S_{\text{C}}$ with sufficient accuracy is a formidable task. Therefore, we estimate $\Delta^{\text{Binding-theory*}}S_{\text{C}}$ by considering the elimination of the fluctuations in the protein side chains, which are located within the binding interface (we assume that the backbones exhibit essentially no flexibility even before the binding, because isolated actin and isolated myosin possess well-defined structures). We count the number of interface residues N_{r} in stable contact pairs as follows. Two residues are assumed to be in contact with each other if their shortest interatomic distance is <4.5 Å. The contact of an intermolecular residue pair is considered to be stable when it is maintained in >95% of the snapshot structures along the MD trajectory for actomyosin. We find that N_{r} is 50. By setting the contribution from the side chains at $1.7k_{\text{B}}$ per residue in accordance with the suggestion by Doig and Sternberg (1995), we obtain $-T\Delta^{\text{Binding-theory*}}S_{\text{C}} \sim 50$ kcal/mol. Thus, $\Delta^{\text{Binding-theory*}}G\ \ -T\Delta^{\text{Binding-theory*}}S_{\text{C}} \sim -14.12$ kcal/mol, and the discrepancy between the present theoretical result and $\Delta G^{\text{Binding-experiment}}$ is significantly improved.

Table 13.10 Further decomposition of energetic components (in kcal/mol) upon actin–myosin binding in Δ^{δ}, where δ are "Binding-fixed" and "Binding-theory*". $\Delta^{\delta}E_{total} = \Delta^{\delta}E_{B} + \Delta^{\delta}(E_{vdW} + \varepsilon_{vdW}) + \Delta^{\delta}(E_{ES} + \varepsilon_{ES})$

	$\Delta^{\delta}E_{total}$	$\Delta^{\delta}E_{B}$	$\Delta^{\delta}(E_{ES} + \varepsilon_{ES})$	$\Delta^{\delta}(E_{vdW} + \varepsilon_{vdW})$	$\Delta^{\delta}E_{ES}$	$\Delta^{\delta}\varepsilon_{ES}$	$\Delta^{\delta}E_{vdW}$	$\Delta^{\delta}\varepsilon_{vdW}$
δ = Binding–fixed	31.40 ± 0.85	0	52.66 ± 1.26	−21.27 ± 0.89	−82.04 ± 10.44	134.70 ± 10.01	−156.58 ± 0.71	135.31 ± 0.90
δ = Binding-theory*	70.52 ± 17.38	−41.15 ± 17.96	104.05 ± 10.16	7.63 ± 9.60	−98.96 ± 34.12	203.01 ± 32.29	−172.71 ± 9.81	180.33 ± 2.90

Finally, we comment on the weak binding mode of actin and myosin. According to the experimental measurements (Katoh and Morita 1996), the absolute values of the binding entropy and enthalpy for the weak binding are considerably smaller than those for the strong binding are. Our result suggests that the strong binding is mainly driven by the water-entropy gain. Therefore, a possible reason for the smaller binding entropy is the decrease in water-entropy gain upon binding. In the weak binding mode, when ATP is bound to myosin, a tight packing of myosin and ATP occurs and the structure of myosin undergoes a large change. As a result, a cleft is formed near the actin–myosin interface (Coureux et al. 2004). Due to the cleft, the decrease in the total EV for water molecules upon actin–myosin binding could become smaller, leading to a smaller gain of water entropy. It is also possible that the formation of such a cleft results in the recovery of flexibility within the cleft portion, leading to a reduction of the conformational entropy loss upon binding. However, the actin–myosin binding could be driven by the water-entropy gain even in the weak binding mode.

13.6 A Unified Theoretical Framework

The binding of biomolecules results in a large water-entropy gain. The water-entropy gain arises from an overlap of the EVs (a decrease in system EV) of biomolecules upon binding due to the entropic EV effect (see Sect. 13.2 and Fig. 13.1). The properties of the polyatomic structure within the binding interface significantly affect the entropic EV effect; the close packing of the backbone and side-chain atoms causes a significant decrease in system EV. Therefore, **the shape complementarity at the atomic level, which is often observed in the binding interface, is responsible for the binding due to water entropy** (see Fig. 13.3c).

The close packing within the target interface is also accompanied by a large energetic gain due to the direct vdW interaction between biomolecules; however, the gain is largely canceled by the vdW-dehydration energy. Therefore, we conclude that the shape complementarity formation contributes to the binding mainly by water-entropy gain and vdW energy.

The shape complementarity is achieved in a system-dependent manner. For the IDP–target binding, such shape complementarity is achieved by the structure formation of IDP in harmony with the target portions where the IDP binds. For the actin–myosin binding, such shape complementarity could be better for the rigor binding state than the weak binding state (in the former, no nucleotide or ADP is bound to myosin, whereas in the latter, ATP or ADP + Pi is bound to myosin). In the general binding processes (e.g., the lock-key (Fischer 1894) or induced-fit (Koshland 1958) binding model), the shape complementarity between portions of the interface regions at the atomic level definitely plays an essential role, and this provides a significant contribution to the water-entropy gain.

On the other hand, the energetic gain due to the direct electrostatic interaction between biomolecules is canceled by the electrostatic-dehydration energy. Usually, if the positively charged groups of a biomolecule are buried within the binding interface, they are in contact with the negatively charged groups. **The electrostatic complementarity, which is essential for compensating the dehydration energy, is thus formed** (Fig. 13.4). However, the electrostatic complementarity is imperfect for the three cases discussed in this chapter; therefore, the net change in the electrostatic energy of the system is positive, and it hinders the binding (it is not obvious whether the total system energy change is positive or negative).

In summary, the binding of biomolecules is mainly driven by a large water-entropy gain and is characterized by sufficiently high shape complementarity together with as much electrostatic complementarity as possible within the binding interface. The binding is accomplished only when the free energy gain originating from these complementarities overcomes the free energy loss (i.e., conformational entropy loss), which originates from the elimination of the fluctuations in the biomolecules upon binding.

References

Avalos JL, Celic I, Muhammad S, Cosgrove MS, Boeke JD, Wolberger C (2002) Mol Cell 10:523
Beglov D, Roux B (1995) J Chem Phys 103:360
Beglov D, Roux B (1996) J Chem Phys 104:8678
Cann NM, Patey GN (1997) J Chem Phys 106:8165
Chiba S, Harano Y, Roth R, Kinoshita M, Sakurai M (2012) J Comput Chem 33:550
Chodera JD, Mobley DL, Shirts MR, Dixon RW, Branson K, Pande VS (2011) Curr Opin Struct Biol 21:150
Coureux P-D, Sweeney HL, Houdusse A (2004) EMBO J 23:4527
Doig AJ, Sternberg MJE (1995) Protein Sci 4:2247
Fischer E (1894) Ber Dtsch Chem Ges 27:2985
Fitter J (2003) Biophys J 84:3924
Gilson MK, Given JA, Bush BL, McCammon JA (1997) Biophys J 72:1047
Hansen J-P, McDonald LR (2006) Theory of simple liquids, 3rd ed. Academic Press, London
Harano Y, Kinoshita M (2005) Biophys J 89:2701
Harano Y, Yoshidome T, Kinoshita M (2008) J Chem Phys 129:145103
Hayashi T, Kinoshita M (2016) Phys Chem Chem Phys 18:32406
Hayashi T, Oshima H, Mashima T, Nagata T, Katahira M, Kinoshita M (2014) Nucleic Acids Res 42:6861
Hayashi T, Oshima H, Yasuda S, Kinoshita M (2015) J Phys Chem B 119:14120
Hayashi T, Oshima H, Harano Y, Kinoshita M (2016) J Phys Condens Matter 28:344003
Hayashi T, Yasuda S, Škrbić T, Giacometti A, Kinoshita M (2017) J Chem Phys 147, 125102
Hoe KK, Verma CS, Lane DP (2014) Nat Rev Drug Discovery 13:217–236
Hornak V, Abel R, Okur A, Strockbine B, Roitberg A, Simmerling C (2006) Proteins 65:712
Hsu W-L, Oldfield CJ, Xue B, Meng J, Huang F, Romero P, Uversky VN, Dunker AK (2013) Protein Sci 22:258
Huang N, Jacobson MP (2007) Curr. Opin. Drug Discovery Dev. 10:325

Imai T, Harano Y, Kinoshita M, Kovalenko A, Hirata F (2007) J Chem Phys 126:225102
Karagiannis P, Ishii Y, Yanagida T (2014) Chem Rev 114:3318
Katoh T, Morita F (1996) J Biochem 120:189
Keskin O, Tuncbag N, Gursoy A (2016) Chem Rev 116:4884
Kinoshita M (2008) J Chem Phys 128:024507
Kinoshita M (2013) Biophys. Rev. 5:283
Kinoshita M, Bérard DR (1996) J Comput Phys 124:230
Kinoshita M, Yoshidome T (2009) J Chem Phys 130:144705
König P-M, Roth R, Mecke KR (2004) Phys Rev Lett 93:160601
Koshland DE Jr (1958) Proc Natl Acad Sci USA 44:98
Kovalenko A, Hirata F (1999) J Chem Phys 110:10095
Kusalik PG, Patey GN (1988a) J Chem Phys 88:7715
Kusalik PG, Patey GN (1988b) J Mol Phys 65:1105
Leach AR, Shoichet BK, Peishoff CE (2006) J Med Chem 49:5851
Lorenz M, Holmes KC (2010) Proc Natl Acad Sci USA 107:12529
Lowe ED, Tews I, Cheng KY, Brown NR, Gul S, Noble MEM, Gamblin SJ, Johnson LN (2002) Biochemistry 41:15625
Luchko T, Gusarov S, Roe DR, Simmerling C, Case DA, Tuszynski J, Kovalenko A (2010) J Chem Theory Comput 6:607
Mashima T, Matsugami A, Nishikawa F, Nishikawa S, Katahira M (2009) Nucleic Acids Res 37:6249
Mashima T, Nishikawa F, Kamatari YO, Fujiwara H, Saimura M, Nagata T, Kodaki T, Nishikawa S, Kuwata K, Katahira M (2013) Nucleic Acids Res 41:1355
Mujtaba S, He Y, Zeng L, Yan S, Plotnikova O, Sachchidanand R, Sanchez NJ, Zeleznik-Le Z Ronai, Zhou M-M (2004) Mol Cell 13:251
Okazaki K, Sato T, Takano M (2012) J Am Chem Soc 134:8918
Oshima H, Kinoshita M (2015) J Chem Phys 142:145103
Oshima H, Kinoshita M (2016) J Comput Chem 37:712
Oshima H, Yasuda S, Yoshidome T, Ikeguchi M, Kinoshita M (2011) Phys Chem Chem Phys 13:16236
Oshima H, Hayashi T, Kinoshita M (2016) Biophys J 110:2496
Preller M, Holmes KC (2013) Cytoskeleton (Hoboken) 70:651
Ratkova EL, Palmer DS, Fedorov MV (2015) Chem Rev 115:6312
Roth R, Harano Y, Kinoshita M (2006) Phys Rev Lett 97:078101
Rustandi RR, Drohat AC, Baldisseri DM, Wilder PT, Weber DJ (1998) Biochemistry 37:1951
Shirts MR, Mobley DL, Brown SP (2010) Free energy calculations in structure-based drug design. In: Kenneth DR, Merz M, Reynolds CH (eds) Structure based drug design. Cambridge University Press, New York
Sweeney HL, Houdusse A (2010) Annu Rev Biophys 39:539
Takács B, O'Neall-Hennessey E, Hetényi C, Kardos J, Szent-Györgyi AG, Kovács M (2011) Biophys Rev 5:283
Truchon J-F, Pettitt BM, Labute P (2014) J Chem Theory Comput 10:934
Várkuti BH, Yang Z, Kintses B, Erdélyi P, Bárdos-Nagy I, Kovács AL, Hári P, Kellermayer M, Vellai T, Málnási-Csizmadia A (2012) Nat Struct Mol Biol 19:299
Várkuti BH, Yang Z, Málnási-Csizmadia A (2015) J Biol Chem 290:1679
Vogelstein B, Lane D, Levine AJ (2000) Nature 408:307
Yoshidome T, Kinoshita M (2012) Phys Chem Chem Phys 14:14554
Yoshidome T, Kinoshita M, Hirota S, Baden N, Terazima M (2008) J Chem Phys 128:225104
Zhou H-X, Gilson MK (2009) Chem Rev 109:4092

Part III
Functioning Mechanisms of Protein Machinery

Chapter 14
Ratchet Model of Motor Proteins and Its Energetics

Yohei Nakayama and Eiro Muneyuki

Abstract We describe the energetics of motor proteins by taking F_1-ATPase as an example. To clarify the meaning of the mechanical works which are experimentally measured, we first identify the heat, which satisfies the second law of thermodynamics, and next formulate the work through the first law of thermodynamics. The theoretical expressions of the mechanical works reveal that the experimentally measured works have a theoretical basis. Contrary to the mechanical work, the measurement of the heat is still developing. We introduce an attempt to measure a part of the heat, which is related to the experimentally accessible quantities through a nonequilibrium equality.

Keywords Motor protein · Single-molecule experiment · First and second laws of thermodynamics · Nonequilibrium system

14.1 Introduction

It is important to unravel the energy balance of the molecular motor proteins in operation for a better understanding of their functions. The framework of stochastic thermodynamics provides us with a way to extract thermodynamic quantities, such as work and heat (Sekimoto 2010), from a time series obtained by the single-molecule observations (Noji et al. 1997). In this short review, we describe the energetics of the molecular motor proteins using a ratchet model of F_1-ATPase (Elston et al. 1998; Oster and Wang 2000; Liu et al. 2003; Gaspard and Gerritsma 2007) as an example to approach it experimentally. The key concept is to identify the heat, which satisfies the second law of thermodynamics. The work performed on or done by the molecular motor is immediately formulated from consistency with the first law of thermodynamics.

Y. Nakayama · E. Muneyuki (✉)
Faculty of Science and Engineering, Department of Physics Chuo University, Tokyo 112-8551, Japan
e-mail: emuneyuk@phys.chuo-u.ac.jp

© Springer Nature Singapore Pte Ltd. 2018

M. Suzuki (ed.), *The Role of Water in ATP Hydrolysis Energy Transduction by Protein Machinery*, https://doi.org/10.1007/978-981-10-8459-1_14

F_1-ATPase is a rotary molecular motor protein. The complex of F_1-ATPase consists of five kinds of subunits, α, β, γ, δ and ε. The γ subunit acts as a stalk surrounded by a cylinder formed by three $\alpha\beta$ pairs (Fig. 14.1) (Abrahams et al. 1994). The conformational change in the $\alpha_3\beta_3$ cylinder accompanying the hydrolysis of ATP rotates the central γ stalk (with the ε subunit) counterclockwise viewed from the stalk protruding from the cylinder (Abrahams et al. 1994). The first single-molecule observation of F_1-ATPase was realized by fixing its β subunit on a glass surface and attaching a fluorescently labeled actin filament to its γ subunit as a marker (Noji et al. 1997).

The technique of the single-molecule observation paved the way for the elucidation of the efficiency of F_1-ATPase based on the video-recorded time-series. In the next single-molecule experiment (Yasuda et al. 1998), it was attempted to estimate the work done by F_1-ATPase from the experimentally measured rotational rate and the theoretical value of the frictional drag coefficient. However, as we discuss below, what was measured in this experiment is the conversion efficiency from the chemical work to the heat rather than that between the chemical and mechanical works. Keeping in mind that F_1-ATPase is rotated forcibly by another rotary molecular motor protein, F_0, to synthesize ATP from ADP and inorganic phosphate (Pi) in vivo, the latter efficiency is considered to be essential for F_1-ATPase. The milestone toward this direction is introductions of magnetic tweezer (Itoh et al. 2004) and electrorotation method to the single-molecule experiment of F_1-ATPase. These two methods may apply a trapping potential and a controlled external torque, respectively. It was shown that the transduction from the chemical work and the mechanical work is nearly perfect in F_1-ATPase by measuring the work done against the constant torque (Toyabe et al. 2011) and the trapping potential (Saita et al. 2015).

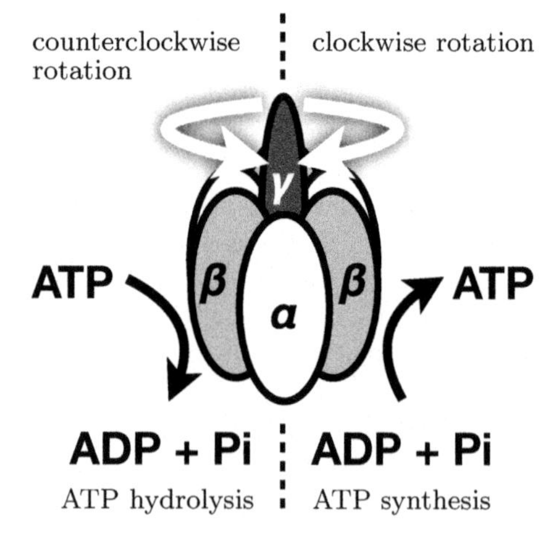

Fig. 14.1 Schematic picture of F_1-ATPase. We here depict $\alpha_3\beta_3\gamma$ subcomplex. The $\alpha_3\beta_3$ part forms a cylinder in which the γ subunit rotates. When F_1-ATPase hydrolyses ATP, the γ subunit rotates counterclockwise. The enforced rotation in the opposite direction synthesizes ATP from ADP and Pi

14.2 Ratchet Model of Molecular Motors

We start with an introduction of a stochastic model of F_1-ATPase. In most of the single-molecule experiments, the rotation of F_1-ATPase was measured by attaching some probes such as an actin filament (Noji et al. 1997), a polystyrene particle (Noji et al. 2001), a gold bead (Yasuda et al. 2001), or a magnetic bead (Itoh et al. 2004) to the γ subunit. Therefore, in a theoretical model, we should obviously describe the motion of the probe. Here, we represent the azimuthal angle of the γ subunit and hence the probe as θ. In addition, we take into account a label of internal states of F_1-ATPase, n. When the chemical events (binding, release, or reaction of the substrate) are rate-determining, F_1-ATPase shows the step-wise rotation with corresponding pauses (Yasuda et al. 1998, 2001). The internal degree of freedom, n, is necessary to describe these kinds of motion.

Next, let us consider the dynamics of θ. Since the size of the probes used for the single-molecule experiments of F_1-ATPase ranges from ~40 nm (Yasuda et al. 2001) to ~μm (Yasuda et al. 1998), the Brownian motion of the probes is noticeable. In addition, the smallness of the probes makes its moment of inertia negligible. Based on these considerations, we describe the motion of the probe by an overdamped Langevin equation with the motor-probe interaction potential, $U_{\text{motor}}(\theta, n)$, the external torque, $f_{\text{ext}}(\theta)$, and the external potential, $U_{\text{ext}}(\theta; \nu)$:

$$\Gamma\dot{\theta} = -\frac{\partial U_{\text{motor}}(\theta, n)}{\partial \theta} + f_{\text{ext}}(\theta) - \frac{\partial U_{\text{ext}}(\theta; \nu)}{\partial \theta} + \sqrt{2\Gamma k_B T}\xi. \tag{14.1}$$

Here, Γ is the frictional coefficient, ν is a controllable parameter, k_B is the Boltzmann constant, and T is the temperature of the environment. $\sqrt{2\Gamma k_B T}\xi$ represents the Gaussian thermal random torque characterized by $\langle \xi \rangle = 0$ and $\langle \xi(t)\xi(t') \rangle = \delta(t - t')$, where $\langle \cdot \rangle$ is the ensemble average of $\cdot$. The internal states of F_1-ATPase, n, affects the motion of the probe through U_{motor}. We suppose that $f_{\text{ext}}(\theta)$ and $U_{\text{ext}}(\theta; \nu)$ are applied externally by the electrorotation method (Watanabe-Nakayama et al. 2008) and the magnetic tweezer (Itoh et al. 2004), respectively. In this case, ν corresponds to the position and the strength of the magnetic tweezer. We represent the set of parameters ν and T as α.

In order to complete the description, we specify the dynamics of the internal states, n. Since we consider the changes in n to be triggered by the chemical events, we assume that such changes occur with the reaction rate coefficients, $k_{n\to m}(\theta)$. Here, $k_{n\to m}(\theta)$ generally depends on the angle, θ. This means that the rates of the chemical events are regulated by the mechanical motion. Therefore, n regulates the change in θ through $U_{\text{motor}}(\theta, n)$, and vice versa [through $k_{n\to m}(\theta)$]. This feature is known as the mechanochemical coupling (Harada et al. 1990) and was verified experimentally (Watanabe et al. 2012).

Before closing this section, we introduce another description of this ratchet model. Although we consider, in the above, a realization of θ and n which fluctuate stochastically, we may consider a collection of many realizations, an ensemble (Fig. 14.2).

The ensemble is characterized by various probability distributions. Since the model has the Markov property, all probability distributions are essentially characterized by knowing the change in the probability distribution over time t (and the initial probability distribution) (van Kampen 1992). The time evolution of the joint probability distribution of θ and n at time t, $P(\theta, n, t)$, is expressed by the master equation as

$$\frac{\partial P(\theta,n,t)}{\partial t} = -\frac{\partial J_\theta(\theta,n,t)}{\partial \theta} + \sum_{m \neq n}[-k_{n\to m}(\theta)P(\theta,n,t) + k_{m\to n}(\theta)P(\theta,m,t)], \quad (14.2)$$

where

$$J_\theta(\theta,n,t) = \frac{1}{\Gamma}\left(-\frac{\partial U_{\mathrm{motor}}(\theta,n)}{\partial \theta} + f_{\mathrm{ext}}(\theta) - \frac{\partial U_{\mathrm{ext}}(\theta;\nu)}{\partial \theta}\right)P(\theta,n,t) - \frac{k_B T}{\Gamma}\frac{\partial P(\theta,n,t)}{\partial \theta} \quad (14.3)$$

is a probability flux along θ. The first term of Eq. (14.2) represents the change in $P(\theta, n, t)$ with the motion of the probe. The second and third terms of Eq. (14.2)

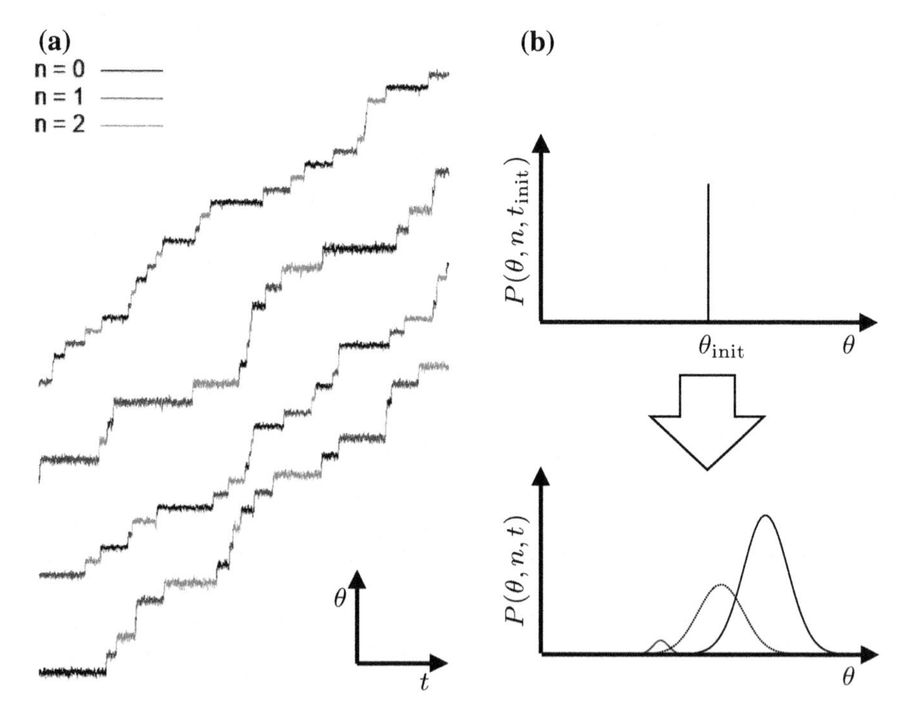

Fig. 14.2 Two theoretical descriptions. **a** A collection of many time series. We represent the value of n by the line color. **b** The behavior of the joint probability distribution, $P(\theta, n, t)$. An initial state $P(\theta, n, t_{\mathrm{init}}) = \delta(\theta - \theta_{\mathrm{init}})\delta_{n,n_{\mathrm{init}}}$ prepared at t_{init} evolves into $P(\theta, n, t)$ according to Eq. (14.2)

correspond the loss and gain of the internal state, n, respectively. The derivation of $J_\theta(\theta, n, t)$ starting from Eq. (14.1) is found in (Sekimoto 2010).

14.3 Second Law of Thermodynamics and Heat

In this section, we identify the heat, which satisfies an inequality corresponding to the second law of thermodynamics.

First, we see how equilibrium states appear in the ratchet model of F_1-ATPase. This is because the second law of thermodynamics claims that the change in the entropy of the system is larger than the integral of the heat divided by the temperature of the environment in transitions between equilibrium states. Since the external torque, the time-dependent external potential and the chemical reaction may drive the system out of equilibrium, we consider a case, where $f_{\text{ext}}(\theta) = 0$, α is fixed constant, and the hydrolysis and synthesis of ATP is in equilibrium. The last condition imposes detailed balance, $k_{n\to m}(\theta)/k_{m\to n}(\theta) = \exp\{-\beta[U_{\text{motor}}(\theta, m) - U_{\text{motor}}(\theta, n)]\}$. Here, $\beta = 1/k_B T$ is the inverse temperature. It may be shown under these conditions that $P(\theta, n, t)$ approaches the equilibrium distribution

$$P_{\text{eq}}(\theta, n; \alpha) = \frac{1}{Z(\alpha)} \exp\left[-\frac{U_{\text{motor}}(\theta, n) + U_{\text{ext}}(\theta; \nu)}{k_B T}\right], \tag{14.4}$$

where $Z(\alpha)$ is a partition function. Since this equilibrium distribution is equivalent to the canonical distribution, the thermodynamic state functions of this state are considered to be given by equilibrium statistical mechanics. Therefore, the entropy of F_1-ATPase is given through its free energy, $\mathcal{F}(\alpha) = -k_B T \ln Z(\alpha)$, as

$$S(\alpha) = -\frac{\partial \mathcal{F}(\alpha)}{\partial T} = -k_B \int d\theta \sum_n P_{\text{eq}}(\theta, n; \alpha) \ln P_{\text{eq}}(\theta, n; \alpha). \tag{14.5}$$

The entropy, S, is called the Gibbs entropy.

The second law of thermodynamics that the heat (per unit time), Q, satisfies may be written as

$$S(T; \nu_{\text{f}}) - S(T; \nu_{\text{i}}) \geq -\frac{1}{T} \int_{t_i}^{t_f} Q \, dt, \tag{14.6}$$

when ν changes from ν_{i} to ν_{f} during the time interval $[t_i, t_f]$. However, we impose a stronger condition

$$k_B \frac{d}{dt} H[P(\theta, n, t)] \geq -\frac{Q}{T}, \tag{14.7}$$

where $H[P(\theta, n, t)]$ is the Shannon entropy given as

$$H[P(\theta, n, t)] = -\int d\theta \sum_{n} P(\theta, n, t) \ln P(\theta, n, t). \tag{14.8}$$

The equality of Eq. (14.7) holds for the equilibrium states. Since $S(\alpha) = k_B H[P_{eq}(\theta, n; \alpha)]$, Eq. (14.6) immediately follows from the time integral of Eq. (14.7).

Next, we introduce an essential assumption to identify the expression of Q. The assumption is that Q may be written in the form of

$$\begin{aligned} \frac{Q}{T} =& k_B \int d\theta \sum_{n} \left[\mathcal{A}_\theta \left(\frac{F(\theta, n)}{\Gamma}, \frac{1}{\Gamma} \frac{\partial F(\theta, n)}{\partial \theta}, \frac{k_B T}{\Gamma} \right) \right. \\ & \left. + \sum_{m \neq n} \mathcal{A}_{n,m}(k_{n \to m}(\theta), k_{m \to n}(\theta)) \right] P(\theta, n, t), \end{aligned} \tag{14.9}$$

where $\mathcal{A}_\theta$ is the heat assigned to Brownian motion, $\mathcal{A}_{n,m}$ is the heat assigned to the switch of the internal states from n to m, and $F(\theta, n) := -\partial U_{\text{motor}}(\theta, n)/\partial\theta + f_{\text{ext}}(\theta) - \partial U_{\text{ext}}(\theta; \nu)/\partial\theta$ is total mechanical torque. This assumption means that Q may be decomposed into the contributions of the motions in the different directions, which depend on θ, n and t through only the quantities which determine the time evolution of $P(\theta, n, t)$ in each direction at that space-time point. Through dimensional analysis, the general forms of $\mathcal{A}_\theta$ and $\mathcal{A}_{n,m}$ are given as

$$\mathcal{A}_\theta = a \frac{1}{\Gamma} \frac{\partial F(\theta, n)}{\partial \theta} + b \frac{\Gamma}{k_B T} \left(\frac{F(\theta, n)}{\Gamma} \right)^2, \tag{14.10}$$

$$\mathcal{A}_{n,m} = k_{n \to m}(\theta) \left[\ln \frac{k_{n \to m}(\theta)}{k_{m \to n}(\theta)} + c \left(\frac{k_{n \to m}(\theta)}{k_{m \to n}(\theta)} \right) \right], \tag{14.11}$$

where a and b are dimensionless constants, and c is a dimensionless function of the ratio $k_{n \to m}(\theta)/k_{m \to n}(\theta)$. Here, we take into consideration that $\mathcal{A}_\theta$ is a polynomial with respect to $F(\theta, n)/\Gamma$ and $[\partial F(\theta, n)/\partial\theta]/\Gamma$, since they may become zero. Substituting Eqs. (14.10, 14.11) into Eq. (14.7), we obtain

$$k_B \frac{\mathrm{d}}{\mathrm{d}t} H[P(\theta,n,t)] + \frac{Q}{T} = -k_B \int \mathrm{d}\theta \sum_n \frac{\partial P(\theta,n,t)}{\partial t} \ln P(\theta,n,t) + \frac{Q}{T}$$

$$= \int \mathrm{d}\theta \sum_n \left\{ \frac{\Gamma}{T} \left(\frac{J_\theta(\theta,n,t)}{P(\theta,n,t)} \right)^2 P(\theta,n,t) \right.$$

$$+ \frac{F(\theta,n)}{T} \left((a-1) J_\theta(\theta,n,t) + (b-a) \frac{F(\theta,n)}{\Gamma} P(\theta,n,t) \right)$$

$$\left. + k_B \sum_{m \neq n} k_{n\to m}(\theta) P(\theta,n,t) \left[\ln \frac{k_{n\to m}(\theta) P(\theta,n,t)}{k_{m\to n}(\theta) P(\theta,m,t)} + c \left(\frac{k_{n\to m}(\theta)}{k_{m\to n}(\theta)} \right) \right] \right\}. \quad (14.12)$$

Since $J_\theta(\theta,n,t) = 0$ and $k_{n\to m}(\theta)P(\theta,n,t) = k_{m\to n}(\theta)P(\theta,m,t)$ in the equilibrium states, it follows from the equality and the inequality of Eq. (14.7) in the linearly nonequilibrium states around the equilibrium states that $a = b = 1$ and $c = 0$. Finally, we arrive an expression of the heat

$$Q = \int \mathrm{d}\theta \sum_n \left[F(\theta,n) J_\theta(\theta,n,t) + k_B T \sum_{m \neq n} k_{n\to m}(\theta) P(\theta,n,t) \ln \frac{k_{n\to m}(\theta)}{k_{m\to n}(\theta)} \right]. \quad (14.13)$$

Although not explained in detail in this review, the heat (Eq. (14.13)) is equivalent to the ensemble average of heat formulated for each realization (Sekimoto 2010), and such stochastic heat is known to satisfy several nonequilibrium equalities which are called fluctuation theorem (Jarzynski 1997; Kurchan 1998; Crooks 1999).

14.4 First Law of Thermodynamics and Work

Since we identify the heat in the previous section, we may formulate the work by relying on the first law of thermodynamics. The first law of thermodynamics claims that the change in the internal energy is equal to the sum of the work and the heat. By choosing $U(\theta,n;\nu) = U_{\mathrm{motor}}(\theta,n) + U_{\mathrm{ext}}(\theta;\nu)$ as the internal energy of F_1-ATPase, the change in the ensemble average of $U(\theta,n;\nu)$ becomes

$$\frac{\mathrm{d}\langle U(\theta,n;\nu)\rangle}{\mathrm{d}t} = \frac{\mathrm{d}}{\mathrm{d}t} \int \mathrm{d}\theta \sum_n U(\theta,n;\nu) P(\theta,n,t)$$

$$= \int \mathrm{d}\theta \sum_n \left[\frac{\mathrm{d}\nu}{\mathrm{d}t} \frac{\partial U(\theta,n;\nu)}{\partial \nu} P(\theta,n,t) + U(\theta,n;\nu) \frac{\partial P(\theta,n,t)}{\partial t} \right]$$

$$= \int \mathrm{d}\theta \sum_n \left[\frac{\mathrm{d}\nu}{\mathrm{d}t} \frac{\partial U(\theta,n;\nu)}{\partial \nu} P(\theta,n,t) + \frac{\partial U(\theta,n;\nu)}{\partial \theta} J_\theta(\theta,n,t) \right.$$

$$\left. \sum_{m \neq n} [-U(\theta,n;\nu) + U(\theta,m;\nu)] k_{n\to m}(\theta) P(\theta,n,t) \right], \quad (14.14)$$

where we take into consideration the dependence of ν on t and use Eq. (14.2). Comparing Eq. (14.14) with the expression of heat (Eq. (14.13)), we obtain the work (per unit time) as

$$W = \frac{\mathrm{d}\langle U(\theta, n; \nu)\rangle}{\mathrm{d}t} + Q = \int \mathrm{d}\theta \sum_n \left[\frac{\mathrm{d}\nu}{\mathrm{d}t} \frac{\partial U(\theta, n; \nu)}{\partial \nu} P(\theta, n, t) \right.$$

$$\left. + f_{\mathrm{ext}}(\theta) J_\theta(\theta, n, t) + k_B T \sum_{m \neq n} k_{n\to m}(\theta) P(\theta, n, t) \ln \frac{e^{-\beta U(\theta, n; \nu)} k_{n\to m}(\theta)}{e^{-\beta U(\theta, m; \nu)} k_{m\to n}(\theta)} \right]. \quad (14.15)$$

Since the first, second, and third terms in the right-hand side of Eq. (14.15) result from the change in ν, the nonequilibrium torque, $f_{\mathrm{ext}}(\theta)$, and the chemical reaction, respectively, we define

$$W_{\mathrm{potential}} := \frac{\mathrm{d}\nu}{\mathrm{d}t} \int \mathrm{d}\theta \sum_n \frac{\partial U(\theta, n; \nu)}{\partial \nu} P(\theta, n, t), \quad (14.16)$$

$$W_{\mathrm{torque}} := \int \mathrm{d}\theta \sum_n f_{\mathrm{ext}}(\theta) J_\theta(\theta, n, t), \quad (14.17)$$

$$W_{\mathrm{chemical}} := k_B T \int \mathrm{d}\theta \sum_n \sum_{m \neq n} k_{n\to m}(\theta) P(\theta, n, t) \ln \frac{e^{-\beta U(\theta, n; \nu)} k_{n\to m}(\theta)}{e^{-\beta U(\theta, m; \nu)} k_{m\to n}(\theta)}. \quad (14.18)$$

$W_{\mathrm{potential}}$ is the ensemble average of work formulated for each realization (Sekimoto 2010).

14.5 Experimental Measurement of Work

The expressions of the works (Eqs. (14.16)–(14.18)) clarify that we can never discuss the conversion between the different kinds of the works without applying torque or potential externally. The first attempt to measure the work done by a single-F_1-ATPase molecule was made by using the electrorotation method (Watanabe-Nakayama et al. 2008). Seven years later, the work against the potential applied by the magnetic tweezer was measured (Saita et al. 2015). In this section, we explain that the measured quantities are equivalent to the definitions of the mechanical works (Eq. (14.17) and Eq. (14.16), respectively).

In (Watanabe-Nakayama et al. 2008; Toyabe et al. 2011; Toyabe and Muneyuki 2015), a spatially constant torque, $f_{\mathrm{ext}}(\theta) = -f$, was applied to a single F_1-ATPase molecule. In these experiments, they measured the responses of the average rotational rate, $\langle\dot{\theta}\rangle$, to the constant torque, f, and determined a stall torque, f_{stall}, where the average rotational rate vanishes, $\langle\dot{\theta}\rangle = 0$. In the absence of ADP and Pi which are the products of hydrolysis of ATP, it was shown that $\langle\dot{\theta}\rangle$ is decreased with the increase of f, but does not substantially vanish unless the F_1-ATPase is broken (Watanabe-

Nakayama et al. 2008). This result means $f_{\mathrm{stall}} = \infty$ on the other hand, in the presence of ADP and Pi, the stall torque remained finite and was compared with the chemical potential difference between ATP and the products, $\Delta\mu$. By assuming that F_1-ATPase rotates 120° per one ATP, the maximum work per one ATP, $f_{\mathrm{stall}} \times 120°$, was shown to be nearly equal to $\Delta\mu$ at various experimental conditions (Toyabe et al. 2011). In terms of the ratchet model, for the constant torque, W_{torque} and $W_{\mathrm{potential}}$ are rewritten as

$$W_{\mathrm{torque}} = -f \int \mathrm{d}\theta \sum_n J_\theta(\theta, n, t) = -f\langle\dot{\theta}\rangle, \qquad W_{\mathrm{potential}} = 0. \tag{14.19}$$

Therefore, what they measured is equivalent to W_{torque}. Here, W_{torque} is defined not per one ATP but per unit time. Their results indicate a nearly perfect transduction from the free energy to the mechanical work (Toyabe et al. 2011). Note that the chemical potential difference was determined from the equilibrium constant between ATP, ADP, and Pi (Rosing and Slater 1972).

In (Saita et al. 2015), the work done against the external potential was measured. They estimated the potential applied by the magnetic tweezer as

$$U_{\mathrm{ext}}(\theta; \theta_{\mathrm{TC}}) = \frac{k}{4} \cos 2(\theta - \theta_{\mathrm{TC}}), \tag{14.20}$$

where θ_{TC} and k represent the center angle and stiffness of the magnetic trap, respectively. It results in

$$W_{\mathrm{torque}} = 0, \qquad W_{\mathrm{potential}} = \frac{\mathrm{d}\theta_{\mathrm{TC}}}{\mathrm{d}t} \int \mathrm{d}\theta \sum_n \frac{k}{2} \sin 2(\theta - \theta_{\mathrm{TC}}) P(\theta, n, t). \tag{14.21}$$

In their experiment, a quantity essentially equivalent to $W_{\mathrm{potential}}$ was measured, although the ensemble average in the definition of $W_{\mathrm{potential}}$ was replaced with the temporal average. Comparing with $\Delta\mu$ again, they showed that the transduction from the free energy to the mechanical work against the external potential is nearly perfect, when θ_{TC} changes very slowly (at 0.04 Hz).

14.6 In the Case of Perfect Coupling

We may simplify the energetics of the ratchet model based on the experimental results explained in the previous section. It was clarified that the maximum work per one ATP is equal to $\Delta\mu$. These results strongly suggest that the hydrolysis of one ATP is tightly coupled with the 120° rotation even under the external torque or potential. Assuming the perfect coupling between one chemical reaction and 120° rotation, we may write a net number of ATP hydrolyzed as a function of θ and n, $N(\theta, n)$. Here, the periodic boundary condition is not imposed on θ. Furthermore, we may introduce the free energy of F_1-ATPase and the surrounding environment including ATP, ADP, and Pi, as

$$V(\theta, n; \nu) = U(\theta, n; \nu) - N(\theta, n)\Delta\mu. \tag{14.22}$$

The introduction of $V(\theta, n; \nu)$ imposes restrictions on the reaction rate coefficients, $k_{n\to m}(\theta)$. According to the reaction rate theory (Kramers 1940), $k_{n\to m}(\theta)$ satisfies local detailed balance condition (Crooks 1999)

$$\frac{k_{n\to m}(\theta)}{k_{m\to n}(\theta)} = \exp\{-\beta[V(\theta, m; \nu) - V(\theta, n; \nu)]\} \tag{14.23}$$

By substituting Eq. (14.23) into Eq. (14.18), we obtain the simplified expression of W_{chemical} [Eq. (14.18)] as

$$W_{\text{chemical}} = \Delta\mu \int d\theta \sum_n \sum_{m\neq n} k_{n\to m}(\theta)P(\theta, n, t)[N(\theta, m) - N(\theta, n)]. \tag{14.24}$$

With an appropriate choice of $N(\theta, n)$, W_{chemical} reproduces the estimation of the free-energy consumption rate (work rate done by the particle reservoirs) in the previous section.

The local detailed balance condition [Eq. (14.23)] provides a way to describe the energetics of the ratchet model in a different manner. If we regard the composite system which consists of F_1-ATPase and the environment as the system of interest, the discussion in Sect. 14.4 should be reconsidered. By replacing $U(\theta, n; \nu)$ with $V(\theta, n; \nu)$, we obtain a new expression of the work as

$$\begin{aligned}\mathcal{W} &= \int d\theta \sum_n \left[\frac{d\nu}{dt}\frac{\partial U(\theta, n; \nu)}{\partial \nu}P(\theta, n, t) + f_{\text{ext}}(\theta)J_\theta(\theta, n, t)\right] \\ &= W_{\text{potential}} + W_{\text{torque}}.\end{aligned} \tag{14.25}$$

While, in the original description, F_1-ATPase converts the different kinds of the works each other, in this description, it consumes the free energy to perform the mechanical work (Fig. 14.3). This description is used in (Toyabe et al. 2012). In the same manner, we may also include $f_{\text{ext}}(\theta)$ into the potential of F_1-ATPase.

14.7 Experimental Measurement of Heat

As can be seen by comparing the expressions of the heat (Eq. (14.13)) and the work (Eqs. (14.16), (14.17), and (14.24)), it appears more difficult to measure the heat than the work. The reason is that Eq. (14.13) includes the motor-probe interaction potential, $U_{\text{motor}}(\theta, n)$, explicitly. Since, in the typical single molecule experiments,

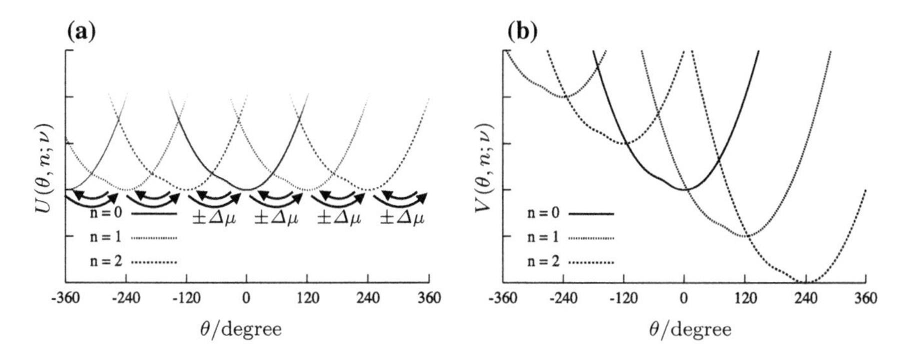

Fig. 14.3 $U(\theta, n; \nu)$ and $V(\theta, n; \nu)$. **a** The original description. F_1-ATPase performs on or is done the chemical work, $\pm\Delta\mu$, per forward and backward step, respectively. **b** Another description. F_1-ATPase converts the free energy, $V(\theta, n; \nu)$, to the mechanical work, and vice versa

we may only measure a time-series of θ, the measurement of $U_{\text{motor}}(\theta, n)$ is a difficult task.

Harada and Sasa proposed an equality which connects a part of the heat in the steady state with the experimentally accessible quantities (Harada and Sasa 2005, 2006). Here, we explain the equality called Harada–Sasa equality without the derivation. The quantities which should be experimentally measured are the correlation function of velocity fluctuations, $C(t)$, and the linear response function of velocity to a perturbation torque, $R(t)$. The explicit definition of $C(t)$ is given as

$$C(t) = \left\langle (\dot{\theta}_{\tau+t} - \langle\dot{\theta}\rangle_{\text{st}})(\dot{\theta}_{\tau} - \langle\dot{\theta}\rangle)_{\text{st}} \right\rangle_{\text{st}}, \tag{14.26}$$

where we show the time dependence of $\dot{\theta}$ by the subscript and $\langle\cdot\rangle_{\text{st}}$ represents the ensemble average in the steady state. It is necessary to apply a small time-dependent perturbative torque, $\epsilon(t)$, in order to measure $R(t)$. When the ensemble average under the perturbative torque is represented by a subscript ϵ, $R(t)$ may be written as

$$\langle\dot{\theta}_{\tau}\rangle_{\epsilon} - \langle\dot{\theta}_{\tau}\rangle_{\text{st}} = \int_{-\infty}^{\tau} d\tau' R(\tau - \tau')\epsilon(\tau') + O[\epsilon(t)^2] \tag{14.27}$$

Harada–Sasa equality relates the heat exchanged through the degree of the probe, Q_{probe} with $C(0)$ and $R(+0) = \lim_{t\to+0} R(t)$ as

$$\begin{aligned} Q_{\text{probe}} &:= \int d\theta \sum_{n} F(\theta, n) J_{\theta}^{\text{st}}(\theta, n) = \Gamma[\langle\dot{\theta}\rangle_{\text{st}}^2 + C(0) - k_B T R(+0)] \\ &= \Gamma\langle\dot{\theta}\rangle_{\text{st}}^2 + \frac{\Gamma}{2\pi}\int d\omega \left\{\tilde{C}(\omega) - 2k_B T \Re\left[\tilde{R}(\omega)\right]\right\}, \end{aligned} \tag{14.28}$$

where $J^{\rm st}_{\theta}(\theta, n)$ is a steady probability flux along θ; $\tilde{C}(\omega)$ is the Fourier transform of $C(t)$ and $\mathfrak{R}\,[\tilde{R}(\omega)]$ is a real part of the Fourier transform of $R(t)$. $Q_{\rm probe}$ is the first term of the right-hand side of Eq. (14.13) in the steady state. Since the response function is easily measured in Fourier space, the expression of the second line in Eq. (14.28) is used in practice.

In F_1-ATPase, $Q_{\rm probe}$ which is measured by Harada–Sasa equality was shown to be equal with the difference between the free-energy consumption rate, $3\,\Delta\mu\langle\frac{\dot{\theta}}{2\pi}\rangle$, and the work rate against the external constant torque, $-W_{\rm torque}$ (Toyabe et al. 2010). Considering together with the first law of thermodynamics, this result means that $Q = Q_{\rm probe}$, and the switch of the internal states occurs at intersection points of $V(\theta, n; \nu)$ on average (Fig. 14.3).

We finally emphasize that the framework of Harada–Sasa equality strongly relies on the assumption that Γ is constant. The energetics discussed in Sects. 14.3–14.4 is quite robust in the sense that it may be extended at least to the case where Γ depends on θ and n. In contrast, we cannot extend Harada–Sasa equality to such cases with an ad hoc fix. It is an open question whether we can develop a method to measure the heat without the help of a specific model.

References

Abrahams JP, Andrew GW, Leslie AG, Lutter R, Walker JE (1994) Structure at 2.8 Å resolution of F_1-ATPase from bovine heart mitochondria. Nature 370(6491):621–628

Crooks GE (1999) Entropy production fluctuation theorem and the nonequilibrium work relation for free energy differences. Phys Rev E 60(3):2721–2726

Elston T, Wang H, Oster G (1998) Energy transduction in ATP synthase. Nature 391(6666):510–513

Gaspard P, Gerritsma E (2007) The stochastic chemomechanics of the F_1-ATPase molecular motor. J Theor Biology 247(4):672–686

Harada T, Sasa SI (2005) Equality connecting energy dissipation with a violation of the fluctuation-response relation. Phys Rev Lett 95(13)

Harada T, Sasa SI, (2006) Energy dissipation and violation of the fluctuation-response relation in nonequilibrium Langevin systems. Phys Rev E 73(2)

Harada Y, Sakurada K, Aoki T, Thomas DD, Yanagida T (1990) Mechanochemical coupling in actomyosin energy transduction studied by in vitro movement assay. J Mol Biol 216(1):49–68

Itoh H, Takahashi A, Adachi K, Noji H, Yasuda R, Yoshida M, Kinosita K (2004) Mechanically driven ATP synthesis by F_1-ATPase. Nature 427(6973):465–468

Jarzynski C (1997) Nonequilibrium equality for free energy differences. Phys Rev Lett 78(14):2690–2693

Kramers HA (1940) Brownian motion in a field of force and the diffusion model of chemical reactions. Physica 7(4):284–304

Kullback S, Leibler RA (1951) On information and sufficiency. Ann Math Stat 22(1):79–86

Kurchan J (1998) Fluctuation theorem for stochastic dynamics. J Phys A: Math Gen 31(16):3719–3729

Liu MS, Todd BD, Sadus RJ (2003) Kinetics and chemomechanical properties of the F_1-ATPase molecular motor. J Chem Phys 118(21):9890–9898

Noji H, Yasuda R, Yoshida M, Kinosita K (1997) Direct observation of the rotation of F_1-ATPase. Nature 386(6622):299–302

Noji H, Bald D, Yasuda R, Itoh H, Yoshida M, Kinosita K (2001) Purine but Not Pyrimidine Nucleotides Support Rotation of F_1-ATPase. J Biol Chem 276(27):25480–25486
Oster G, Wang H (2000)Reverse engineering a protein: the mechanochemistry of ATP synthase. Biochimica et Biophysica Acta (BBA)—Bioenergetics 1458(2):482–510
Rosing J, Slater EC (1972) The value of $\Delta G°$ for the hydrolysis of ATP. Biochimica et Biophysica Acta (BBA)—Bioenergetics 267(2):275–290
Saita E, Suzuki T, Kinosita K, Yoshida M (2015) Simple mechanism whereby the F_1-ATPase motor rotates with near-perfect chemomechanical energy conversion. Proc Natl Acad Sci 112(31):9626–9631
Sekimoto K (2010) Stochastic energetics. In: Lecture notes in physics, vol 799. Springer, Berlin. ISBN 978-3-642-05410-5
Toyabe S, Muneyuki E (2015) Single molecule thermodynamics of ATP synthesis by F_1-ATPase. New J Phys 17(1):015008
Toyabe S, Okamoto T, Watanabe-Nakayama T, Taketani H, Kudo S, Muneyuki E (2010) Nonequilibrium energetics of a single F_1-ATPase molecule. Phys Rev Lett 104(19)
Toyabe S, Watanabe-Nakayama T, Okamoto T, Kudo S, Muneyuki E (2011) Thermodynamic efficiency and mechanochemical coupling of F_1-ATPase. Proc Natl Acad Sci 108(44):17951–17956
Toyabe S, Ueno H, Muneyuki E (2012) Recovery of state-specific potential of molecular motor from single-molecule trajectory. EPL (Europhysics Letters) 97(4):40004
van Kampen NG (1992) Stochastic Processes in Physics and Chemistry. North-Holland personal library. North-Holland, Amsterdam Tokyo. ISBN 0-444-89349-0
Watanabe R, Okuno D, Sakakihara S, Shimabukuro K, Iino R, Yoshida M, Noji H (2012) Mechanical modulation of catalytic power on F_1-ATPase. Nat Chem Biol 8(1):86–92
Watanabe-Nakayama T, Toyabe S, Kudo S, Sugiyama S, Yoshida M, Muneyuki E (2008) Effect of external torque on the ATP-driven rotation of F_1-ATPase. Biochem Biophys Res Commun 366(4):951–957
Yasuda R, Noji H, Kinosita K, Yoshida M (1998) F_1-ATPase Is a Highly Efficient Molecular Motor that Rotates with Discrete 120° Steps. Cell 93(7):1117–1124
Yasuda R, Noji H, Yoshida M, Kinosita K, Itoh H (2001) Resolution of distinct rotational substeps by submillisecond kinetic analysis of F_1-ATPase. Nature 410(6831):898–904

Chapter 15
Single-Molecule Analysis of Actomyosin in the Presence of Osmolyte

Mitsuhiro Iwaki, Kohji Ito and Keisuke Fujita

Abstract Actomyosin is a protein complex composed of myosin and actin, which is well known for being the minimal contractile unit of muscle. The chemical free energy of ATP is converted into mechanical work by the complex, and the single-molecule mechanical properties of myosin are well characterized in vitro. However, the aqueous solution environment in in vitro assay is far from that in cells, where biomolecules are crowded, which influences osmotic pressure, and processes such as folding, and association and diffusion of proteins. Here, to bridge the gap between in vitro and in-cell environment, we observed mechanical motion of actomyosin-V in the presence of the osmolyte sucrose, as a model system. Single-molecule observation of myosin-V motor domains (heads) on actin filament at varying sucrose concentration revealed modulated mechanical elementary processes suggesting increased affinity of heads with actin and more robust force generation possibly accompanied by a sliding motion of myosin head along actin.

Keywords Actomyosin · Osmolyte · Single-molecule observation
Myosin-V · Force generation mechanism

M. Iwaki (✉) · K. Fujita
Quantitative Biology Center, RIKEN/Graduate School of Frontier Biosciences, Osaka University, Osaka, Japan
e-mail: iwaki@riken.jp

K. Fujita
e-mail: keisuke.fujita@riken.jp

K. Ito
Graduate School of Science, Chiba University, Chiba, Japan
e-mail: k-ito@faculty.chiba-u.jp

© Springer Nature Singapore Pte Ltd. 2018

M. Suzuki (ed.), *The Role of Water in ATP Hydrolysis Energy Transduction by Protein Machinery*, https://doi.org/10.1007/978-981-10-8459-1_15

15.1 Introduction

Biological molecular machines (e.g., proteins) emerge physiological functions in aqueous solution. Actomyosin is one such protein complex, consisting of myosin and actin, and known to be a minimal contractile unit of muscle. It hydrolyzes ATP and converts its chemical free energy into mechanical work. During this chemo-mechanical transduction, myosin and actin in water undergo Brownian motion, interact with each other, and move with in one direction (Fujita et al. 2012). Accompanied by the dynamic transduction processes, the surrounding bulk water molecules are agitated and should contribute to the energy transduction; however, little is understood about the role of water.

Clarification of role of water on the chemomechanical transduction is especially important to bridge the gap between in vitro and in-cell experiments. In cells, biomolecules are crowded, influencing osmotic pressure, and processes such as folding, and association and diffusion of proteins (Boersma et al. 2015), in which water also plays an important role. Here, we discuss the effect of water on chemomechanical transduction of actomyosin. We examine mechanical motion of both ensemble and single myosin-V on actin filament in the presence of the osmolyte sucrose, as a model system to perturb water dynamics. Sucrose cannot completely mimic macromolecules in cells; however, sucrose is a stable and reversible agent which is useful for studying actomyosin (Ando and Asai 1977). Myosin constitutes a superfamily of over 24 classes (Foth et al. 2006). Among these myosins, myosin-V has ideal features suitable for single-molecule studies including observation of mechanical elementary processes, enabling quantification of the effect of sucrose on the mechanical processes in detail. We hope our results will be of use to predict energy transduction of actomyosin in a physiological cellular environment.

15.2 The Effect of Sucrose on Ensemble Myosin-V Motility

The in vitro motility assay is a powerful assay system to analyze unidirectional motion of actomyosin (Harada et al. 1987). In the system, we observe fluorescently labeled actin filaments sliding over a coverslip that has been covered with purified myosin molecules in the presence of ATP. Because myosin molecules are densely adsorbed onto a glass surface, many myosin molecules drive a single actin filament. Figure 15.1 shows dependency of the sliding velocity of actin filaments on the sucrose concentration. Without sucrose, the velocity was 210 nms^{-1} at saturating ATP concentration (2 mM), consistent with previous reports using myosin-V (Mehta et al. 1999), whereas the velocity monotonically decreased with increasing sucrose concentration and became zero at ~2.0 M. Although sucrose increases solution viscosity and drag force against sliding actin filaments, it is negligible

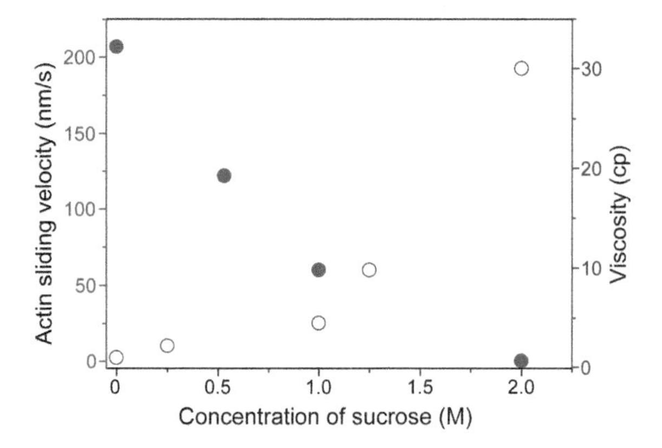

Fig. 15.1 Dependency of sucrose on the sliding velocity of actin filaments. Red circles indicate mean actin sliding velocity and blue open circles indicate viscosity of solution

compared with force generated by ensemble myosins (the order of 10^1–10^2 pN) because the drag force (F) without sucrose is calculated as

$$F = c_{\|} L v \sim 2.4 fN,$$

where L is the mean length of actin filament (5 μm), v is the sliding velocity (210 nms^{-1}), and the drag coefficient $c_{\|}$ is defined as

$$c_{\|} \cong \frac{2\pi\eta}{\ln(2h/r)},$$

where η is the viscosity of water (0.89×10^{-6} pN ms nm^{-2} at 25 °C), h is the height from the glass surface (~25 nm), and r is the radius of an actin filament (3 nm) (Howard 2001). Therefore, even ~30 times larger viscosity at 2.0 M sucrose shown in Fig. 15.1 cannot stall actin filaments sliding motion and is not a dominant factor for the stall.

15.3 The Effect of Sucrose on Single Myosin-V Motility

In vitro motility assay revealed sucrose brought actomyosin-V sliding to a halt, and possibly, sucrose affected coordinated motion between myosin-V molecules and/or chemomechanical transduction of individual myosin-V. To fully comprehend the mechanism, we observed unidirectional motion of single myosin-V in the presence of sucrose.

Myosin-V is an intracellular vesicle transporter that is particularly popular in single-molecule studies because single myosin-V can processively move along actin filaments without dissociating away from actin (Mehta et al. 1999), enabling the observation of mechanical motion in detail at the single-molecule level. During the processive motion, two motor domains (heads) span the actin half-helical pitch of 36 nm as a waiting state and walk along actin by a hand-over-hand mechanism (Yildiz et al. 2003) (Fig. 15.2). Bright fluorescent probes such as quantum dots (QDs) and total internal reflection microscopy (TIRFM) enable motion capture of single myosin-V at a few nanometer accuracy. We observed the processive motion both with and without sucrose (Fig. 15.3a). TIRFM observations of QDs attached to the center-of-mass position of myosin-V showed regular 36 nm processive steps, consistent with previous reports (Mehta et al. 1999), and 36 nm step sizes were not affected by sucrose (Fig. 15.3b). However, stepping rate (inverse of dwell time between adjacent 36 nm steps) linearly decreased as increasing sucrose (Fig. 15.3c), and the result suggested processive motion stalled near 2.0 M condition, consistent with in vitro motility assay. Therefore, we concluded stalls observed in in vitro motility assay at 2.0 M sucrose were caused by a direct effect of sucrose on individual myosin-V motility.

It is well known that dwell time includes ADP release time from the rear head, ATP binding time to the rear head followed by a detachment of the ATP-bound

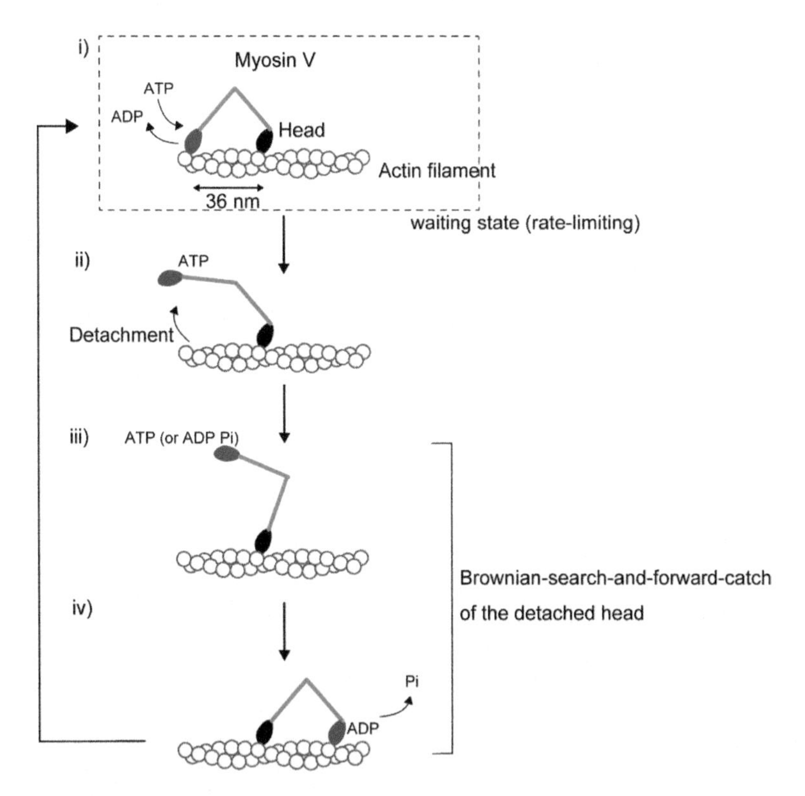

Fig. 15.2 Hand-over-hand mechanism of myosin-V. **i** Myosin heads strongly attach to actin filament and span the half-helical pitch. ADP is released from the rear head (red) followed by an ATP binding. **ii–iii** Upon ATP binding, the rear head detaches from actin and searches for a forward landing site on actin. **iv** Inorganic phosphate release occurs soon after landing on actin and transitions to strong binding state with actin

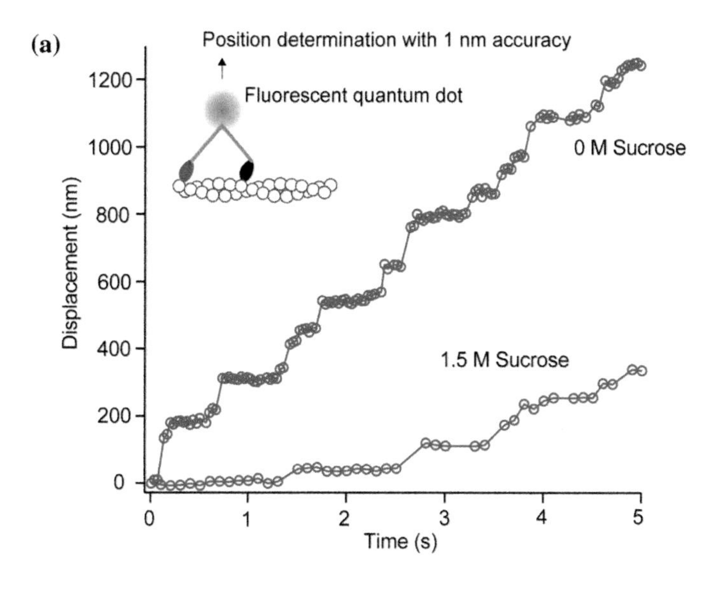
(a)
Position determination with 1 nm accuracy
Fluorescent quantum dot
0 M Sucrose
1.5 M Sucrose
Displacement (nm)
Time (s)
0
200
400
600
800
1000
1200
0
1
2
3
4
5

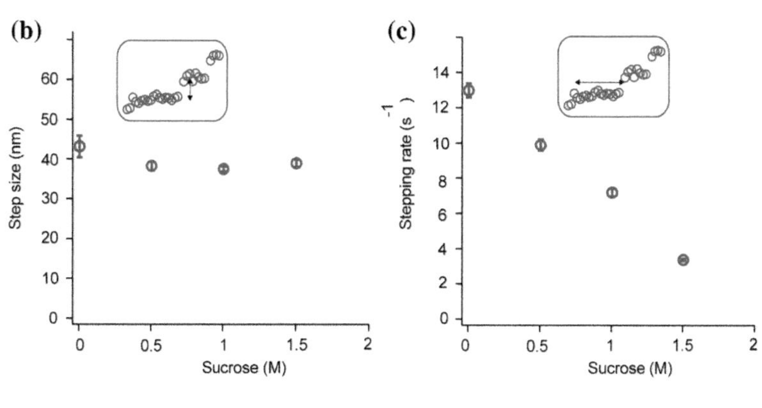
(b)
Step size (nm)
Sucrose (M)
(c)
Stepping rate (s^{-1})
Sucrose (M)

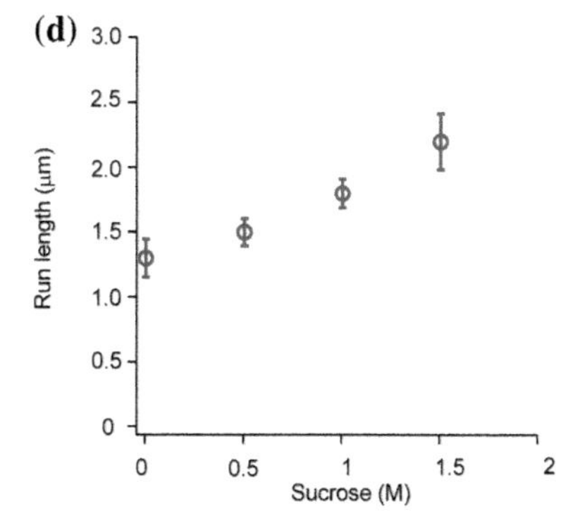
(d)
Run length (μm)
Sucrose (M)

◀**Fig. 15.3** Single-molecule imaging of quantum dot attached to myosin-V in the presence of sucrose. **a** Typical trajectories of processive motion of myosin-V recorded at 30 fps. (inset) A fluorescent quantum dot was attached to the center-of-mass position of dimeric myosin-V, and the fluorescent spot position was determined with 1 nm accuracy. **b** Mean step sizes at various sucrose concentrations. **c** Mean stepping rates at various sucrose concentrations. **d** Mean run length at various sucrose concentrations

head from actin (Fig. 15.2 state i–ii) and Brownian search-and-forward-catch of the head (Iwaki et al. 2009; Rief et al. 2000; Shiroguchi and Kinosita 2007) (Fig. 15.2 state iii–iv). Normally, the duration of the Brownian search-and-forward-catch is short (10–20 ms) (Dunn and Spudich 2007) and undetectable at our time resolution (33 ms); hence, if sucrose slowed the process to explain the decreased stepping rate, we should observe it as a large fluctuation of QDs and 36 nm steps should be divided into two phases (Dunn and Spudich 2007; Fujita et al. 2012). We could not detect such trajectories; therefore, state i and/or ii in Fig. 15.2 was slowed by sucrose. Regarding processive run length (travel distance of myosin-V before dissociating away from actin), sucrose increased the length (Fig. 15.3d), implying myosin heads stably bound to actin during the processive motion.

15.4 The Effect of Sucrose on Transient Kinetics of ATP Hydrolysis Cycle

To discover a process sensitive to sucrose, we performed kinetic analyzes of myosin both in the presence and absence of sucrose. In the kinetic studies, we used *Arabidopsis* myosin-III ATM1 (ATM1) because the inhibition effect of the sliding velocity of actin filaments by ATM1 was similar to that by myosin-V, and high amounts of protein are needed for kinetic studies, and high yields are obtained for ATM1. Similar to myosin-V, the sliding velocity by ATM1 decreased linearly when sucrose concentration increased and became to zero at 2.0 M sucrose. In general, the actin sliding velocities of myosins depend on the time when they are strongly bound to actin (t_s). t_s is primarily determined by the ADP dissociation rate from actomyosin and the actomyosin dissociation rate upon ATP binding (Millar and Geeves 1983; Siemankowski et al. 1985). Therefore, we determined these rates both without and with 2.0 M sucrose.

The actomyosin dissociation reaction upon ATP binding can be described as a two-step process, and the rate constant (k_{obs}) is expressed as $K_1k_{+2}[ATP]/(1 + K_1[ATP])$, (Millar and Geeves 1983), where A and M represent actin and myosin, respectively (Fig. 15.4 kinetic scheme). The first step in Fig. 15.4 kinetic scheme is the formation of a collision complex between actomyosin and ATP, which are in rapid equilibrium (equilibrium constant, K_1). The second step is the dissociation of myosin–ATP from actin following the isomerization of myosin (rate constant, k_{+2}). To determine K_1 and k_{+2}, a pyrene–actin–ATM1 complex was dissociated by mixing with 10–8000 μM ATP using a stopped-flow apparatus, and

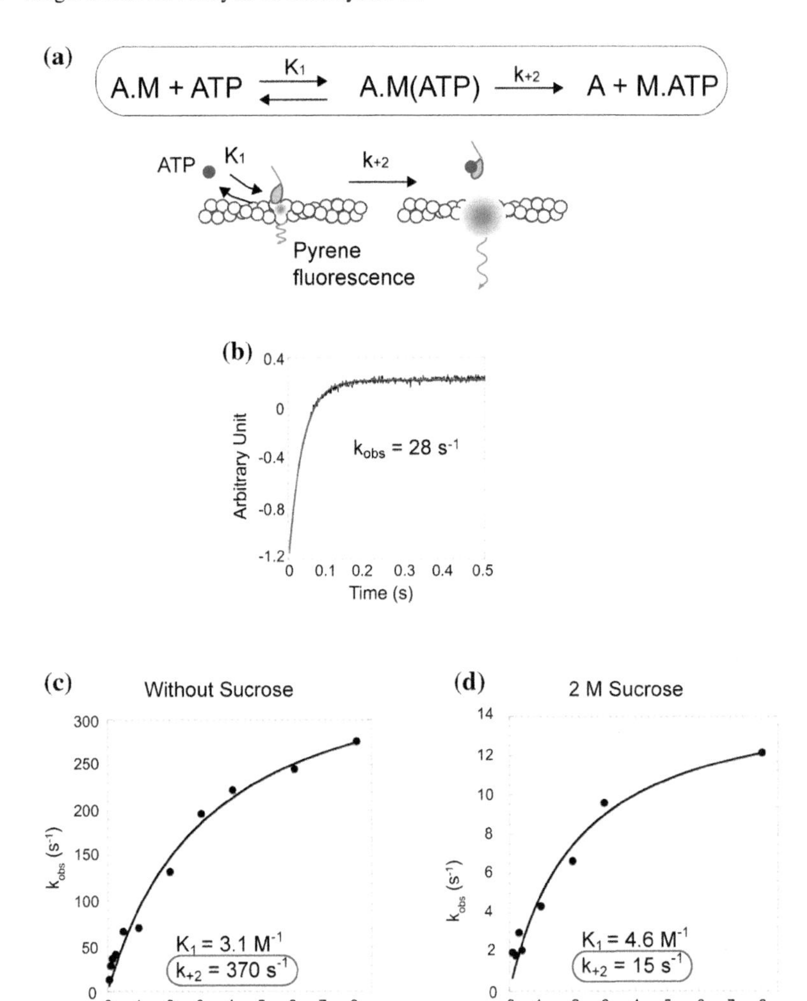

Fig. 15.4 Kinetic analyses of myosin ATM1 using stopped-flow apparatus both with and without sucrose. ATP-induced dissociation of the pyrene–acto-ATM1 complex. **a** The actomyosin dissociation reaction upon ATP binding can be described as a two-step process. The first step is the formation of a collision complex between actomyosin and ATP, which are in rapid equilibrium. The second step is the dissociation of myosin–ATP from actin following the isomerization of myosin. **b** Pyrene–acto-ATM1 (0.5 μM) was mixed with 100 μM ATP final concentration in the absence of sucrose. The transient is the average of five separate recordings, and the red line is a single-exponential fit, which yielded a rate constant of 28 s^{-1} in the example shown. **c–d** The observed rate constant of ATP-induced dissociation of the acto-ATM1complex (k_{obs}) was plotted against ATP concentrations. Values of K_1 and k_{+2} were obtained by fitting the data to K_1k_{+2} [ATP]/(1 + K_1 [ATP]). (c, without sucrose; d, with 2.0 M sucrose). The values of K_1 and k_{+2} without sucrose were 3.1 M^{-1} and 370 s^{-1}, respectively, and those with 2.0 M sucrose were 4.6 M^{-1} and 15 s^{-1}, respectively

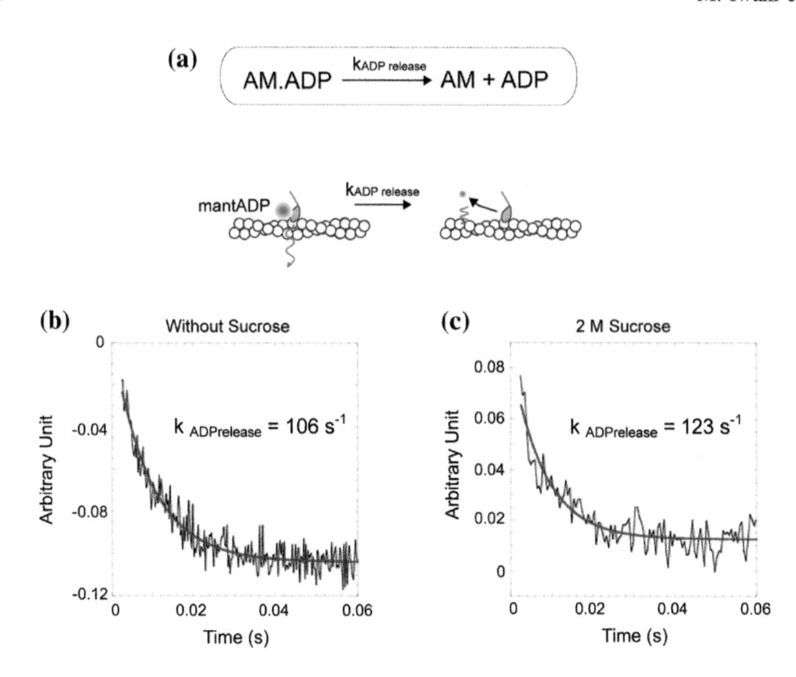

Fig. 15.5 Kinetic analyses of myosin ATM1 using stopped-flow apparatus both with and without sucrose. **a** MantADP dissociation from acto-ATM1. A mixture of 0.2 μM ATM1-MD, 5 μM actin, and 5 μM mantADP was mixed with 2 mM ATP (final concentration). The dissociation of mantADP from acto-ATM1 was measured using fluorescence energy transfer between tryptophans of ATM1 and mantADP. **b–c** The trajectory is the average of five separate recordings, and the red line is a single-exponential fit, which yielded rate constants of 106 s^{-1} (**b**) and 123^{-1} (**c**)

the acto-ATM1 dissociation was monitored by the increase in pyrene fluorescence (Geeves and Jeffries 1988). The values of K_1 and k_{+2} were obtained by fitting the data to $K_1k_{+2}[ATP]/(1 + K_1[ATP])$. The values of K_1 and k_{+2} without sucrose were 3.1 M^{-1} and 370 s^{-1}, respectively, and those with 2.0 M sucrose were 4.6 M^{-1} and 15 s^{-1}, respectively, (Fig. 15.4). Thus, the K_1 value was almost unchanged, but the k_{+2} value decreased to only 4% by adding 2.0 M sucrose.

The rate of ADP dissociation from acto-ATM1 was determined by measuring the decrease in the fluorescence intensity of mantADP (Fig. 15.5) (De La Cruz et al. 1999). Time course of the change in fluorescence followed a single exponential. The rate constants in the absence of sucrose and the presence of 2.0 sucrose were almost the same (106 s^{-1} in the absence of sucrose, 123 s^{-1} in the presence of 2.0 M sucrose).

In conclusion, it is strongly suggested that sucrose increased the affinity between myosin heads and actin, resulting in an inhibition of dissociation of myosin heads from actin (Fig. 15.2 state ii) and a decrease of myosin motility as observed in single-molecule assays.

15.5 Force Measurement in the Presence of Sucrose

We finally performed force measurement of single myosin-V both with and without sucrose using optical tweezers. Optical tweezers are a conventional biophysical tool to quantify generated force by single motor protein (Iwaki et al. 2006; Svoboda et al. 1993). In our assay system, myosin heads were labeled with a short (~60 nm in length) double-stranded DNA molecule (DNA handle) and the end of the DNA handle was attached with polystyrene bead (200 nm in diameter) to optically capture the bead by an focused infrared beam (Fujita et al. 2012). Figure 15.6a shows a typical trajectory of an optically trapped bead tagged with single myosin-V. Single myosin-V pulled the bead and reached stall and then detached from actin and returned to optical trapping center. The process was repeated. The trapping force is proportional to the displacement from the trap center (set to zero displacement), and we estimated the stall force as 0.9 ± 0.3 pN without sucrose and 0.86 ± 0.4 pN with 1.0 M sucrose, respectively (Fig. 15.6b–d). Considered from step sizes near stall force of 77 nm (Fujita et al. 2012), output work can be

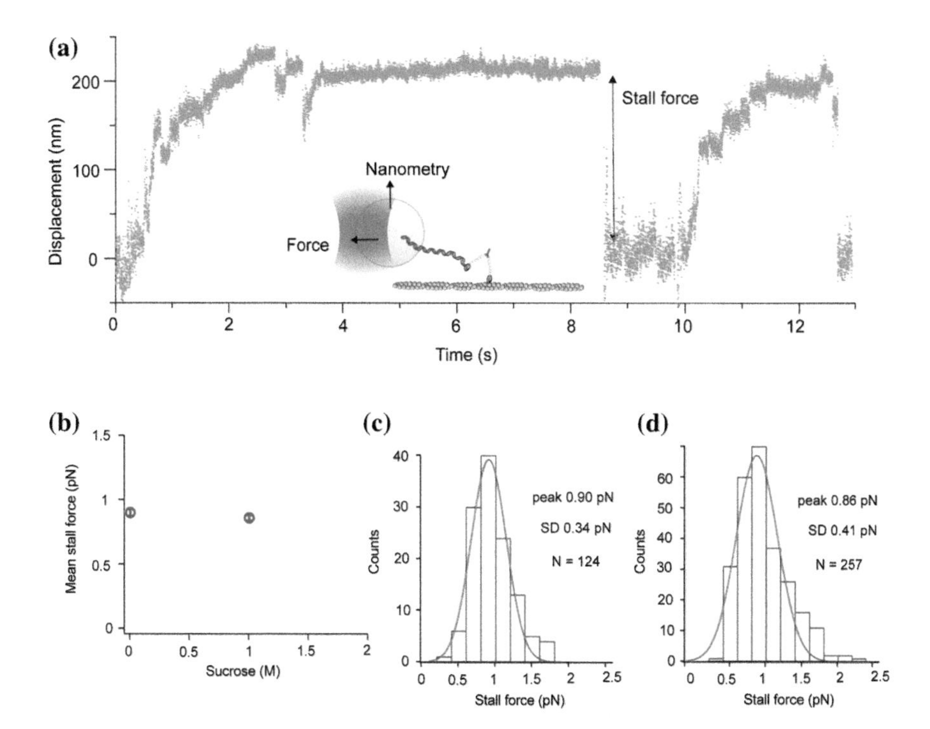

Fig. 15.6 Force measurement of single myosin-V in the presence of sucrose by optical tweezers. **a** A typical trajectory of myosin-V under load. Optically trapped bead position was determined by a quadrant photodiode with several nanometer accuracy at 24 kHz recording rate. Stall force was determined by displacement multiplied by trap stiffness. **b** Mean stall force at 0 M and 1.0 M sucrose. **c** and **d** Histogram of stall force for 0 M sucrose (**c**) and 1.0 M sucrose (**d**)

calculated to be 69 pNnm (77 nm $\times$ 0.9 pN), which is interpreted as 69 $\sim$ 86% of input energy (80 $\sim$ 100 pNnm) of ATP was converted to mechanical work regardless of sucrose concentration.

15.6 Conclusion

In this study, we examined the effect of sucrose on single myosin-V motility and our results strongly suggested that sucrose strengthened the affinity between myosin head and actin. This effect causes prolonged processive runs and possibly stabilizes force generation by reducing a risk of detachment from actin.

It is phenomenologically known that saccharide and polyol prevent denaturation of proteins and stabilize protein structure (possibly because saccharide and polyol have high affinity with water and suppress hydration accompanied by protein unfolding); therefore, our results here are reasonable in this viewpoint. The quantitative mechanism for the stabilization of actomyosin complex is still unclear, but Highsmith et al. (1996) argue that osmolytes induce dehydration of interface between myosin head and actin, increasing the free energy difference for actomyosin binding. In addition to that possibility, Amano et al. (2010) argue that entropic effects arising from the translational displacement of water molecules will be strengthened by medium size particles like sucrose (Kinoshita, personal communication). In their concept (see also Chap. 18), when myosin and actin interact with each other, the excluded volumes overlap, leading to an increase in the total volume available to the translational displacement of water and entropy gain occurs. Then, sucrose strengthens the effect.

The cell is highly crowded with various macromolecules (50–400 $mgml^{-1}$) (Cayley et al. 1991; Fulton 1982). Co-solvent, sucrose used in our study is different from these macromolecules, however, the scenarios proposed by Highsmith or Kinoshita will be applicable to the in-cell environment, and our results will help to compensate for the gap between in vitro and in-cell experiments. Especially in muscle, sarcomeric proteins (e.g., myosin, actin) are densely packed into sarcomere structure. The surrounding macromolecules may increase affinity between myosin heads and actin, resulting in a robust force generation. Also, actomyosin forms a contractile ring at the last step of cell division. In this step, cytoplasmic protein concentration increases (Cookson et al. 2010), which may stabilize the contraction through the effect of water discussed here and accelerate cell division. We are currently endeavoring to develop a force sensor (Iwaki et al. 2016) to measure contractile force in vivo, which will enable us to investigate our hypothesis.

In this review, we did not show an effect of sucrose on actomyosin force generation mechanism (or mechanical elementary processes); however, we have observed small (<10 nm) successive steps in the rising phase of the displacement by optical tweezers assay in the presence of sucrose. Although this is still a preliminary result, the trajectory implies sliding motion of myosin heads along actin filament as proposed by Kitamura et al. (1999) by using a rather unique setup. Because we could not observe such sliding motions in the absence of sucrose, increased affinity between myosin head and actin by sucrose might induce the sliding. If so, actomyosin force generation mechanism depends on environment and myosin may generate force by sliding along actin in-cell environment. Further investigation is necessary to confirm this speculation.

References

Amano K, Yoshidome T, Iwaki M, Suzuki M, Kinoshita M (2010) Entropic potential field formed for a linear-motor protein near a filament: Statistical-mechanical analyses using simple models. J Chem Phys 133:045103

Ando T, Asai H (1977) The effects of solvent viscosity on the kinetic parameters of myosin and heavy meromyosin ATPase. J Bioenerg Biomembr 9:283–288

Boersma AJ, Zuhorn IS, Poolman B (2015) A sensor for quantification of macromolecular crowding in living cells. Nature methods 12: 227-229, 221 p following 229

Cayley S, Lewis BA, Guttman HJ, Record MT Jr (1991) Characterization of the cytoplasm of Escherichia coli K-12 as a function of external osmolarity. Implications for protein-DNA interactions in vivo. J Mol Biol 222:281–300

Cookson NA, Cookson SW, Tsimring LS, Hasty J (2010) Cell cycle-dependent variations in protein concentration. Nucleic Acids Res 38:2676–2681

De La Cruz EM, Wells AL, Rosenfeld SS, Ostap EM, Sweeney HL (1999) The kinetic mechanism of myosin V. Proc Natl Acad Sci USA 96:13726–13731

Dunn AR, Spudich JA (2007) Dynamics of the unbound head during myosin V processive translocation. Nat Struct Mol Biol 14:246–248

Foth BJ, Goedecke MC, Soldati D (2006) New insights into myosin evolution and classification. Proc Natl Acad Sci USA 103:3681–3686

Fujita K, Iwaki M, Iwane AH, Marcucci L, Yanagida T (2012) Switching of myosin-V motion between the lever-arm swing and brownian search-and-catch. Nature communications 3:956

Fulton AB (1982) How crowded is the cytoplasm? Cell 30:345–347

Geeves MA, Jeffries TE (1988) The effect of nucleotide upon a specific isomerization of actomyosin subfragment 1. Biochem J 256:41–46

Harada Y, Noguchi A, Kishino A, Yanagida T (1987) Sliding movement of single actin filaments on one-headed myosin filaments. Nature 326:805–808

Highsmith S, Duignan K, Cooke R, Cohen J (1996) Osmotic pressure probe of actin-myosin hydration changes during ATP hydrolysis. Biophys J 70:2830–2837

Howard J (2001) Mechanics of motor proteins and the cytoskeleton. Sinauer Associates, MA

Iwaki M, Iwane AH, Shimokawa T, Cooke R, Yanagida T (2009) Brownian search-and-catch mechanism for myosin-VI steps. Nat Chem Biol 5:403–405

Iwaki M, Tanaka H, Iwane AH, Katayama E, Ikebe M, Yanagida T (2006) Cargo-binding makes a wild-type single-headed myosin-VI move processively. Biophys J 90:3643–3652

Iwaki M, Wickham SF, Ikezaki K, Yanagida T, Shih WM (2016) A programmable DNA origami nanospring that reveals force-induced adjacent binding of myosin VI heads. Nature communications 7:13715

Kitamura K, Tokunaga M, Iwane AH, Yanagida T (1999) A single myosin head moves along an actin filament with regular steps of 5.3 nanometres. Nature 397:129–134

Mehta AD, Rock RS, Rief M, Spudich JA, Mooseker MS, Cheney RE (1999) Myosin-V is a processive actin-based motor. Nature 400:590–593

Millar NC, Geeves MA (1983) The limiting rate of the ATP-mediated dissociation of actin from rabbit skeletal muscle myosin subfragment 1. FEBS Lett 160:141–148

Rief M, Rock RS, Mehta AD, Mooseker MS, Cheney RE, Spudich JA (2000) Myosin-V stepping kinetics: a molecular model for processivity. Proc Natl Acad Sci USA 97:9482–9486
Shiroguchi K, Kinosita K Jr (2007) Myosin V walks by lever action and Brownian motion. Science 316:1208–1212
Siemankowski RF, Wiseman MO, White HD (1985) ADP dissociation from actomyosin subfragment 1 is sufficiently slow to limit the unloaded shortening velocity in vertebrate muscle. Proc Natl Acad Sci USA 82:658–662
Svoboda K, Schmidt CF, Schnapp BJ, Block SM (1993) Direct observation of kinesin stepping by optical trapping interferometry. Nature 365:721–727
Yildiz A, Forkey JN, McKinney SA, Ha T, Goldman YE, Selvin PR (2003) Myosin V walks hand-over-hand: single fluorophore imaging with 1.5-nm localization. Science 300:2061–2065

Chapter 16
Novel Intermolecular Surface Force Unveils the Driving Force of the Actomyosin System

Makoto Suzuki, George Mogami, Takahiro Watanabe and Nobuyuki Matubayasi

Abstract In this chapter, we discuss the role of water in actomyosin-force generation. We have been investigating the hydration properties of ions, organic molecules, and proteins. These studies revealed that actin filaments (F-actin) are surrounded by a hyper-mobile water (HMW) layer and restrained water layer, while myosin subfragment 1 (S1) has only a typical restrained hydration layer. The understanding of the physicochemical properties of HMW has been greatly advanced by recent theoretical studies on statistical mechanics and solution chemistry. To explain the mechanism of force generation of actomyosin using ATP hydrolysis, we propose a driving force hypothesis based on novel intermolecular surface force. This hypothesis is consistent with the reported biochemical kinetics and thermodynamic parameters for the primary reaction steps. The gradient field of solvation free energy of S1 is generated in close proximity to F-actin.

M. Suzuki (✉)
Biological and Molecular Dynamics, Institute of Multidisciplinary Research for Advanced Materials (IMRAM), Tohoku University, Katahira 2-1-1, Aoba-ku, Sendai 980-8577, Japan
e-mail: makoto.suzuki.c5@tohoku.ac.jp

M. Suzuki
Department of Biomolecular Engineering, Graduate School of Engineering, Tohoku University, 6-6-07 Aoba, Aramaki, Aoba-ku, Sendai 980-8579, Japan

G. Mogami · T. Watanabe
Department of Materials Processing, Graduate School of Engineering, Tohoku University, 6-6-02 Aoba, Aramaki, Aoba-Ku, Sendai 980-8579, Japan
e-mail: mogami-g@material.tohoku.ac.jp

N. Matubayasi
Division of Chemical Engineering, Graduate School of Engineering Science, Osaka University, Toyonaka, Osaka 560-8531, Japan
e-mail: nobuyuki@cheng.es.osaka-u.ac.jp

N. Matubayasi
Elements Strategy Initiative for Catalysts and Batteries, Kyoto University, Katsura, Kyoto 615-8520, Japan

© Springer Nature Singapore Pte Ltd. 2018

M. Suzuki (ed.), *The Role of Water in ATP Hydrolysis Energy Transduction by Protein Machinery*, https://doi.org/10.1007/978-981-10-8459-1_16

Keywords ATP hydrolysis • Electric field effect • Hydration free energy Motor protein • Protein–water interaction

16.1 Introduction

What is the direct energy source driving the function of actomyosin, the motor protein system of muscle? Szent-Györgyi pointed out the energetic importance of water in muscle contraction (Szent-Györgyi 1951, 1956). In an electric motor, electric energy is converted into magnetic energy and then into mechanical work, where the direct energy source of force generation is magnetic energy. In a combustion engine, chemical energy is converted into thermal energy by combustion and then into mechanical work by high-pressure gas in the cylinder, where the direct energy source is thermal energy. In the actomyosin motor system, the direct energy source generating the driving force is unknown. Very recently, the authors proposed a physical basis of the actomyosin driving force (Suzuki et al. 2017).

First, we introduce the basic idea of intermolecular surface force. Let us consider two solutes 1 and 2 in water: solute 1 is a long cylindrical highly charged macromolecule with radius R_1 and solute 2 is a spherical macromolecule with radius R_2. The cylindrical macromolecule is a model of an actin filament (F-actin) or double-strand DNA, both of which are negatively charged under a physiological condition. The spherical macromolecule is a model of a myosin subfragment 1 (myosin S1) or DNA binding protein. The two solutes are placed at a separation distance r as shown in Fig. 16.1. Let us assume that the surrounding water layer of solute 1 of thickness h is a hyper-mobile water (HMW, Suzuki 2014) layer formed around F-actin. HMW detected around F-actin in water has a shorter dielectric relaxation time, or a higher polarization rate, than bulk water according to dielectric relaxation spectroscopy (DRS) studies (Kabir et al. 2003; Wazawa et al. 2011; Suzuki et al. 2016). Here h is set to be a few-fold larger than the solvent molecule diameter d_s. In contrast, solute 2 is not surrounded by HMW. For simplicity, the properties of the HMW are considered as uniform in the layer. Although both solutes 1 and 2 restrain one water layer, we omit them here for simplicity. The role of solute 1 is to form the HMW region in the solvent space of solute 2.

In such a non-uniform solvent space, let us consider the solvation free energy of solute 2, $\Delta\mu_2(r)$, at position r as shown in Fig. 16.1. If $\Delta\mu_2(r) < \Delta\mu_{2w}$, solute 2 is attracted to solute 1, where $\Delta\mu_{2w}$ is the solvation free energy of solute 2 in bulk water. We now provide a rough picture for the emergence of attractive interaction between the two solutes. When solute 2 is immersed in the HMW layer around solute 1 ($0 < r < h$), the change in $\Delta\mu_2(r)$, $\Delta\Delta\mu_2(r)$, should be roughly proportional to the immersed surface area ΔS_2 because solvation occurs at the solute surface as follows, using the solvation free energy of HMW-covered solute 1 $\Delta\mu_{1w}$.

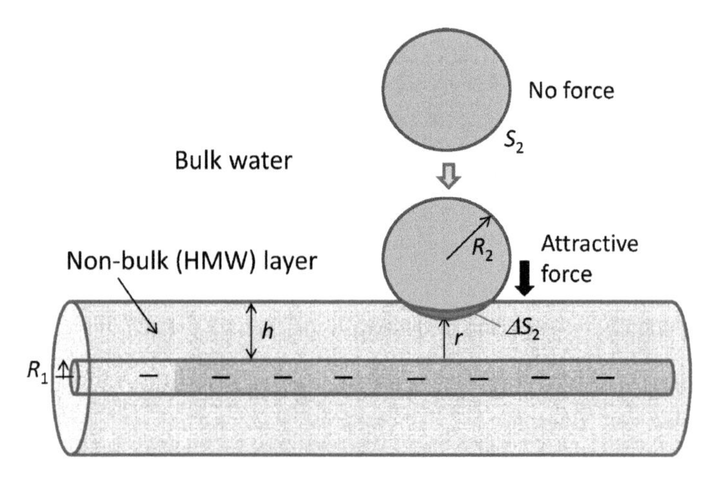

Fig. 16.1 Schematic image of intermolecular surface force

$$\Delta\Delta\mu_2(r) = \Delta\mu_2(r) - (\Delta\mu_{1\mathrm{w}} + \Delta\mu_{2\mathrm{w}}) \tag{16.1}$$

$$\approx \left[-\gamma_{\mathrm{ph}}\Delta S_2 - \gamma_{\mathrm{wh}}(S_1 - \Delta S_2) - \gamma_{\mathrm{pw}}(S_2 - \Delta S_2)\right] - \left(-\gamma_{\mathrm{wh}}S_1 - \gamma_{\mathrm{pw}}S_2\right) \tag{16.2}$$

$$= -\gamma\Delta S_2. \tag{16.3}$$

Here $\gamma = \gamma_{\mathrm{ph}} - (\gamma_{\mathrm{wh}} + \gamma_{\mathrm{pw}})$. γ_{ph}, γ_{wh}, and γ_{pw} are the interfacial tensions of solute 2-HMW, bulk water-HMW, and solute 2-bulk water, respectively.

Additionally, ΔS_2 was used as an approximation of the surface-area reduction of the cylindrical HMW layer when solute 2 was immersed in the layer. In this case, radial force F_r acting on solute 2 is calculated by Eq. 16.4.

$$F_r = -\partial/\partial r(\Delta\Delta\mu_2(r)). \tag{16.4}$$

$$\approx -2\pi R_2\gamma \qquad (2d_{\mathrm{s}} < r \text{ and } 0 \le h - r \ll R_2), \tag{16.5}$$

where the condition $2d_{\mathrm{s}} < r$ is introduced for the first water layers of solutes 1 and 2 to ensure that they do not overlap each other. Otherwise, an additional attractive force may appear between solutes 1 and 2 because of the excluded volume effect (Asakura and Oosawa 1954; Kinoshita 2009). It should be noted that the interface between the bulk water and HMW contains a few water layers over which the water property changes from bulk water to HMW. Thus, if $\gamma > 0$, the force between solutes 1 and 2 is attractive. Interfacial tension γ can be directly estimated as previously described (Suzuki et al. 2017) and is summarized in Sect. 16.3.

The driving mechanism of actomyosin motility was successfully explained utilizing this intermolecular surface force (Suzuki et al. 2017). As shown in Fig. 16.1, the force is in the radial direction of the cylindrical coordinate. In contrast, the longitudinal force along F-actin, $F_z = -\partial/\partial z(\Delta\Delta\mu_2(r))$, was discussed in the model and a myosin-induced asymmetrical modification of the periodic electric field was assumed along F-actin. As shown in Fig. 16.2, the water of ~1 nm thick in close proximity of F-actin is hyper-mobile water (HMW, colored in green), while the first water layer restrained by F-actin is not shown here. HMW has a higher response rate of polarization than bulk water (Kabir et al. 2003; Miyazaki et al. 2008; Mogami et al. 2011, 2013; Suzuki et al. 2016). a Free M.ATP rapidly reaches equilibrium with M.ADP.Pi before binding with F-actin. The red circle represents Pi and the dark blue circle denotes ADP. Since the Pi-release rate from M.ADP.Pi is slow (~0.04 s^{-1}) (Taylor 1977), the M.ADP.Pi state is expressed as M.ADP.Pi(c), where (c) represents the closed Pi-release tunnel. b Upon association of M.ADP.Pi (c) with F-actin AM.ADP.Pi(c) is formed. Then the isomerization reaction from b to c occurs. This model is consistent with reported biochemical and single-molecule experimental data (Bagshaw and Trentham 1974; Stein et al. 1981, 1984; Hibberd et al. 1985; Webb et al. 1986; Kawai and Halvorson 1991; Lawson et al. 2004; Takagi et al. 2004; Muretta et al. 2013, 2015). c AM.ADP.Pi(o), the Pi-release-tunnel open state, is formed upon interaction with F-actin, while the structure of F-actin is modified over several actin protomers (colored in pink) at the backside of bound myosin S1 and the state of water surrounding F-actin also

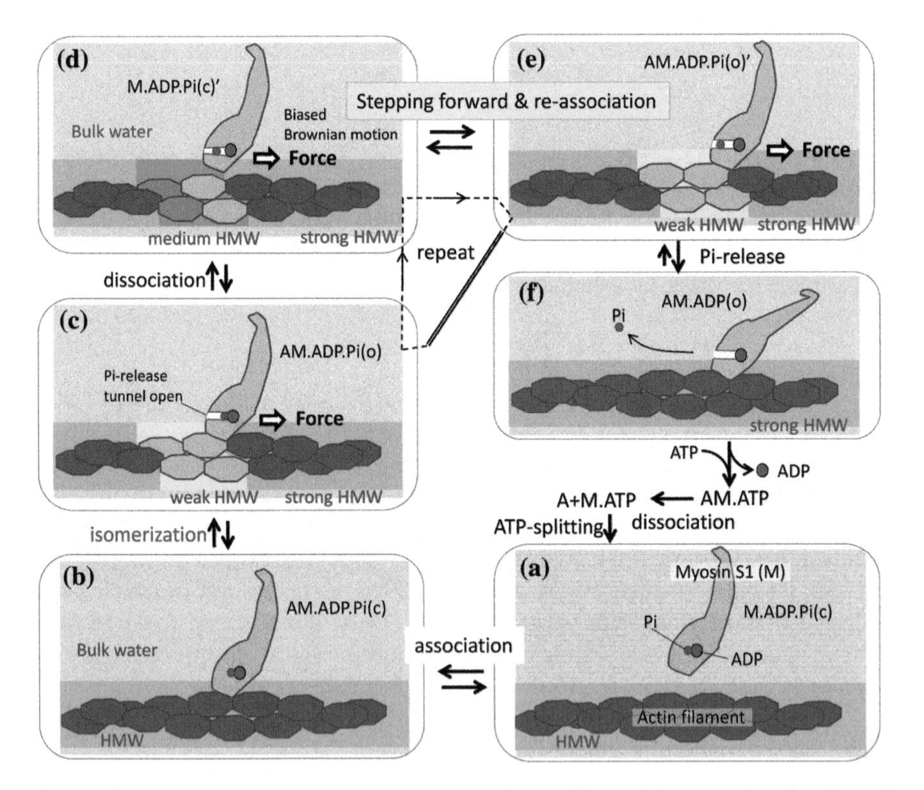

Fig. 16.2 Proposed driving mechanism of actomyosin (modified with permission from Suzuki et al. 2017. Copyright 2017 Wiley.)

changes into a weak HMW state. The myosin S1 is pulled to the right by the force exerted by the solvation free-energy gradient along F-actin. Inside myosin S1, the Pi tends to diffuse in the tunnel. d M.ADP.Pi(o) is readily dissociated from F-actin and drifts to the right under the influence of the driving force.

Subsequently, the structures of F-actin and M.ADP.Pi(o) should change toward the states of free F-actin and M.ADP.Pi(c), respectively, and the water changes with respect to the F-actin state. The changes are indicated by alterations in color from pink to light red for actin, light blue to blue for the surrounding water and from M. ADP.Pi(o) to M.ADP.Pi(c)' where the Pi moves in the tunnel. The Brownian displacement (Chauwin et al. 1994) of M.ADP.Pi(c)' can be affected by the applied load on myosin S1 and by the length of the lever arm which has a flexible joint at the end of the myosin S1. e M.ADP.Pi(c)' re-associates with F-actin at the more right position than c and becomes AM.ADP.Pi(o)' with the open tunnel. Simultaneously, F-actin changes its structure again similar to c, where the Pi comes closer to the exit. Until the Pi is released from AM.ADP.Pi or M.ADP.Pi, this cycle c-d-e-c accompanying displacement of myosin S1 toward the right can repeat as indicated by the dashed cycle. This model is consistent with experimental evidence of significant myofibril contraction before starting turn-over of ATP hydrolysis for 50 ms after addition of calcium (Ohno and Kodama 1991) and multi-step movement per ATP (Yanagida et al. 1985; Kitamura et al. 1999). f The AM.ADP state is a strong binding state, where the lever arm becomes straight after Pi-release (Volkmann et al. 2003) which may induce additional displacement between F-actin and myosin. Clockwise processes are dominant in the figure under the high-ATP and low-ADP conditions.

In this hypothesis, two assumptions were set. First, the periodic electric field distribution along an actin filament (F-actin) is unidirectionally modified upon binding of myosin subfragment 1 (myosin S1) with ADP and inorganic phosphate Pi (M.ADP.Pi complex). Second, the solvation free energy of myosin S1 depends on the external electric field strength and the solvation free energy of myosin S1 in close proximity to F-actin can become the potential force driving myosin S1 along F-actin. The first assumption is supported by a recent study (Ngo et al. 2015; Hirakawa et al. 2017), which demonstrated unidirectional cooperative binding of heavy meromyosin on F-actin and by the electric field and hydration analyses on F-actin, whose structure was altered by subtle stimulations such as exchanging Mg^{2+} and Ca^{2+} ions (Suzuki et al. 2016). The second assumption is supported by solvation free energy calculations on glycine and two small proteins utilizing molecular dynamics (MD) and the energy representation method (Karino and Matubayasi 2011; Sakuraba and Matubayasi 2014). More direct examination is required to determine the orientation dependence of the solute dipole and side-chain species on the electric field effect of $\Delta\mu$.

According to $\Delta\Delta\mu$ analysis for the 10^8 V m^{-1}-field effect (Suzuki et al. 2017), the interfacial tension γ in Eq. 16.3 can be estimated for horse heart cytochrome c (hh cyt c) to be +0.014 J m^{-2} which is 19% of the interfacial tension of water and air and 27% of that of water and hexane. Therefore, hh cyt c is attracted to the HMW layer by this interfacial tension, as shown in Fig. 16.1. The value of γ should change depending on the side-chain species and their arrangement on the protein surface. Assuming $\gamma = 0.014$ J m^{-2} and $R_2 = 2.5$ nm for the myosin head in Eq. 16.5, the radial attractive force reaches 220 pN. The electric field strength changes in the range of 10^7–10^8 V m^{-1} in the HMW layer. Assuming a proportional relationship between $\Delta\Delta\mu$ and electric field strength, the attractive force is 22–220 pN. The relationship between $\Delta\Delta\mu$ and electric field strength is described in the following section.

The occurrence of HMW in the proximity of F-actin was first detected in our dielectric relaxation spectroscopy (DRS) study (Kabir et al. 2003). The electric field strength around F-actin was estimated to be 10^7–10^8 V m^{-1}, which is in accordance with the theoretical conditions to form HMW around a charged particle (Kinoshita and Suzuki 2009; Kubota et al. 2012; Mogami et al. 2013, 2016). However, whether such a high field gradient in the longitudinal direction along F-actin is generated upon binding with myosin S1 with ADP and Pi (M.ADP.Pi) was not determined. In a previous study, the binding effect of myosin S1 with adenylyl-imidodiphosphate (M.AMPPNP) (Greene and Eisenberg 1980; Inoue and Tonomura 1980) on the F-actin hydration state was examined by DRS (Suzuki et al. 2017). Under low ionic conditions of 25 mM KCl, it was observed that the average HMW intensity showed a reduction of 20% per actin for the M.AMPPNP binding compared to that of free F-actin at a molar ratio of [myosin S1]/[actin] = 2/13. If four actin molecules and the surrounding HMW are modified by one myosin S1 binding as reported previously (Siddique et al. 2005), the reduction in the HMW intensity in proximity to the modified actin is estimated to be 33% of the value for actin before myosin binding. The cooperative structural change and polymorphic feature of F-actin are now widely known (Prochniewicz and Thomas 1997; Egelman and Orlova 1995; Galkin et al. 2010; Oda and Maeda 2010). A correlation between the F-actin structure and hydration state was also demonstrated (Suzuki et al. 2016). Based on the observed unidirectional structural changes in F-actin upon heavy meromyosin binding (Hirakawa et al. 2017), the first assumption is plausible.

The driving mechanism shown in Fig. 16.2 can be examined as the number dependence of bound myosin S1 on the hydration state of F-actin. From the cross-sectional view of myofilament with a hexagonal arrangement of myosin and actin filaments, three myosin filaments exist as the nearest neighbors of an F-actin (Woledge et al. 1985). The number of myosin S1 that can bind F-actin is estimated to be at most 2.6 per 36.5 nm of half-helical pitch of F-actin, assuming only one S1 of a double-headed heavy meromyosin can bind F-actin in the presence of ATP. Thus, five actin protomers are available for one myosin S1. According to the model shown in Fig. 16.2, the generated force per actin filament is proportional to the number of bound M.ADP.Pi unless the modified actin region (colored in pink) is overlapped by other bound M.ADP.Pi. The reported number of four actin protomers cooperatively modified per myosin S1 (Siddique et al. 2005) is consistent with the maximum S1 number of 2.6 per half-helical pitch of F-actin. Therefore, it is important to examine how the HMW around F-actin depends on the number of myosin S1 bound to F-actin and how the temperature dependence of the HMW state of F-actin is linked to force generation in the body. In this chapter, we examine these points.

16.2 Hydration State of Actomyosin

16.2.1 Preparation of Actomyosin for Hydration Measurements

Myosin S1 was prepared from chicken breast muscle using papain following the Margossian and Lowey method (1982). Myosin S1 was purified utilizing the hydrophobic column TOYOPEARL Butyl-650S (Tosoh, Tokyo, Japan) followed by further purification with the gel-permeation column Sephacryl 5-200 HR(GE Healthcare, Little Chalfont, UK). The purity of the myosin S1 solution was determined to be 87% using SDS-PAGE. The common solution conditions were 50 mM KCl, 10 mM HEPES–KOH pH 7.8, 2.0 mM $MgCl_2$, and 0.01 wt%NaN_3. The ATPase activity of myosin S1 was 0.04 s^{-1} at 20 °C. The concentration of myosin S1 in the range of 5–10 mg/mL was precisely determined using the absorption coefficient of 0.75 Abs/(mg/mL/cm) at 280 nm. The partial specific volumes (v_B) of M and M.ADP in solution were determined to be 0.742 ± 0.002 and 0.740 ± 0.008, respectively, utilizing an Anton Paar density meter DMA5000M (Graz, Austria). DRS measurements were performed using the microwave network analyzer 8364C (Keysight Technologies, Santa Rose, CA, USA) followed by analysis described previously (Suzuki et al. 2016).

Actin was prepared from chicken breast muscle by the Spudich and Watt method (1971). The purity of G-actin was observed to be 99% on SDS-PAGE results analyzed using the ImageJ software (Bethesda, MD, USA). G-actin solutions were prepared in a buffer containing 2 mM HEPES–KOH pH 7.8, 0.1 mM ATP, 0.1 mM $CaCl_2$, and 1 mM DTT. The polymerization of G-actin to F-actin was performed by increasing the KCl, $MgCl_2$, and HEPES-KOH (pH 7.6) to 50 mM (or 25 mM), 2 mM, and 10 mM, respectively, in the DRS measurement cell with gentle stirring for 10 min to avoid introducing air bubbles into the highly viscous solution. The measured actin and myosin S1 mixed solutions in the presence of ADP were mixtures of AM, AM.ADP, F-actin, M.ADP, and negligible free myosin S1. In the ADP case, the ratio of AM:AM.ADP:M.ADP was 16.5%:82.1%:1.4%, based on the binding constant, 1×10^6 M^{-1} for myosin S1 and ADP (Kodama 1985).

Although the sample solutions were not pure, the effect of the nucleotide species on the hydration state of the actin–myosin S1 complex could be deduced by combining the results with M.ADP and F-actin at a fixed ADP concentration of 1 mM. Subsequently, DRS measurements were performed at 20 °C. DRS parameters were utilized according to a previous report (Suzuki et al. 2016).

Acto-S1 solutions for DRS were prepared in a two-step procedure. First, actin in 10 mM HEPES-KOH (pH 7.5–7.6) containing 0.1 mM ATP, 0.1 mM $CaCl_2$, and 1 mM DTT was polymerized by adding KCl and $MgCl_2$ to 50 and 2 mM, respectively. S1 in 10 mM HEPES (pH 7.4) containing 0.1 mM $CaCl_2$, 50 mM KCl, and 2 mM $MgCl_2$ was then added and mixed. To ensure thorough mixing of the proteins at such high concentrations, the ATP concentration was raised to 0.2 mM just prior to mixing. The G-actin unit/S1 molar ratios adopted were 13:1, 13:2, and 13:3 which approximately corresponded to the state of an actin filament fully overlapping with myosin in the muscle.

16.2.2 Hydration Properties of Actomyosin S1 with ADP

The hydration state of F-actin was studied using high-resolution microwave DRS to determine the rotational response (DR frequency) of water hydrating proteins and the volume of hydration shells. The results indicated that both the HMW shell ($f_1 > f_w$) and restrained water shell ($f_2 < f_w$) were present around F-actin (Kabir et al. 2003), where f_w is the DR frequency of pure water. The molecular surface of actin is rich in negative charges, which along with its filamentous structure provides a structural basis for the induction of a hyper-mobile state of water.

The volume of the HMW component reportedly increased when myosin S1 was bound to F-actin (Suzuki et al. 2004), while no hyper-mobile component exists in the hydration shell of S1 itself or many other globular proteins (Suzuki et al. 1996; Yokoyama et al. 2001).

We performed hydration measurements of actomyosin by DRS to investigate the hydration state dependence of F-actin on the bound myosin S1 number to F-actin. The results are summarized in Table 16.1. The parameters were calculated after excluding the effect of bulk water by the volume fraction of 1- ϕ utilizing a mixture theory, where ϕ was set to 0.07 which was sufficiently larger than the volume fraction of water modified by the solute proteins of 10 mg/mL, as described previously (Suzuki et al. 2016). In all cases of actomyosin complexes, HMW was detected. The DR frequency of HMW decreased with an increasing number of myosin S1 bound to F-actin N_m, while the dispersion amplitude $\delta_1$increased with N_m, as shown in Fig. 16.3. The HMW intensity D_{hmw}, defined as $D_{hmw} = (f_1 - f_w)\delta_1/(f_w\delta_w)$, decreased with increasing N_m, as shown in the inset of Fig. 16.3. Because these data were estimated for a total protein concentration of 10 mg/mL, D_{hmw}/R indicates HMW intensity per actin molecule, where R is the weight ratio of actin in the actomyosin complexes. The 13:1 and 13:2 actomyosin complexes exhibited $D_{hmw}/R > 0.08$, while that of the 13:3 actomyosin complex decreased markedly, which is consistent with the cooperative structural change in F-actin

Table 16.1 Hydration properties of actomyosin (dielectric relaxation parameters at 20 °C, KCl 50 mM, and ϕ/c = 0.007 mL/mg)

Sample	Molar ratio (actin:S1)	f_2 GHz	δ_2	f_1 GHz	δ_1	δ_{bulk}	D^{a}_{hmw}
Myosin S1	0:1	5.1	6.1			55.7	0
SE		0.1	0.1	–	–	0.1	
F-actin	1:0	6.0	9.5	24.6	22.2	39.8	0.132
SE		0.4	0.6	0.4	2.0	1.5	0.012
Acto-S1.ADP	13:1	6.6	6.2	20.5	25.6	41.2	0.0704
SE		0.2	0.3	0.2	1.1	0.5	0.0030
Acto-S1.ADP	13:2	6.8	4.9	19.3	31.8	36.1	0.0574
SE		0.2	0.1	0.1	0.8	0.9	0.0014
Acto-S1.ADP	13:3	6.2	7.3	18.4	35.5	31.0	0.0390
SE		0.4	0.8	0.1	1.1	1.5	0.0012
F-actin K100[b]	1:0	5.4	7.3	24.2	17.8	45.0	0.099
SE		0.4	0.6	0.4	2.0	1.5	0.010
F-actin K25[c]	1:0	6.7	7.6	24.6	18.3	46.1	0.1091
SE		0.4	0.6	0.4	2.0	1.5	0.0092
Acto-S1.ADP K25[c]	13:2	3.9	4.3	20.8	25.6	41.3	0.0763
SE		0.2	0.1	0.3	2.4	2.6	0.0010
Acto-S1. AMPPNP K25[c]	13:2	4.9	3.1	22.1	13.9	53.3	0.0556
SE		0.5	0.5	2.2	4.7	5.1	0.0040

[a]$D_{hmw} = (f_1 - f_w)\ \delta_1/(f_w\delta_w)$. f_w: dielectric relaxation frequency of the bulk water (17 GHz at 20 °C), δ_w: dielectric dispersion amplitude of the bulk water (75 at 20 °C)
[b]KCl 100 mM
[c]KCl 25 mM (Suzuki et al. 2017)

found in a previous study (Siddique et al. 2005) and suggests overlap of the modified actin region, as shown in Fig. 16.1c–e.

In addition, the 50 mM KCl F-actin was found to form the highest D_{hmw} compared to the 25 and 100 mM KCl F-actin systems. F-actin, AM.ADP, and AM. AMPPNP (AMPPNP: adenylyl-imidodiphosphate) in the 25 mM KCl, 10 mM HEPES (pH 7.8), 0.1 mM $CaCl_2$, and 2 mM $MgCl_2$ buffer solution were displayed for comparison, where AM.AMPPNP exhibited the lowest D_{hmw} (Suzuki et al. 2017).

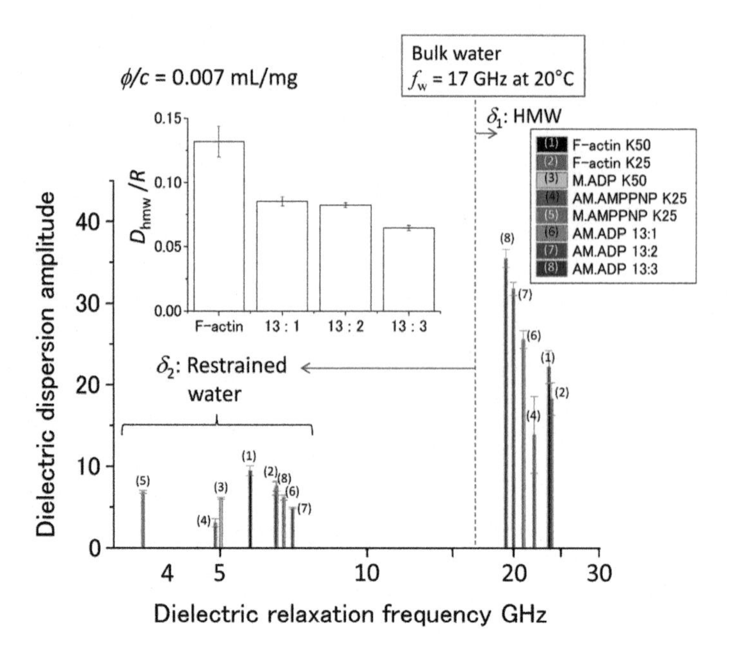

Fig. 16.3 Dielectric dispersion amplitudes of the hydration layer of actomyosin. δ_1: dispersion amplitude of HMW, δ_2: dispersion amplitude of restrained water. K25: KCl 25 mM, K50: KCl 50 mM. The inset shows the hyper-mobile water intensity D_{hmw} divided by the weight ratio R of actin in the actomyosin complexes depending on the number of bound myosin S1.ADP per 36.5 nm of F-actin on the average

16.3 Calculation of Solvation Free Energy of Glycine and Small Proteins Under an External Electric Field

16.3.1 Method of the Solvation Free Energy Calculation

Here, we review an MD study for the solvation free energies of glycine as a small polar organic molecule and bovine pancreatic trypsin inhibitor (BPTI, Mw 6513) and horse heart cytochrome *c* (hh cyt *c*, Mw 12,384) as two small proteins, in water under an external electric field as described previously (Suzuki et al. 2017). We performed an additional study of glycine in different orientations to the electric field for this chapter. Each solute and 1000–20,000 water molecules were placed in a cubic unit cell with the periodic boundary condition, and MD simulation was conducted over 5 ns at a sampling interval of 100 fs in the isothermal–isobaric (NPT) ensemble at 20 °C and 1 bar using the MD program package NAMD 2.11 (Phillips et al. 2005). The TIP3P model was used for water (Jorgensen et al. 1983), CHARMM22 (Brooks et al. 1983; MacKerell et al. 1998) for glycine and BPTI, and CHARMM36 (Best et al. 2012) for hh cyt *c*. Hh cyt *c* was described utilizing with a snapshot structure and the partial charges obtained in a previous MD simulation study (Yamamori et al. 2016) based on the Amber99sb force field (Hornak et al. 2006) and the RESP procedure. The electrostatic interaction was handled by the particle-mesh Ewald (PME) method (Essmann et al. 1995) with a real-space cutoff of 12 Å, a spline order of 4, an inverse decay length of 0.258 $Å^{-1}$, and a reciprocal-space mesh size of 96 for each of the x, y, and z directions. The Lennard–Jones (LJ) interaction was truncated by applying the switching function (Brooks et al. 1983) with the switching range of 10–12 Å. The Langevin dynamics was employed for temperature control at a damping coefficient of 5 ps^{-1} with the Brünger–Brooks–Karplus algorithm at a time step of 2 fs (Brünger et al. 1984; Van Gunsteren and Berendsen 1988). The pressure was maintained by the Langevin piston Nosé–Hoover method with barostat oscillation and damping time constants of 200 and 100 fs, respectively (Martyna et al. 1994; Feller et al. 1995). The water molecules were kept rigid with SETTLE (Miyamoto et al. 1992). MD simulations were run under application of a uniform static electric field with a strength range from 0 to 1×10^8 V m^{-1}. We utilized the "collective variable calculations" (colvars) module (Fiorin et al. 2013) of NAMD, where each solute was fixed to the initial structure and the dipole moment of each solute was aligned to the direction of the applied electric field.

16.3.2 Solvation Free Energy of a Polar Organic Molecule Under the Electric Field

The solvation free energy ($\Delta\mu$) of glycine was calculated under the external electric field E in the range 0–1×10^8 V m^{-1}. The non-dissociated glycine molecule was placed in the following four conditions: a fixed structure glycine with a dipole moment of 3.05 D is placed in parallel, anti-parallel, and perpendicular to the electric field, and a free structure glycine with an average dipole moment of 2.41 ± 0.71 D is placed allowing free rotation in the electric field. For a dipole moment p of 3D, the orientation energy ($-pE$) is −0.144 kcal/mol at 1×10^8 V m^{-1}, which is 25% of thermal energy, and the polarization energy ($-\alpha E^2/2$), where $\alpha = p^2/3kT$, is −0.006 kcal/mol, which is 1% of thermal energy.

Electric field dependence of solvation free energy $\Delta\mu$ of glycine was analyzed as shown in Fig. 16.4. $\Delta\mu$ was divided into two terms: the solute–solvent (two body) interaction energy $\Delta\mu_{\text{two}}$ and the solvent–solvent (many body) interaction energy $\Delta\mu_{\text{many}}$, where $\Delta\mu_{\text{two}}$ was calculated directly and $\Delta\mu_{\text{many}}$ was obtained as $\Delta\mu - \Delta\mu_{\text{two}}$ as described previously (Mogami et al. 2016). $\Delta\mu_{\text{two}}$ was largely negative for all cases at $|E| = 0$ and increased with increasing $|E|$, while $\Delta\mu_{\text{many}}$ was positive for all cases at $|E| = 0$ and decreased with $|E|$. There appeared a compensation between $\Delta\mu_{\text{two}}$ and $\Delta\mu_{\text{many}}$, which caused almost constant or very weak electric field dependence of $\Delta\mu$. Is this compensation observed in protein solutions? We performed similar analysis on the $\Delta\mu$ of horse heart cytochrome c (hh cyt c) in the next section in addition to a review of our previous study (Suzuki et al. 2017).

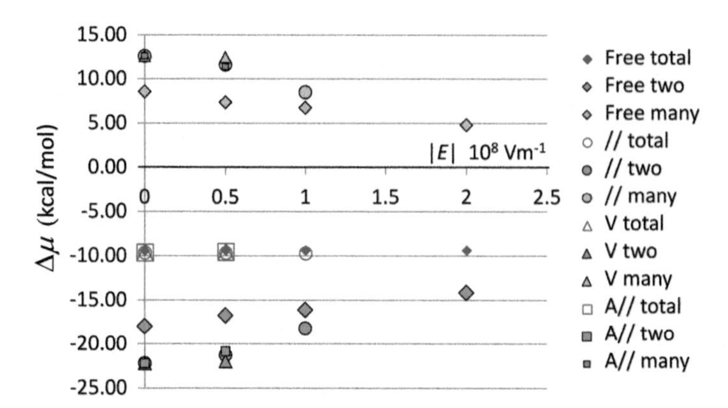

Fig. 16.4 Electric field dependence of constituent of solvation free energy of glycine. Total: total solvation free energy, two: solute–solvent interaction energy, many: solvent–solvent interaction energy, where "many" = "total" – "two". "Free" means rotationally not-restrained. The glycine dipole was fixed in parallel (//), perpendicular (V), and anti-parallel (A//) to *E*

16.3.3 Solvation Free Energy of Small Proteins Under the Electric Field

The solvation free energies ($\Delta\mu$) of small proteins were calculated under the two external electric field conditions ($|E| = 1 \times 10^8$ V m^{-1} or 0) (Suzuki et al. 2017). The results of the difference ($-\Delta\Delta\mu = -(\Delta\mu\ (E) - \Delta\mu\ (0))$ are shown in Fig. 16.5. The $\Delta\Delta\mu$ was found to be roughly proportional to the SASA of solute molecule, consistent with the understanding of the solvation free energy of a protein molecule

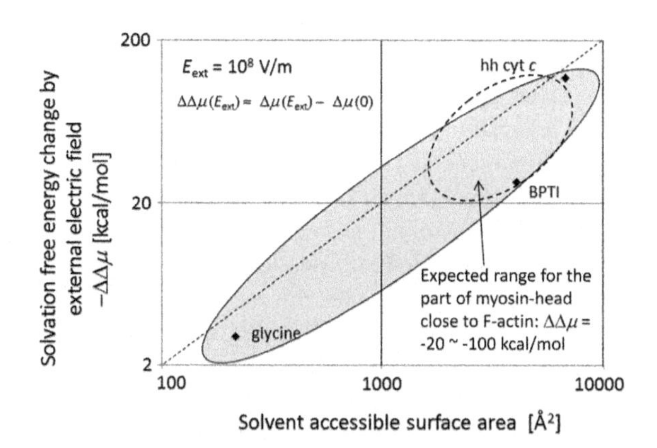

Fig. 16.5 The solvation free energy decrement vs. solvent accessible surface area (SASA) of small proteins and an amino acid when applying electric field of 1×10^8 V m^{-1}. Three solutes, glycine, BPTI (Mw 6513), and hh cyt *c* (Mw 12,384) were chosen to estimate the $\Delta\Delta\mu$ of unknown proteins. The expected proportional relationship between $-\Delta\Delta\mu$ and SASA was roughly confirmed. The dashed-circle indicates the expected range of $-\Delta\Delta\mu$ of the immersed part of the myosin S1 in the HMW layer in close proximity of F-actin. (Reprinted with permission from Suzuki et al. 2017. Copyright 2017 Wiley.)

that is roughly proportional to the SASA (Oobatake and Ooi 1988, 1993). The interfacial tension γ for hh cyt *c* between the bulk and high-electric-field (HMW) phases can be estimated to be 12 mN m^{-1} calculated as $-\Delta\Delta\mu$ (118 kcal/mol) divided by 6753 Å^2. By Eq. (16.5), the attraction force strength is roughly 110 pN with R_2 of 1.5 nm in Fig. 16.1.

In Fig. 16.6, the SASA of the HMW-immersed part of myosin S1 bound to F-actin is roughly estimated to a range from 1600 to 6500 Å^2, which is 1/20th to 1/5th of the SASA of S1dC.ADP (32460 Å^2; pdb:1mma) analyzed using the UCSF Chimera software package (2017). The SASA of glycine is ~200 Å^2. The $\Delta\Delta\mu$ of myosin S1, induced by a difference in the electric field strength, (e.g., a variation of 0.3×10^8 V m^{-1}) between the front- and backsides of myosin S1 along F-actin, becomes 1.2–4.8 × 10^{-20} J (Suzuki et al. 2017), as shown in Fig. 16.6.

The driving force acting on myosin S1 was estimated from $\Delta\Delta\mu$ divided by the diameter of myosin S1 ~ 5 nm to generate 2.4–9.6 pN. The rationale for using the diameter (*d*) of myosin S1 is that the free energy transfer of myosin S1 from the low-HMW to the high-HMW can be calculated from the difference of $\Delta\mu$ at $-d/2$ and $+d/2$ from the center of myosin S1 along F-actin as shown in Fig. 16.6. Since the electric field distribution around F-actin is produced by the surface charges of F-actin, the reaction force acts on F-actin itself in the reverse direction through the electric field. The force due to the change in the solvation free energy is a kind of surface force. Thus, the previous surface force models (Suzuki 1994, 2004) could simulate the double-hyperbolic force-velocity relation of muscle contraction (Edmann 1988). Indeed, the estimated force range agrees with the experimental value of approximately 3 pN (Ishijima et al. 1991).

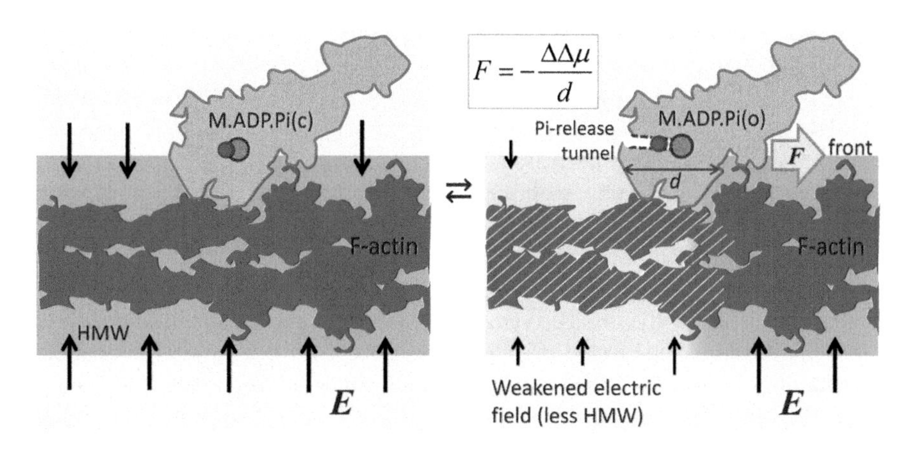

Fig. 16.6 Force generation model of actomyosin utilizing the gradient field of the solvation free energy of myosin S1. Negatively charged F-actin produces an electric field high enough to generate HMW (green) in proximity to F-actin. Upon binding of myosin S1 with ADP and Pi the HMW intensity decreases, producing the $\Delta\mu$ difference of myosin S1 between the front and backsides of myosin S1 bound to F-actin. (Reprinted with permission from Suzuki et al. 2017. Copyright 2017 Wiley.) Actomyosin structure was modified from Fujii and Namba 2017

Table 16.2 Electric field dependence of solvation free energy of hh cyt *c* at $|E| = 1 \times 10^8$ V m^{-1}, utilizing the method of Fig. 16.4. (in kcal/mol)

	$\Delta\mu_{total}(0)$	$\Delta\mu_{two}(0)$	$\Delta\mu_{many}(0)$	$\Delta\mu_{total}(E)$	$\Delta\mu_{two}(E)$	$\Delta\mu_{many}(E)$
Glycine	−9.8	68.8	−78.6	−12.6	58.6	−71.2
hh cyt *c*	−1460.4	−4001.1	2540.7	−1578.1	−4012.3	2434.1

	SASA Å^2	*p* D	$\Delta\Delta\mu^{a}_{total}$	$\Delta\Delta\mu^{a}_{two}$	$\Delta\Delta\mu^{a}_{many}$
Glycine	216	3.05	−2.9	−10.2	7.3
hh cyt *c*	6753	~240[b]	117.8	−11.2	−106.6

[a] $\Delta\Delta\mu_x = \Delta\mu_x(E) - \Delta\mu_x(0)$, (x = total, two, and many)
[b] from (Takashima 2002)

In Table 16.2, we analyzed the electric field dependence of solvation free energy of hh cyt *c*. $\Delta\mu_{two}$ was largely negative at $|E| = 0$ and decreased with increasing $|E|$, while $\Delta\mu_{many}$ was positive at $|E| = 0$ and further decreased with $|E|$ compared to $\Delta\mu_{two}$. Although there is some compensation for $\Delta\mu_{total}$ between $\Delta\mu_{two}$ and $\Delta\mu_{many}$, the marked decrease in $\Delta\mu_{many}$ with $|E|$ reduces the compensation and $\Delta\mu_{total}$ largely. This mechanism of electric field dependence of the solvation free energy of proteins is an attractive issue for the next stage.

16.4 Conclusion

In this chapter, we reviewed our recently proposed model of the driving force for actomyosin based on a novel intermolecular surface force and described the aspects of this force that had remained unclear so far. First, the number dependence of myosin S1 bound to F-actin per 13 actin protomers (half the helical pitch of F-actin) on the hydration state of F-actin was examined by DRS. The result showed that a high-HMW level was maintained upon two myosins S1 binding per 13 actin protomers and was generally decreased by three myosins S1 binding per 13 actin protomers. This is compatible with the internal structural arrangement of actin and myosin arrangement in a myofilament, giving a maximum binding number of 2.6 per 13 actin protomers. Second, the orientation effect of the solute dipole on the solvation free energy in an external field of 10^7–10^8 V m^{-1} was examined for the small polar molecule glycine. The vertical orientation of glycine to the field showed a non-significant change in the solvation free energy, while a parallel/anti-parallel orientation showed a decrease/an increase in the solvation free energy.

On the other hand for the small proteins, the solvation free energy of protein molecules clearly decreased when an electric field was applied. The difference between a protein and small polar molecule will be studied in future studies.

The intermolecular surface force due to the electric field effect on solvation free energy is strong and can be applied to understand actomyosin motility and versatile biological protein machineries.

References

Asakura S, Oosawa F (1954) On interaction between two bodies immersed in a solution of macromolecules. J. Chem. Phys. 22:1255–1256
Bagshaw CR, Trentham DR (1974) The characterization of Myosin-product complexes and of product-release steps during the magnesium ion-dependent adenosine triphosphatase reaction. Biochem J 141:331–349
Best RB, Zhu X, Shim J, Lopes PEM, Mittal J, Feig M, MacKerell Jr AD (2012) Optimization of the additive CHARMM all-atom protein force field targeting improved sampling of the backbone φ, ψ and side-chain $\chi 1$ and $\chi 2$ dihedral angles. J Chem Theory Comput 8:3257–3273
Brooks BR, Bruccoleri RE, Olafson BD, States DJ, Swaminathan S, Karplus M (1983) CHARMM: a program for macromolecular energy, minimization, and dynamics calculations. J Comput Chem 4:187–217
Brünger A, Brooks IIICL, Karplus M (1984) Stochastic boundary conditions for molecular dynamics simulations of ST2 water. Chem Phys Lett 105:495–500
Chauwin JF, Ajdari A, Prost J (1994) Force-free motion in asymmetric structures: a mechanism without diffusive steps. Europhys Lett 27:421–426
Edmann KAP (1988) Double hyperbolic force-velocity relation in frog muscle fibers. J Physiol 404:301–321
Egelman EH, Orlova A (1995) New insights into actin filament dynamics. Curr Opin Struct Biol 5:172–180
Essmann U, Perera L, Berkowitz ML, Darden T, Lee H, Pedersen JG (1995) A smooth particle mesh Ewald method. J Chem Phys. 103:8577–8593
Feller SE, Zhang Y, Pastor RW, Brooks BR (1995) Constant pressure molecular dynamics simulation: The Langevin Piston method. J Chem Phys 103:4613–4621
Fiorin G, Klein ML, Hénin J (2013) Using collective variables to drive molecular dynamics simulations. Mol Phys 111:3345–3362
Fujii T, Namba K (2017) Structure of actomyosin rigour complex at 5.2 Å resolution and insights into the ATPase cycle mechanism. Nat Commun 8 (1–11):13969
Galkin VE, Orloval A, Schröder GF, Egelman EH (2010) Structural polymorphism in F-actin. Nat Struct Mol Biol 17:1318–1324
Greene LE, Eisenberg E (1980) Dissociation of the Actin ~ Subflagment1 Complex by Adeny1-5'-yl Imidodiphosphate, ADP, and PPi. J Biol Chem 255:543–548
Hibberd MG, Dantzig JA, Trentham DR, Goldman YE (1985) Phosphate release and force generation in skeletal muscle fibers. Science 228:1317–1319
Hirakawa R, Nishikawa Y, Uyeda TQP, Tokuraku K (2017) Unidirectional growth of HMM clusters along actin filaments revealed by real time fluorescence microscopy. Cytoskeleton 74:482–489. https://doi.org/10.1002/cm.21408
Hornak V, Abel R, Okur A, Strockbine B, Roitberg A, Simmerling C (2006) Comparison of multiple Amber force fields and development of improved protein backbone parameters. Proteins Struct Funct, Bioinf 65:712–725
Inoue A, Tonomura Y (1980) Dissociation of Acto ~ H-Meromyosin and That of Acto-Subfragment-l Induced by Adenyl-5'-yl-imidodiphosphate: Evidence for a Ternary Complex of F-Actin, Myosin Head, and Substrate. J Biochem 88:1643–1651
Ishijima A, Doi T, Sakurada K, Yanagida T (1991) Sub-piconewton force fluctuations of actomyosin in vitro. Nature 352:301–306
Jorgensen WL, Chandrasekhar J, Madura JD, Impey RW, Klein ML (1983) Comparison of simple potential functions for simulating liquid water. J Chem Phys 79:926–935
Kabir SR, Yokoyama K, Mihashi K, Kodama T, Suzuki M (2003) Hyper-mobile water is induced around actin filaments. Biophys J 85:3154–3161
Karino Y, Matubayasi N (2011) Communication: free-energy analysis of hydration effect on protein with explicit solvent: equilibrium fluctuation of cytochrome c. J Chem Phys 134(1–11): 041105

Kawai M, Halvorson HR (1991) Two step mechanism of phosphate release and the mechanism of force generation in chemically skinned fibers of rabbit psoas muscle. Biophys J 59:329–342
Kinoshita M (2009) Importance of translational entropy of water in biological self-assembly processes like protein folding. Int J Mol Sci 10:1064–1080
Kinoshita M, Suzuki M (2009) A statistical-mechanical analysis on the hypermobile water around a large solute with high surface charge density. J Chem Phys 130(1–11):014707
Kitamura K, Tokunaga M, Iwane AH, Yanagida T (1999) A single myosin head moves along an actin filament with regular steps of 5.3 nanometres. Nature 397(6715):129–134
Kodama T (1985) Thermodynamic analysis on muscle ATPase mechanisms. Physiol Rev 65: 467–551
Kubota Y, Yoshimori A, Matubayasi N, Suzuki M, Akiyama R (2012) Molecular dynamics study of fast dielectric relaxation of water around a molecular-sized ion. J Chem Phys 137(1–4): 224502
Lawson JD, Pate E, Rayment I, Yount RG (2004) Molecular dynamics analysis of structural factors influencing the Pi-release tunnel in myosin. Biophys J 86:3794–3803
MacKerell AD Jr, Bashford D, Bellott M, Dunbrack RL Jr, Evanseck JD, Field MJ, Fischer S, Gao J, Guo H, Ha S et al (1998) All-atom empirical potential for molecular modeling and dynamics studies of proteins. J Phys Chem B 102:3586–3616
Margossian SS, Lowey S (1982) Preparation of myosin and its subfragments from rabbit skeletal muscle. San Diego, California: Elsevier, Inc. Methods Enzymol 85 Pt B:55–71
Martyna GJ, Tobias DJ, Klein ML (1994) Constant pressure molecular dynamics algorithms. J Chem Phys 101:4177–4189
Miyamoto S, Kollman PA (1992) SETTLE: an analytical version of the SHAKE and RATTLE algorithm for rigid water models. J Comput Chem 13:952–962
Miyazaki T, Wazawa T, Mogami G, Kodama T, Suzuki M (2008) Measurement of the dielectric relaxation property of water-ion loose complex in aqueous solutions of salt at low concentrations. J Phys Chem A 112:10801–10806
Mogami G, Wazawa T, Morimoto N, Kodama T, Suzuki M (2011) Hydration properties of adenosine phosphate series as studied by microwave dielectric spectroscopy. Biophys Chem 154:1–7
Mogami G, Miyazaki T, Wazawa T, Matubayasi N, Suzuki M (2013) Anion-dependence of fast relaxation component in Na-, K-halide solutions at low concentrations measured by high-resolution microwave dielectric spectroscopy. J Phys Chem A 117:4851–4862
Mogami G, Suzuki M, Matubayasi N (2016) Spatial-decomposition analysis of energetics of ionic hydration. J Phys Chem B 120:1813–1821
Muretta JM, Petersen KJ, Thomas DD (2013) Direct real-time detection of the actin-activated power stroke within the myosin catalytic domain. Proc Natl Acad Sci USA 110:7211–7216
Muretta JM, Rohde JA, Johnsrud DO, Cornea S, Thomas DD (2015) Direct real-time detection of the structural and biochemical events in the myosin powerstroke. Proc Natl Acad Sci USA 112:14272–14277
Ngo KX, Kodera N, Katayama E, Ando T, Uyeda TQP (2015) Cofilin-induced unidirectional cooperative conformational changes in actin filaments revealed by high-speed atomic force microscopy. eLife(1–22):4:e04806
Oda T, Maeda Y (2010) Multiple conformations of F-actin. Structure 18:761–767
Oobatake M, Ooi T (1988) Characteristic thermodynamic properties of hydrated water for 20 amino acid residues in globular proteins. J Biochem 104:433–439
Oobatake M, Ooi T (1993) Hydration and heat stability effects on protein unfolding. Prog Biophys Mol Biol 59:237–284
Ohno T, Kodama T (1991) Kinetics of adenosine triphosphate hydrolysis by shortening myofibrils from rabbit psoas muscle. J Physiol 441:685–702
Phillips JC, Braun R, Wang W, Gumbart J, Tajkhorshid E, Villa E, Chipot C, Skeel RD, Kale L, Schulten K (2005) Scalable molecular dynamics with NAMD. J Comput Chem 26:1781–1802

Prochniewicz E, Thomas DD (1997) Perturbations of functional interactions with myosin induce long-range allosteric and cooperative structural changes in actin. Biochemistry 36:12845–12853

Sakuraba S, Matubayasi N (2014) ERmod: fast and versatile computation software for solvation free energy with approximate theory of solutions. J Comp Chem 35:1592–1608

Siddique MSP, Mogami G, Miyazaki T, Katayama E, Uyeda TQP, Suzuki M (2005) Cooperative structural change of actin filaments interacting with activated myosin motor domain, detected with copolymers of pyrene-labeled actin and acto-S1 chimera protein. Biochem Biophys Res Commun 337:1185–1191

Spudich JA, Watt S (1971) The regulation of rabbit skeletal muscle contraction. 1. Biochemical studies of the interaction of the tropomyosin-troponin complex with actin and the proteolytic fragments of myosin. J Biol Chem 246:4866–4871

Stein LA, Chock PB, Eisenberg E (1981) Mechanism of the actomyosin ATPase: effect of actin on ATP hydrolysis step. Proc Natl Acad Sci USA 78:1346–1350

Stein LA, Chock PB, Eisenberg E (1984) The rate-limiting step in the actomyosin adenosinetriphosphate cycle. Biochemistry 23:1555–1563

Suzuki M (1994) New concept of a hydrophobicity motor based on local hydrophobicity transition of functional polymer substrate for micro/nano machines. Polym Gels Networks 2:279–287

Suzuki M, Shigematsu J, Kodama T (1996) Hydration study of proteins in solution by microwave dielectric analysis. J Phys Chem 100:7279–7282

Suzuki M (2004) Actomyosin motor mechanism: affinity gradient surface force model. Prog Coll Polym Sci 125:38–41

Suzuki M, Kabir SR, Siddique MSP, Nazia US, Miyazaki T, Kodama T (2004) Myosin-induced volume increase of the hyper-mobile water surrounding actin filaments. Biochem Biophys Res Commun 322:340–346

Suzuki M (2014) What is "hypermobile" water?: detected in alkali halide, adenosine phosphate and F-actin solutions by high-resolution microwave dielectric spectroscopy. Pure Appl Chem 86:181–189

Suzuki M, Imao A, Mogami G, Chishima R, Watanabe T, Yamaguchi T, Morimoto N, Wazawa T (2016) Strong dependence of hydration state of F-actin on the bound $Mg2^{+}/Ca2^{+}$ ions. J Phys Chem B 120:6917–6928

Suzuki M, Mogami G, Ohsugi H, Watanabe T, Matubayasi N (2017) Physical driving force of actomyosin motility based on the hydration effect. Cytoskeleton 74:512–527. https://doi.org/10.1002/cm.21417

Szent-Gyorgyi A (1951) Nature of the contraction of muscle. Nature 4245(1951):380–381

Szent-Gyorgyi A (1956) Bioenerg Sci 124:873–875

Takagi Y, Shuman H, Goldman YE (2004) Coupling between phosphate release and force generation in muscle actomyosin. Philos Trans R Soc Lond B Biol Sci 359:1913–1920

Takashima S (2002) Electric dipole moments of globular proteins: measurement and calculation with NMR and X-ray databases. J Non-Cryst Solids 305:303–310

Taylor EW (1977) Transient phase of adenosine triphosphate hydrolysis by myosin, heavy meromyosin, and head. Biochemistry 17:732–740

UCSF Chimera. http://www.cgl.ucsf.edu/chimera/. Accessed 18 July 2017

Van Gunsteren WF, Berendsen HJC (1988) A Leap-frog algorithm for stochastic dynamics. Mol Simul 1:173–185

Volkmann N, Ouyang G, Trybus KM, DeRosier DJ, Lowey S, Hanein D (2003) Myosin isoforms show unique conformations in the actin-bound state. Proc Nat Acad Sci USA 100:3227–3232

Wazawa T, Sagawa T, Ogawa T, Morimoto N, Suzuki M (2011) Hyper-mobility of water around actin filaments revealed using pulse-field gradient spin-echo 1H-NMR and fluorescence spectroscopy. Biochem Biophys Res Commun 404:985–990

Webb MR, Hibberd MG, Goldman YE, Trentham DR (1986) Oxygene exchange between Pi in the medium and water during ATP hydrolysis mediated by skinned fibers from rabbit skeletal muscle: evidence for Pi binding to a force-generating state. J Biol Chem 261:15557–15564

Woledge RC, Curtin NA, Homsher E (1985) Energetic aspects of muscle contraction, Chap 3. Academic Press, London, pp 119–165

Yamamori Y, Ishizuka R, Karino Y, Sakuraba S, Matubayasi N (2016) Interaction-component analysis of the hydration and urea effects on cytochrome c. J Chem Phys 144:085102

Yanagida T, Arata T, Oosawa F (1985) Sliding distance of actin filament induced by a myosin crossbridge during one ATP hydrolysis cycle. Nature 316:366

Yokoyama K, Kamei T, Minami H, Suzuki H (2001) Hydration study of globular proteins by microwave dielectric spectroscopy. J Phys Chem B 105:12622–12627

Chapter 17
Extremophilic Enzymes Related to Energy Conversion

Satoshi Wakai and Yoshihiro Sambongi

Abstract Across the Earth, a variety of organisms inhabit both mild and extreme environments wherever liquid water is available. Among these, extremophilic microorganisms, termed extremophiles, favorably live in extreme environments by adapting their physiological properties. Such extremophiles must acquire energy in order to maintain their cell homeostasis, which is functionally similar to organisms living in mild environments. Numerous enzyme proteins from extremophiles such as thermophiles, psychrophiles, piezophiles, and halophiles have been investigated to date, revealing both unity and diversity in their biochemical and structural biological features through comparison with their homologous counterpart enzymes from organisms living in mild environments. In this chapter, we aim to summarize the biochemical and thermodynamic aspects of enzymes related to the energy conversion that occurs in extremophiles. The obtained insights into extremophilic enzymes related to energy conversion thereby allow us to decipher the mechanistic fundamentals of these protein machineries.

Keywords Extremophile · Enzyme · Energy conversion · Biochemistry Thermodynamics

17.1 Introduction

Water in liquid form is essential for living organisms. A variety of organisms inhabit almost all environments across the Earth where liquid water is available. Organisms living in extreme environments (e.g., high and low temperature, high pressure, high salinity, acidic or alkaline pH conditions) are termed extremophiles.

S. Wakai (✉)
Graduate School of Science, Technology, and Innovation, Kobe University, Kobe, Japan
e-mail: wakaists@pegasus.kobe-u.ac.jp

Y. Sambongi
Graduate School of Biosphere Science, Hiroshima University, Hiroshima, Japan

© Springer Nature Singapore Pte Ltd. 2018

M. Suzuki (ed.), *The Role of Water in ATP Hydrolysis Energy Transduction by Protein Machinery*, https://doi.org/10.1007/978-981-10-8459-1_17

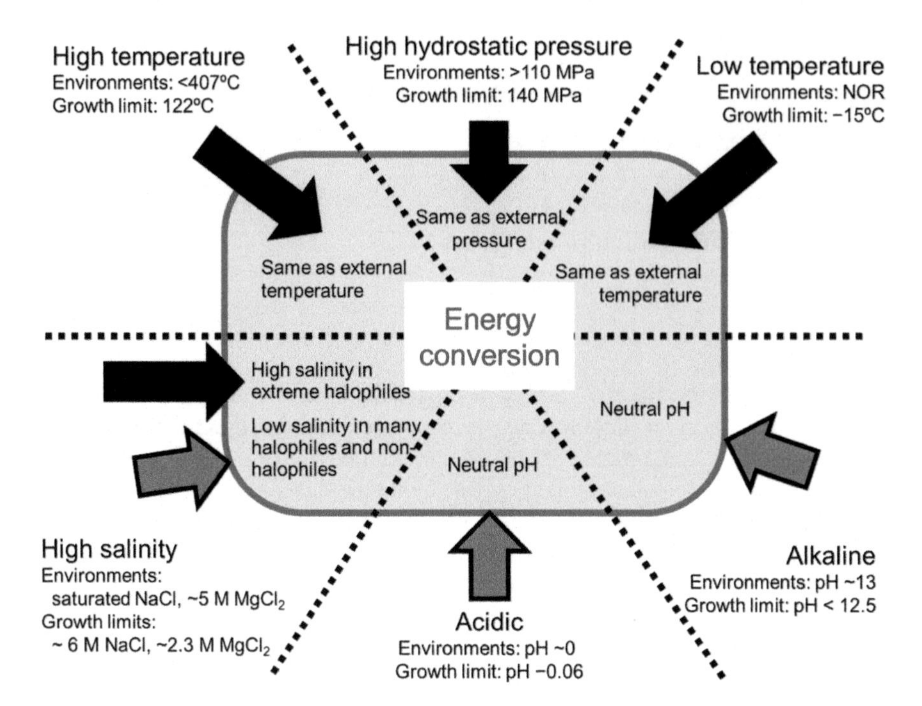

Fig. 17.1 Influences of extreme physicochemical factors on biological energy conversion. Black arrows represent physicochemical parameters which transfer across the cellular membrane. Gray arrows represent chemical parameters which block by the cellular membrane. Each value of "Environments" and "Growth limit" represents maximal value observed in each extreme environment where liquid water exists and upper or lower limit value which each extremophile can grow, respectively; NOR represents non-official record

Physicochemical properties of extreme environments, such as high temperature and hydrostatic pressure, may influence the cellular components of extremophiles not only exposed to the external environment but also existing inside the cell (Fig. 17.1). Extremophiles must acquire energy to maintain cellular homeostasis by means of enzymes related to energy conversion, which must adapt to such harsh conditions. Through deeper understanding of the energy conversion enzymes isolated from extreme environments, of which for some the availability of liquid water is a limitation, we can thereby decipher the mechanistic fundamentals of such protein machineries.

Adaptation strategies of extremophilic enzymes have been studied extensively (Adams et al. 1995; Stetter 1999; Madern et al. 2000; Demirjian et al. 2001; Feller and Gerday 2003; Dubnovitsky et al. 2005; Zhang and Ge 2013; Graziano and Merlino 2014). For example, enzymes from thermophiles and piezophiles exhibit higher stability against heat and pressure than those from mesophiles and pressure-sensitive organisms, and those from halophiles have adapted to be active under high salt conditions. These studies have primarily been carried out in view of biochemistry, protein engineering, and structural biology. Recently, understanding of the adaptation strategies of extremophilic enzymes has also progressed through mutagenesis experiments in conjunction with crystal structure analysis.

For further understanding of the functioning mechanisms of such protein machinery, the relationships among the enzyme protein itself, substrate, and solvent must be investigated in solution. Substrate dissolved in solvent binds to the reaction center of the enzyme protein, and its product is in turn re-dissolved into the solvent after catalytic reaction. Such phenomena occurring in solution can be evaluated by means of thermodynamic analysis, such as calorimetric measurement. However, little is known regarding the thermodynamic aspect of energy conversion reactions that occur under the extreme conditions in which extremophilic enzymes generally function.

In this chapter, we first describe extremophilic environments and microorganisms in order to share the latest knowledge regarding the limitations of life. We next discuss the adaptation mechanisms of extremophilic enzymes to each extreme environment. Finally, we focus on the thermophilic and halophilic enzymes related to energy conversion, which illuminates fundamentals of the mechanisms of such protein machineries.

17.2 Extremophilic Environments and Microorganisms

An environment suitable for human life is mild and limited to narrow regions. In contrast, extreme environments where humans cannot live unprotected are found across broad areas, in which various types of extremophiles reside if water is available in liquid form (Table 17.1). In this section, extreme environments are described with regard to three physical aspects, high and low temperatures and high pressure, and three chemical aspects, salinity, acidic and alkaline pH, with provision of maximal recorded conditions and probabilities of the presence of liquid water.

17.2.1 Extreme Environments on the Earth

High temperature—Maximal temperature described to date in regions where liquid water exists on the Earth is 407 °C (transient 464 °C) in deep-sea hydrothermal vents (Bischoff and Rosenbauer 1988; Koschinsky et al. 2008); such environments are also high pressure to maintain water in the liquid form at atmospheric pressure. High-temperature environments are also found in hot springs.

Low temperature—Lowest temperature recorded at atmospheric pressure is −89.2 °C where water is in the ice form (Feller 2010). Although an official record of lowest temperature in liquid water is unknown, the presence of liquid water at subzero degrees Celsius has been observed in various environments, such as seawater in polar sites.

Table 17.1 Extreme environments across the Earth where liquid water is available and extremophiles

Physicochemical parameters	Environments		Organisms		References
	Examples	Limits	Examples	Growth limits	
High temperature	Deep-sea hydrothermal vents and hot springs	407 °C in the area at 5° S on Mid-Atlantic Ridge (transient 464 °C)	*Methanopyrus kandleri*	122 °C	Bischoff and Rosenbauer (1988), Koschinsky et al. (2008), Takai et al. (2008)
Low temperature	Polar sites	Non-official record (−89.2 °C official records at atmosphere in Antarctica)	*Planococcus halocryophilus*	−15 °C	Feller (2010), Mykytczuk et al. (2013)
High pressure	Deep-sea and the Earth's deep crust	110 MPa in the Challenger Deep of the Mariana Trench	*Colwellia marinimaniae*	140 MPa	Kato et al. (1997), Kusube et al. (2017)
High salinity	Salt lakes, saltern soils, and deep-sea basin	Saturated NaCl 5 M $MgCl_2$ at the discovery basin in the Eastern Mediterranean Sea	Many extremely halophilic archaea (in saturated NaCl) and *Halobacterium sodomense* (in $MgCl_2$)	Saturate (NaCl) 2.3 M $MgCl_2$	van der Wielen et al. (2005), Hallsworth et al. (2007), Oren (1983)
Acidic pH	Acidic hot springs, acidic mines, and solfataric fields	~0 at the solfataric spring, sulfurous deposits, dry, hot soil in the solfataric field in Japan	*Picrophilus oshimae* KAW2/2	−0.06	Schleper et al. (1995a, b)
Alkaline pH	Alkaline lakes, ground water, and gold mines	~13 at the Gorka pit lake in Poland	*Alkaliphilus transvaalensis* SAGM1	12.5	Czop et al. (2011), Takai et al. (2001)

High pressure—Highest hydrostatic pressure is 110 MPa in the Challenger Deep of the Mariana Trench (Kato et al. 1997). The Earth's deep crust infiltrated with water also constitutes a high-hydrostatic-pressure environment, although the pressure values are not recorded. It is likely that liquid water exists in such high hydrostatic pressure crust, considering that huge numbers of microorganisms live in the oceans and terrestrial subsurfaces (Whitman et al. 1998).

High salinity—Among hyper-saline environments, saturated NaCl environments are observed in, e.g., salt lakes and saltern soils. It should be noted that $MgCl_2$-rich environments also exist on the Earth. For example, the concentration of $MgCl_2$ in the water of the Discovery basin in the Eastern Mediterranean Sea reaches to approximately 5 M (van der Wielen et al. 2005).

Acidic pH—Strongly acidic environments (<pH 3) are observed in mining sites of coal and metal ores. Extremely acidic environments (around zero pH) are also observed in acidic hot springs and solfataric fields, which are formed by high concentrations of sulfuric acid (Schleper et al. 1995a). Many of these extremely acidic environments are also high-temperature environments.

Alkaline pH—Alkaline waters in natural environments are frequently observed in alkaline lakes and groundwater discharging from ultramafic rocks, whereas the strongest alkaline water is derived from mining pit lakes formed by human economic activity. A strongly alkaline environment, with leachate around pH 13, is the hyper-alkaline Górka pit Lake in the Chrzanow region in the south of Poland (Czop et al. 2011).

Such extreme environments are observed worldwide, with some simultaneously exhibiting multiple extreme parameters, e.g., high temperature–high pressure, high temperature–high salinity, high salinity–alkaline pH, or high temperature–high pressure–high-salinity. Extreme environments having single or multiple extreme parameters are apparently lethal for the most living organisms; conversely, many extremophiles "live" and acquire energy and maintain cellular homeostasis under the harsh conditions.

17.2.2 Extremophiles Living in Extreme Environments

Thermophiles—Many thermophiles live in hot springs, hydrothermal vents, and similar environments. Of these, the archaeon *Methanopyrus kandleri* belonging to phylum *Euryarchaeota* can grow at 122 °C, which is the maximal temperature ever reported (Takai et al. 2008). This value will be eventually updated depending on the determination of cultivation conditions, because some viable but non-culturable microorganisms were detected in hydrothermal vents at 365 °C (Takai et al. 2004). The maximal values of temperature at which organisms can live are most probably restricted by damage to biomolecules by heat.

Psychrophiles—Many psychrophiles are observed in circa- or subzero degree Celsius environments such as deep-sea and polar regions. In contrast to high temperature, low temperature does not cause lethal effects in many non-psychrophilic organisms. The bacterium *Planococcus halocryophilus* belonging to phylum *Firmicutes* can grow at −15 °C (Mykytczuk et al. 2013). The limit of living microorganisms in low temperature is restricted by whether energy can be acquired and cellular homeostasis can be maintained, because low temperature reduces enzyme catalysis activity, enzyme stability, membrane fluidity, and availability of liquid water (Feller 2007).

Piezophiles—Deep sea also constitutes a high-hydrostatic-pressure environment. The bacterium *Colwellia marinimaniae* belonging to class *Gammaproteobacteria* can grow at 140 MPa (Kusube et al. 2017), which is the highest hydrostatic pressure value for piezophiles ever reported. High hydrostatic pressure acts as a bacteriostatic factor rather than as a lethal one; thus, non-piezophilic microorganisms such as *Escherichia coli* can also survive after experiencing high hydrostatic pressure of about 100 MPa, although extremely high pressure at over 300 MPa causes lethal effects (Kimura et al. 2017).

Halophiles—On the basis of optimal NaCl concentration for cell growth, halophiles are classified into slight (0.2–0.5 M), moderate (0.5–2.0 M), borderline extreme (2.0–3.0 M), and extreme (3.0–4.0) (Kushner 1993). Many extremely halophilic microorganisms growing under saturated NaCl conditions have been reported, whereas there is no report of extreme halophiles growing at saturated $MgCl_2$ conditions such as occur in the Discovery basin, Mariana Trench. The present limit of life is likely to be 2.3 M $MgCl_2$ (Hallsworth et al. 2007), at which the archaeon *Halorubrum sodomense* belonging to phylum *Euryarchaeota* can grow.

Acidophiles—Many mesophilic and thermophilic acidophiles are isolated from acidic hot springs, acid mine drainages, and solfataric environments (Sharma et al. 2012). Among these, the archaeon *Picrophilus oshimae* belonging to phylum *Euryarchaeota* was isolated from a dry, hot soil in the vicinity of Kawayu in Japan, growing optimally at pH 0.9 and 60 °C. Furthermore, it can also grow at pH −0.06, which is the lower limit of all known organisms (Schleper et al. 1995b).

Alkaliphiles—The bacterium *Alkaliphilus transvaalensis* belonging to phylum *Firmicutes* was isolated from alkaline water in an ultradeep gold mine and can grow under strongly alkaline conditions (pH 12.5) (Takai et al. 2001), representing the highest record. Moreover, many alkaliphiles have been isolated not only from alkaline lakes and soils but also from neutral and acidic environments (Horikoshi 1999).

Polyextremophiles—The existence of polyextremophiles, which adapt to multiple extremophilic parameters, such as thermoalkaliphile, haloalkaliphile, and thermo-halo-alkaliphile, has also been recently reported (Harrison et al. 2013). The great majority of extremophiles that show highest or lowest values for individual physicochemical parameters, except the alkaliphile *A. transvaalensis*, are polyextremophiles, because such strictly extreme environments are established by complex physicochemical parameters.

It is not known whether the maximal values observed in known extremophiles constitute the "limits of life," although the living conditions of extremophiles are likely to be near the edge of these "limits of life." In any extreme environment, extremophiles have a requirement for liquid water and energy for maintaining living activity. Minimal energy potential to support growth has been reported as −20 kJ/mol per reaction (Schink 1997), which indicates that organisms cannot live in thermodynamically cheap environments (>−20 kJ/mol). Values of free energy change for various reactions have been reported at standard conditions, whereas little is known regarding these values at extreme conditions. Accumulation of thermodynamic data under both standard and extreme conditions will be needed to fully understand the fundamentals of life.

17.2.3 The Concept of Habitability and Extremophiles Living on the Edge

The lower limit of energy potential for life was estimated as described above, raising the question of whether an upper limit also exists. The range of energy potential between lower and upper limits would provide an understanding regarding the "limits of life." Hoehler proposed the concept of "habitability as an energy balance" (Hoehler 2007). He described the extremely complicated living energies using two factors, energy potential (voltage) and energy load (power). The biological demand for energy is two-dimensional with discrete power and voltage requirements, and habitability can be calculated by narrowing between the respective upper and lower limits as follows: (1) biological energy quantum, which defines the minimum free energy level for sustaining "Complexity of the system" that is the driving force for biological processes (lower limit of voltage); (2) voltage uptake limit for sustaining "Complexity," above which any additional energy potential goes out of control (upper limit of voltage); (3) minimal maintenance energy for sustaining "Complexity" (lower limit of power); and (4) maximal energy that "Complexity" can control (upper limit of power).

According to this concept, a hyper-temperature environment corresponds to (2) and (4), and extreme oligotrophic conditions and ultralow temperature correspond to (1) and (3). Therefore, the energy metabolism of hyperthermophiles would place them near the edges of (2) and (4). Understanding the limits of energy conversion, which is one of the definitions of living organisms, would thus approach the fundamental of life. Therefore, we focus on energy conversions and their machineries in extremophilic microorganisms.

17.3 Enzymatic Adaptations to Extreme Environments

Extremophiles adapt to individual extreme environments based on strategies at the level of biological molecules. For example, G + C content of nucleic acids, lipid content of cellular membranes, and osmolyte content aim to protect the cell itself or cell components from extreme physicochemical parameters (Morozkina et al. 2010; Siliakus et al. 2017). Enzymes from extremophiles are altered to function in the respective extreme conditions; these are sometimes termed extremozyme (Adams et al. 1995). Such extremozymes adapt to individual physicochemical parameters using various strategies (Table 17.2).

Table 17.2 Adaptation of extremophilic protein to extreme environment

Categories	Strategies	How to	Effects
Thermophilic	Rigidify	• Oligomerization • Increased hydrophobic core • Increased number of disulfide bonds, salt bridges, and surface charge • Decreased surface area	• To decrease the ratio of surface area • To increase the rigidity of the individual subunits • To promote tighter packing of the hydrophobic core • To reduce exposure of hydrophobic residues to solvent
Piezophilic	Rigidify and to reduce the chance for water to penetrate the core of the protein	• Compact and dense hydrophobic core • Prevalence of smaller hydrogen-bonding amino acids (less number of large hydrophobic residues in the core) • Multimerization	• To compact the size of individual monomers • To protect the hydrogen bonds between the subunits
Psychrophilic	Acquire the flexibility and higher turnover number	• Increased glycine residues • Reduced size of nonpolar residues in the core • Reduced proline residues in loop and arginine residues • Increased size of substrate binding site	• To acquire the flexibility • To create weaker hydrophobic interactions • To avoid the formation of salt bridge and hydrogen bond
Halophilic	Acquire the solubility in high salt concentration	• Increased acidic residues on the surface • Less serine • Decreased hydrophobic residues • Acidic peptide insertion	• To maintain hydration shell • To acquire the flexibility of protein • To avoid excessively strong hydrophobic interactions and interfering electrostatic interaction in ion pair
Acidophilic	To avoid insolubilization by aggregation	• Negative surface charge	• To neutralize excess amounts of protons
Alkaliphilic	To avoid insolubilization by aggregation	• Increased hydrophobic residues in the intersubunit contacts	• To minimize accessible surface area against denaturing solvent

17.3.1 Thermophilic Enzymes

Under high temperature conditions, protein is denatured (unfolded), exposing the hydrophobic core to solvent, which causes aggregation. Thermophilic proteins adapt to high temperature through oligomerization and alteration of amino acid residues. The latter includes a decreased ratio of surface area, increased numbers of disulfide bonds, salt bridges, and surface charges, and rigidification of hydrophobic core packing (Table 17.2).

Some thermophilic proteins are oligomerized to increase the rigidity of the individual subunits, promote tighter packing of the hydrophobic core, and reduce exposure of hydrophobic residues to solvent (Vieille and Zeikus 2001). For example, acetyl-CoA synthetase from the hyperthermophilic archaeon *Ignicoccus hospitalis* forms octamers, whereas such enzymes from mesophiles are monomers or dimers (Jetten et al. 1989; Kumari et al. 1995; Mayer et al. 2012). Hydrophobic interactions at subunit–subunit interfaces decrease the overall ratio of accessible surface area to solvent, which results in tighter packing of the hydrophobic core.

In addition, increased numbers of disulfide bonds and salt bridges result in thermal stability. Because disulfide bonds mainly play a role in stabilization of the tertiary structure, a decrease in disulfide bonds in single or double mutants of cysteine residues yields decreased thermal stabilities compared with wild type (Cacciapuoti et al. 1994, 2012). Although in mesophilic proteins the salt bridges play a role as destabilizers, those in thermophilic enzymes act as structurally stabilizing elements when the desolvation penalty and entropic cost associated with ion pairing is overcome at higher temperature (Chan et al. 2011). In addition, an increase in charged residues on the protein surface is observed in thermophilic protein (Fukuchi and Nishikawa 2001). Replacements of thermally labile residues such as asparagine and glutamine to charged residues and charge–charge interaction by increasing charged residues aid in protecting against thermal denaturation (Lee et al. 2005).

There are many studies of thermophilic enzymes. Of these, cytochrome *c*, which plays a role as an electron carrier in energy conversion, has been studied to obtain a systematic understanding of thermophilic proteins. There are many reports about thermal and denaturant stabilities of cytochromes *c* from hyperthermophiles, thermophiles, mesophiles, and psychrophiles (Hasegawa et al. 2000; Oikawa et al. 2005; Nakamura et al. 2006; Ogawa et al. 2007; Hakamada et al. 2008; Obuchi et al. 2009; Yamanaka et al. 2009; Takeda et al. 2009; Takenaka et al. 2010; Oda et al. 2011; Masanari et al. 2011, 2016; Kobayashi et al. 2017). Thermal stabilities of cytochrome *c* proteins from these microorganisms correspond to the growth temperatures of each microorganism (Fig. 17.2). The cytochrome *c* proteins contain a heme in a hydrophobic core, with the hydrophobic interaction between apoprotein and heme playing a key role in the structural organization (Kang and Carey 1999). In the case of hyperthermophilic cytochrome *c* from the hyperthermophilic bacterium *Aquifex aeolicus*, hydrophobic amino acid residues around the heme and an additional α-helix form a densely packing core, and the T_m values, which is the

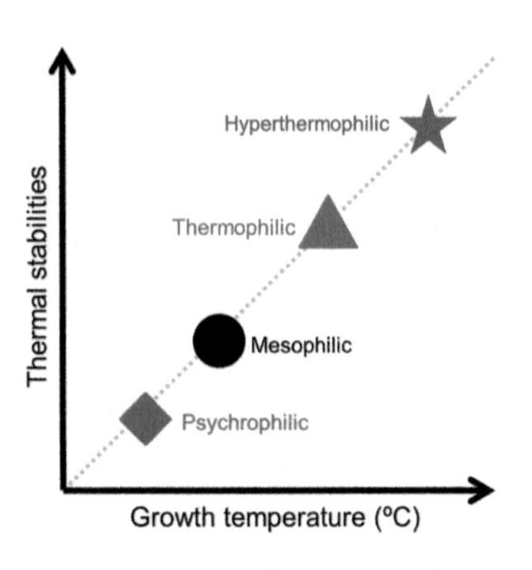

Fig. 17.2 Relationship between growth temperature and thermal stability. Trend in thermal stabilities of cytochrome *c* proteins from hyperthermophile, thermophile, mesophile, and psychrophile was plotted against the optimal growth temperature of each microorganism

temperature at midpoint of the transition during thermal denaturation, of this protein reach to approximately 130 °C (Yamanaka et al. 2011). In the *A. aeolicus* cytochrome *c*, the extremely strong hydrophobic core allows folding at the state of apoprotein without heme (Yamanaka et al. 2009). The importance of a hydrophobic core is also demonstrated by systematic denaturing experiments using different lengths of alkyl urea (Kobayashi et al. 2017). The denaturation of cytochrome *c* depends on the alkyl chain length of each alkyl urea, according to the hydrophobic effect of each reagent. These results clearly indicate that the hydrophobic interaction is important in thermophilic monomer proteins as an adaptation strategy.

In contrast, thermal stabilities of dimer forming cytochrome *c′* proteins clearly depend on the interaction between subunits (Fujii et al. 2017; Kato et al. 2015). Although the interaction between apoprotein and heme is important, the strong subunit–subunit interaction in thermophilic cytochrome *c′* provides higher thermal stability compared to that of mesophilic cytochrome *c′*. Such stabilization strategy corresponds to the general strategy of thermophilic enzymes as described above.

17.3.2 Piezophilic Enzymes

As many piezophiles are also thermophiles as described above, the piezophilic adaptation strategy is difficult to discriminate clearly from the thermophilic one. Therefore, proteins from piezophiles are adapted to high hydrostatic pressure by the formation of a compact, dense hydrophobic core, the prevalence of smaller hydrogen-bonding amino acids, and multimerization. For example, the proteome of the piezophilic hyperthermophile *Pyrococcus abyssi* showed a preference for smaller amino acid residues compared with that of a non-piezophilic thermophile, *Pyrococcus furiosus* (Di Giulio 2005). Such adaptation results in the reduction of large hydrophobic residues, such as tryptophan and tyrosine, in the hydrophobic core. In addition, a DNA-binding protein from *Sulfolobus solfataricus* and a glutamate dehydrogenase from *Thermococcus litoralis* have a compact hydrophobic core (Consonni et al. 1999; Sun et al. 2001). In addition, multimerization, which consists of the compacted individual subunits, reduces the chance for water to penetrate the core of the protein and protects the hydrogen bonding between subunits (Boonyaratanakornkit et al. 2002; Rosenbaum et al. 2012).

In addition to the studies of proteins from thermophilic piezophiles, there are comparative studies of monomeric cytochrome *c* proteins from piezophilic and non-piezophilic psychrophiles. The genera *Shewanella* belonging to class *Gammaproteobacteria* inhabits various terrestrial and water environments (fresh, brackish, and seawaters) (Hau and Gralnick 2007). In seawater environments, *Shewanella* species distribute to not only shallow sea (normal hydrostatic pressure, 0.1 MPa) but also deep sea (higher hydrostatic pressure), and they are separated into piezosensitive and piezophilic (or piezotolerant) types, respectively (Kato and Nogi 2001). In addition, all *Shewanella* species share a common cytochrome *c* as one of the electron carrier proteins in the respiratory chain. Therefore, we can compare the stabilities between cytochrome c_5 proteins from piezophilic and non-piezophilic *Shewanella* spp.

The thermal stabilities of cytochrome c_5 proteins from deep-sea living *S. violacea* and *S. benthica* are higher than those from shallow-sea living *S. amazonensis* and *S. livingstonensis* (Masanari et al. 2014). However, *S. violacea*, *S. benthica*, and *S. livingstonensis* are also psychrophile (Nogi et al. 1998; Kato et al. 1998; Bozal et al. 2002), whereas *S. amazonensis* is mesophile (Venkateswaran et al. 1998). Cytochrome c_5 proteins from piezophilic psychrophiles *S. violacea* and *S. benthica* are clearly stabilized compared with that from the piezosensitive mesophile *S. amazonensis*. As the order of thermal stabilities is almost according to growth temperature of their microorganisms as mentioned in the thermophilic enzymes Sect. 17.3.1 (Fig. 17.2), unexpectedly higher stabilities of the cytochrome c_5 proteins from *S. violacea* and *S. benthica* strongly indicate an adaptation strategy for high hydrostatic pressure.

Crystal structure analysis of *S. violacea* cytochrome c_5 showed an additional hydrogen bond between the peptide chain and heme (Masanari et al. 2016). This additional interaction would strengthen interactions in the hydrophobic core of the globular protein. Accordingly, reciprocal mutagenesis experiments using *S. violacea* and *S. livingstonensis* cytochrome c_5 proteins demonstrated the contribution of this hydrogen bond to higher thermal stability (Masanari et al. 2016). Because piezophilic adaptation strategies aim to reduce the chance for water to penetrate the core of the protein, such additional hydrogen bonds in a monomeric protein would thus contribute as an adaptation strategy.

17.3.3 Psychrophilic Enzymes

Psychrophilic enzymes should be flexible, in contrast to thermophilic and piezophilic enzymes. As psychrophiles live in at below 20 °C, enzymes from psychrophiles must catalyze at low temperatures in which the kinetic energy of a molecule is small. At such low temperatures, the rigid structure of thermophilic and piezophilic enzymes would most likely interfere with local or global conformational changes during the catalytic reaction. Therefore, the protein requires flexibility for catalysis and stability for sustaining the structure (Siddiqui and Cavicchioli 2006; Feller 2010). Some strategies for the adaptation to low temperature have been reported. In order to acquire flexibility, the numbers of glycine residues, which constitute the smallest amino acid, are increased, and the size of nonpolar residues in the protein core is reduced. The former provides greater conformational mobility, and the latter creates weaker hydrophobic interactions. Conversely, proline residues in the loop regions and arginine residues are reduced in order to stabilize the protein by providing conformational rigidity and avoiding the formation of salt bridges and hydrogen bonds.

In addition to such residue selections, psychrophilic enzymes exhibit a unique adaptation strategy for catalysis at low temperatures. The turnover number (k_{cat}) and the K_m value of psychrophilic enzymes are typically higher compared with those of mesophilic enzymes under low temperature (Feller 2010). The higher turnover number at low temperature is due to the increased size of the substrate binding site, and low affinity reduces the energy of activation for the psychrophilic enzymes (D'Amico et al. 2006). Consequently, at low temperature psychrophilic enzymes generally exhibit higher specific activity compared with homologous enzymes from mesophiles. The adaptation strategy of psychrophilic enzymes therefore diametrically differ from those of thermophilic and piezophilic enzymes. Namely, psychrophilic strategies involve the acquisition of flexibility, whereas thermophilic and piezophilic strategies comprise structural stabilization.

17.3.4 Halophilic Enzymes

Halophilic enzymes are active in the presence of salts at a molar order. Many non-halophilic enzymes are inactive in such high salt concentration, because salts strengthen hydrophobic interactions and interfere with the electrostatic interactions between charged amino acids (Mancinelli 2007; Karan et al. 2012). Therefore, halophilic enzymes adapt to avoid these influences from the solvent (Frolow et al. 1996; Mevarech et al. 2000; Soppa 2006; Kastritis et al. 2007; Tadeo et al. 2009). Accordingly, many adaptation strategies are observed in the solvent-exposed surface of each subunit and the subunit–subunit interfaces.

The most well-known feature of extremely halophilic enzymes is the increase in acidic residues on the protein surface. This feature is supported by the crystal structures and bioinformatics analyses (Frolow et al. 1996). In addition, bioinformatics analysis has shown a decrease in the numbers of serine residues, which are preferred when interacting with water (Zhang and Ge 2013). This feature on the surface of halophilic enzymes maintains a hydration shell around the protein by binding hydrated cations, rather than direct hydration of amino acid residues. In addition, halophilic enzymes exhibit decreased numbers of hydrophobic residues, which then form a smaller hydrophobic core (Siglioccolo et al. 2011). This affords increased flexibility to the protein in high salt concentrations by preventing the hydrophobic core from becoming too rigid. Furthermore, in some cases, a peptide insertion containing a large number of acidic amino acid residues provides halophilicity to proteins without generally increasing acidic residues and reducing the hydrophobic core. For example, such insertions have been identified in cysteinyl-tRNA synthase, serinyl-tRNA, and ferredoxin from extremely halophilic archaea (Taupin et al. 1997; Marg et al. 2005; Evilia and Hou 2006).

Although the adaptation mechanisms of thermophilic, piezophilic, and psychrophilic enzymes illustrate how to acquire tolerance for such extreme environments, those of halophilic enzymes are fundamentally different. Specifically, halophilic enzymes, especially extremely halophilic ones, utilize salts (ions) to form a correct folding state.

17.3.5 Acidophilic and Alkaliphilic Enzymes

Little is known regarding the adaptation mechanisms of acidophilic and alkaliphilic enzymes compared with those of thermophilic, piezophilic, and psychrophilic enzymes. Because acidophiles and alkaliphiles maintain neutral intracellular pH (Horikoshi 1999; Sharma et al. 2012), intracellular enzymes do not need to adapt to acidic or alkaline conditions. Conversely, extracellular enzymes must adapt to such pH extremes. Therefore, almost all studies have reported findings regarding enzyme function on extracellular substrates, large molecules such as starch, cellulose, and protein, and insoluble or non-permeable substrates (Sarethy et al. 2011; Sharma et al. 2012).

Uniquely, among these, tetrathionate hydrolase from the acidophilic bacterium *Acidithiobacillus ferrooxidans* requires exposure to acidic solution for protein folding (Kanao et al. 2010). Although it is clear that the acidic condition is essential for enzyme folding, the molecular mechanism has not yet been revealed. In addition, as an adaptation strategy of acidophilic enzymes for acidic conditions, negative surface charge has been reported (Huang et al. 2005).

In comparison, an increase in hydrophobic residues among the intersubunit contacts has been reported as adaptation strategy of alkaliphilic protein (Dubnovitsky et al. 2005; Popinako et al. 2017). Although the acidic and alkaline conditions function as parameters along the same axis as pH, the adaptation strategies of acidophilic and alkaliphilic enzymes clearly differ from each other. Acidophilic enzymes may use negative surface charge for preventing insolubilization by aggregation and neutralizing excess amounts of proton, whereas alkaliphilic enzymes may minimize accessible surface area to escape from the denaturing solvent.

17.4 Energy Conversion of Thermophiles and Their Machineries

Little knowledge of energy conversion based on thermodynamic measurements in thermophilic microorganisms is available compared with the accumulated data from studies on adaptation mechanisms of thermophilic proteins. The basis of energy conversion in biological processes is chemical reactions, which are catalyzed by enzymes. The Gibbs free energy change for catalytic reactions under high temperature is large compared with that under low temperature (Amend and Plyasunov 2001), because the activation energy barrier is reduced by the entropy gain of the $-T\Delta S$ term associated with temperature elevation. Increasing reaction rates based on elevating temperature are according to the Arrhenius equation at a range up to the limiting temperature at which both substrate and product are stable. Although we do not know whether such thermodynamic advantage involves versatile energy conversion, the energy metabolisms of thermophilic microorganisms are extremely versatile (Amend and Shock 2001). In the overall metabolic reactions of thermophiles, both organic (various hydroxyl-carboxylic acids, alcohols, amino acids, and methane) and inorganic (H_2, reduced sulfur compounds, and Fe^{2+}) materials can be used as electron donors and various electron acceptors including O_2, SO_4^{2-}, SO_3^{2-}, $S_2O_3^{2-}$, S^0, NO_3^-, NO_2^-, NO, N_2O, CO_2, CO, and Fe^{3+}. Free energy changes for various combinations thereof have been exhaustively calculated (Amend and Shock 2001). However, the thermodynamic features of each reaction have not yet been linked to functioning mechanisms of catalyzing enzymes.

The thermophilic, facultatively chemolithoautotrophic, hydrogen-oxidizing bacterium *Hydrogenophilus thermoluteolus* can grow heterotrophically on organic acids and autotrophically on hydrogen and carbon dioxide (Goto et al. 1978), and both energy transduction systems are consolidated by oxidative phosphorylation (Wakai et al. 2013a). The inverted membrane vesicles derived from the *H. thermoluteolus* cells autotrophically cultivated with H_2 form a proton gradient with lactate and succinate as substrates for heterotrophic growth. Both energies from substrates for autotrophic and heterotrophic growth are converted to a proton gradient (membrane potential) and then drive ATP synthesis via ATP synthase. Maintaining a heterotrophic energy conversion pathway even when cells are growing with an autotrophic energy conversion pathway may allow the microorganism to live in versatile energy environments, because the origin of molecular hydrogen as an autotrophic substrate is geochemically limited, whereas organic acids, as heterotrophic substrates that are generated by other organisms, exist globally. Accordingly, the thermophilic, facultative chemoautotrophs *Hydrogenophilus* spp. are prevalently distributed in versatile geochemical niches and are also notably detected at non-high temperatures such as the subglacial Lake Vostk, Antarctica as well (Lavire et al. 2006).

ATP synthase is a key enzyme in the energy conversion of *H. thermoluteolus*, because this bacterium grows heterotrophically and autotrophically via oxidative phosphorylation. ATP hydrolysis activity of the *H. thermoluteolus* ATP synthase shows 65 °C as the optimal temperature (Wakai et al. 2013a). Heterologous expression of the *H. thermoluteolus* ATP synthase in *Escherichia coli* showed thermophilic ATP hydrolysis activity, whereas it could not complement the growth of authentic ATP synthase-defective *E. coli* under mesophilic conditions. This thermophilic feature of ATP hydrolysis is likely due to the adaptation mechanism of thermophilic enzymes as described above, whereas the uncoupling of ATP synthesis may suggest that thermophilic enzyme exhibits difficulty in acquiring catalytic ability along with flexibility under mesophilic temperatures. ATP synthase comprises eight kinds of subunits, the proton-translocating Fo portion, which is embedded in the membrane, and the catalytic F1 part, which projects in the cytosol. The correct folding of each subunit, correct organization of all subunits, and flexibility of whole structure are essential to proper functioning.

17.5 Energy Conversion of Halophiles and Their Machineries

Little is known regarding the thermodynamic analysis of energy conversion in thermophiles, with even less known for halophiles. In energy conversion under high temperature, the activation energy barrier is easily overcome by entropy gain, with this event strongly depending on thermal stability of enzymes that provide the reaction field. In contrast, catalytic reactions in high salt conditions involve many negative factors, such as the influences of ionic strength on reaction progression and the influences of dehydration against enzymes, substrates, and cofactors. For examples, the biological energy currency of ATP is highly hydrated, and hydration energy is recognized as one of the origins of the Gibbs free energy change for hydrolysis, together with resonance stabilization and electrostatic repulsion (George et al.1970; de Meis 1989; Saint-Martin et al. 1994; Hong et al. 2012). In hyper-saline environments, an ATP molecule is dehydrated because such environment enforces lower water activity. It has been reported that the free energy change for ATP hydrolysis is reduced under the low water activity conditions established by non-physiological concentrations of dimethyl sulfoxide (Romero and de Meis 1989). Therefore, we discuss issues of salt adaptation and the thermodynamic analysis of high-energy phosphate-compound-hydrolyzing enzymes in halophilic microorganisms.

17.5.1 ATP Hydrolysis by Halophiles

Shewanella spp. live in various salt environments, such as soil, freshwater, brackish water, and seawater (shallow and deep seas), with their halophilicities also being versatile, such as halotolerant, moderately halophilic, and strictly halophilic as described above. Although the study of extracellular electron transfer of *Shewanella* spp. has been focused in the area of electromicrobiology (Nealson and Rowe 2016), in this section we discuss the biochemical features of ATP-hydrolyzing enzymes. *Shewanella* spp. possess a 5′-nucleotidase, which may be related to acquiring energy from substrates in the living environments, and salt sensitivities and adaption strategies of these enzymes from salt-sensitive and halophilic *Shewanella* spp. have been reported (Kuribayashi et al. 2017). These nucleotidases (SVNTase and SANTase) have been purified from the deep-sea living, strictly halophilic *S. violacea* (Nogi et al. 1998) and the brackish water living, halotolerant *S. amazonensis* (Venkateswaran et al. 1998), respectively. ATP hydrolysis activity of SVNTase is highly tolerant to high salt concentrations compared to that of SANTase. The identity of the estimated amino acid sequences is 69.7%, whereas the acidic/basic amino acid ratio and the numbers of salt bridges differ. Among these biochemical and bioinformatics features, we focus on the differences of NaCl tolerance and amino acid composition.

Genome sequences of ten species of *Shewanella* are available, and the NaCl requirement for optimal growth and growth abilities on 0 M NaCl have been also reported (summarized in Kuribayashi et al. 2017). In addition, ten estimated amino acid sequences of 5′NTases have been collected, and model structures were simulated in order to compare the numbers and positions of potential salt bridges. The 5′-NTases from the halophilic *Shewanella* group, which cannot grow on 0 M NaCl, showed higher acidic/basic ratio and lower numbers of salt bridges compared with those from the halotolerant *Shewanella* group, which can grow on 0 M NaCl and at seawater levels. Because the high salt environment, namely strong ionic strength, influences ionic interactions, the finding of fewer salt bridges in 5′-NTases from halophilic *Shewanella* including SVNTase is reasonable. The positions of salt bridges in halophilic *Shewanella* 5′-NTases are also preserved in halotolerant *Shewanella* except for one salt bridge (Fig. 17.3), suggesting that these preserved salt bridges may be required for conformational stability.

Intracellular environments of *Shewanella* spp. regardless of halophilic or halotolerant status do not constitute hyper-saline conditions because they do not adopt a salt-in strategy. Therefore, the energy conversions in the cells would be not influenced by extracellular salt concentration. In contrast, for extremely halophilic microorganisms, which adopt a salt-in strategy as an adaptation mechanism for high salt concentrations (Christian and Waltho1962), both enzyme adaptation and energy conversion must be considered.

17.5.2 Hydrolysis of High-Energy Phosphate Compounds by Extreme Halophiles

Extremely halophilic microorganisms, especially archaea and a few bacteria, accumulate salts in the cells in order to eliminate osmotic pressure via the cellular

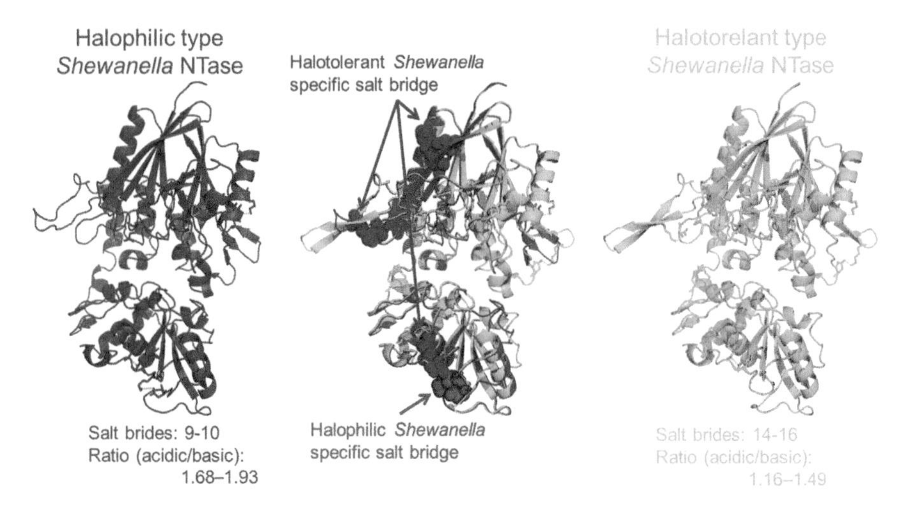

Fig. 17.3 Difference of conserved salt bridges between halophilic and halotolerant *Shewanella* NTases. Model structures of halophilic and halotolerant *Shewanella* NTases were simulated using the *E. coli* NTase crystal structure as a reference. Left and right models represent SVNTase and SANTase, respectively. The numbers of salt bridges and the ratios of acidic per basic amino acid residues were noted in this figure. The conserved positions of salt brides in halophilic and halotolerant NTase represent red and blue color, respectively

membrane (Oren 2013). The intracellular salt concentrations reach to saturation levels of KCl or NaCl depending on the extracellular salt concentrations (Médicis et al. 1986). Almost all non-halophilic and many moderately halophilic enzymes are inactivated in such hyper-saline condition. Furthermore, in such conditions, water activity reduces to approximately 0.8, influencing the hydration states of enzymes, substrates, and cofactors. However, extremely halophilic enzymes must maintain correct folding state and catalyze specific reactions. Although numerous studies describing the effects of salt on enzyme biochemical features and folding state have been reported (Madern et al. 2000; Kastritis et al. 2007; Siglioccolo et al. 2011; Miyashita et al. 2015), to our knowledge, none have been published regarding thermodynamic analysis of enzymatic reactions under high salt conditions.

17.5.2.1 Biochemical Features of Halophilic Enzymes Related to Energy Conversion

The extremely halophilic archaeon *Haloarcula japonica* can grow preferentially at 20% NaCl (3.4 M NaCl) and accumulates large amounts of salt in the cells. Accordingly, there have been many reports regarding intracellular, halophilic enzymes (e.g., cell division protein FtsZ1 (Ozawa et al. 2005); α-amylase (Onodera et al. 2013); dihydrofolate reductase (Miyashita et al. 2015); 2-deoxy-D-ribose-5-phosphate aldolase (Ohshida et al. 2016); and inorganic pyrophosphatase (Wakai et al. 2017a)). In addition, we investigated the halophilicity of ATP hydrolysis activity using cell-free extracts (unpublished), demonstrating that the *H. japonica* cell-free extract preferentially hydrolyzed ATP at 2.5 M Na_2SO_4. Although the activities with NaCl and KCl also showed halophilicities, such activities were only 25% even at 4 M NaCl and 4 M KCl. Similar activities were observed in the membrane fraction, most likely due to AoA1 ATP synthase. The estimated amino acid sequence of each subunit found on the *H. japonica* genome sequence shows high contents of acidic amino acid residues, which corresponds to features of halophilic enzymes and indicates stabilization by interaction between acidic residues and cations. Furthermore, the highest specific activity being observed for Na_2SO_4 may suggest stabilization by anions in addition to cations, as the quaternary structure of malate dehydrogenase from the extremely halophilic archaeon *Haloarcula marismortui* is stabilized by anion (Madern and Ebel 2007). As AoA1 ATP synthase is multicomplex enzyme consisting of ten kinds of subunits, sulfate ions may play an important role in the stability of quaternary structure in the *H. japonica* ATP synthase. We subsequently conducted thermodynamic measurements for ATP hydrolysis in the presence of high concentrations of Na_2SO_4, as accurate measurements have not yet been obtained because Na_2SO_4 readily precipitates in the apparatus.

Because inorganic pyrophosphate (PPi) has the smallest size among high-energy phosphate compounds and ATP analogues, we carried out biochemical and thermodynamic measurements for PPi hydrolysis. In addition to ATP hydrolysis activity, pyrophosphate hydrolysis activity is also observed in the *H. japonica* cell-free extract (Wakai et al. 2013b). Salt dependency of PPi hydrolysis activity clearly differed from that of ATP hydrolysis activity, and higher specific activities were observed not only with Na_2SO_4 but also with NaCl and KCl. Optimal concentrations for Na_2SO_4 and NaCl were approximately 0.75 M and 1.5 M, respectively. In contrast, PPi hydrolysis activity was not observed at any concentration of $(NH_4)_2SO_4$. Thus, PPi hydrolysis activity in the *H. japonica* cell extracts depends on cations (Na^+ or K^+ albeit not $NH_4{}^+$) but not anions. A cation-dependent property is a common feature in extremely halophilic enzymes. A sole gene encoding for inorganic pyrophosphatase (PPase) is found in the *H. japonica* genome sequence, with the estimated amino acid sequence from this gene showing a high content of acidic residues. Crystal structures of bacterial and archaeal PPases have been reported as hexamers (Teplyakov et al. 1994; Kankare et al. 1996; Leppänen et al. 1999; Liu et al. 2004); the model structure of the *H. japonica* PPase was simulated from the *E. coli* PPase. In the hexameric structure of the *H. japonica* PPase, accessible surface areas were occupied by acidic residues, whereas the subunit–subunit interfaces were not (Fig. 17.4). A similar model feature has been reported in the PPase of the extremely halophilic archaeon *Haloferax volcanii* (McMillan et al. 2015).

The PPi hydrolysis activity in *H. japonica* exhibits obligatory halophilic activity, which is inactive in 0 M NaCl at the low Mg^{2+} concentration, whereas this activity is evident even at 0 M NaCl at 100–300 mM Mg^{2+} concentrations (Wakai et al. 2017a). This NaCl-independent activity is higher than the NaCl-dependent activity.

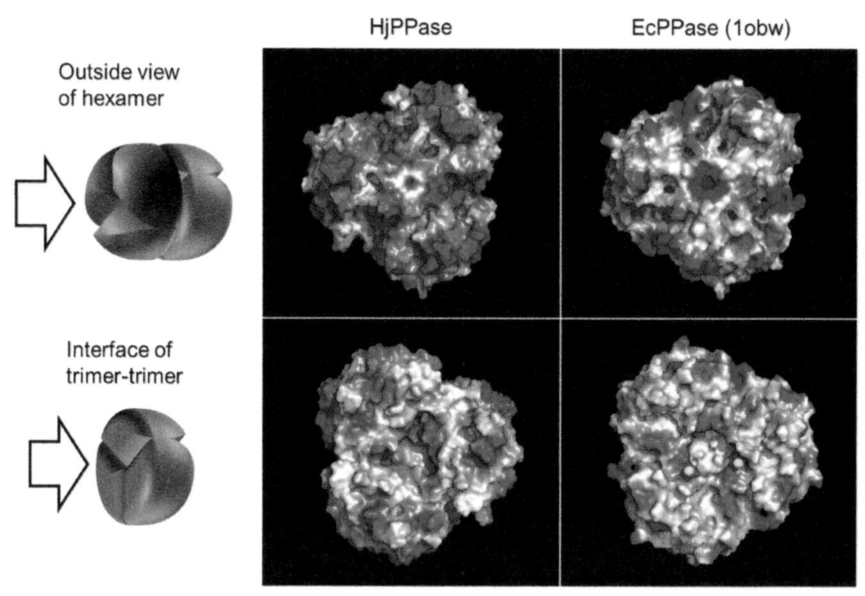

Fig. 17.4 Model structure and distribution of acidic amino acid residues of the *H. japonica* PPase. Model structures of *H. japonica* PPase was simulated using the *E. coli* PPase crystal structure as a reference. Top and bottom pictures represent outside external viewing of hexamer and the interface of trimer–trimer, respectively

Unexpectedly, this NaCl-independent activity was not due to the ionic strength-dependent activity as this activity is inactive at $(NH_4)_2SO_4$. Therefore, the NaCl-independent activity comprises the Mg^{2+}-dependent activity. This activity may have biological significance against intracellular low salinity stress. Notably, extremely halophilic archaea require high concentrations of Mg^{2+} in addition to NaCl for cell growth, and the intracellular Mg^{2+} concentration is also higher than those of non-halophilic archaea (Médicis et al. 1986; Boujelben et al. 2012). Although PPi is produced by intracellular biological reactions, such as synthesis of DNA and RNA, its accumulation causes toxicity. Therefore, the PPi must be hydrolyzed immediately, with PPase playing a key role in PPi detoxification.

17.5.2.2 Thermodynamic Measurement of Enzymatic Reactions by Halophilic Enzymes

It was found that the adaptation strategies of high-energy phosphate-compound-dehydrolyzing enzymes from extreme halophiles correspond to those of other enzymes from extreme halophiles, whereas little is known regarding the influence of salt on the energy state of substrates under such hyper-saline conditions. Free energy changes for hydrolysis of ATP and PPi under gas phase are extremely low compared with those under liquid phase (George et al. 1970; de Meis 1989; Saint-Martin et al.1994; Hong et al. 2012). Thus, the hydration states of ATP and PPi are important for large free energy changed. Hyper-saline conditions such as intracellular conditions of extreme halophiles decrease the water activity and likely alter the hydration state. Because hydration of ATP and PPi is enthalpically driven (Wolfenden 2006), we estimated the values of enthalpy change for enzymatic hydration of PPi under a broad range of salt concentrations using isothermal titration microcalorimetry (Wakai et al. 2013b).

Net enthalpy change value for a reaction must be estimated from the experimentally determined enthalpy change (ΔH_{EXP}) and ionization enthalpy change (ΔH_{ION}) (Fig. 17.5), because heat from all events such as reaction, dilution, and ionization is measured. In order to calculate the ΔH_{EXP}, the apparent enthalpy change ($\Delta H_{\mathrm{Whole,PPase}}$) for PPi hydrolysis and the dilution heat ($\Delta H_{\mathrm{Dilute,PPi}}$) of substrate must be measured in individual solution systems. In addition, the ΔH_{ION} values are calculated from the apparent enthalpy change ($\Delta H_{\mathrm{Whole,Protonation}}$) for buffer protonation and the dilution heat ($\Delta H_{\mathrm{Dilute,HCl}}$) of acid. The correlation between the resulting ΔH_{EXP} and ΔH_{ION} values in some buffers can be expressed as a linear function with the following equation:

$$\Delta H_{\mathrm{EXP}} = \mathrm{n}\,\Delta H_{\mathrm{ION}} + \Delta H_{\mathrm{Whole}}$$

where n represents the slope of the line giving the experimentally determined number of protons released owing to the PPi hydrolysis reaction (and absorbed by the buffer). Linear regression of plots gives the whole catalytic enthalpy change

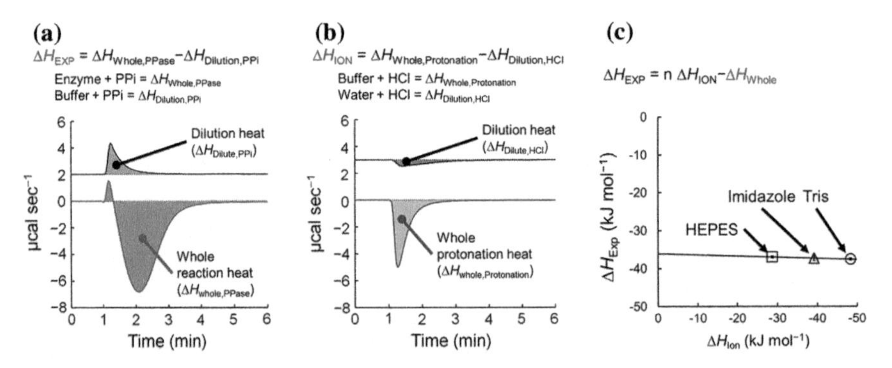

Fig. 17.5 Estimation of the whole catalytic enthalpy change ($\Delta H_{\mathrm{Whole}}$). Each plot is measured by isothermal titration calorimeter. The heat of each event is calculated as the integrated area of each injection. **a** Estimation of the experimentally determined enthalpy change (ΔH_{EXP}). The $\Delta H_{\mathrm{Whole,PPase}}$ and $\Delta H_{\mathrm{Dilute,PPi}}$ are measured by injecting PPi solution into reaction solution with and without PPase, respectively. **b** Estimation of ionization enthalpy change (ΔH_{ION}). The $\Delta H_{\mathrm{Whole,HCl}}$ and $\Delta H_{\mathrm{Dilute,HCl}}$ are measured by injecting HCl solution into buffer solution and pure water (or salt solution), respectively. **c** Estimation of the $\Delta H_{\mathrm{Whole}}$. The correlation between the resulting ΔH_{EXP} and ΔH_{EXP} values in different buffer systems is plotted

(ΔH_{Whole}) through extrapolation of the line to zero heat of buffer, as defined previously (Morin and Freire 1991; Todd and Gomez 2001; Bianconi 2003).

By multitudes of measurements using the above method, the ΔH_{Whole} values for PPi hydrolysis by extremely halophilic *H. japonica* cell-free extract, non-halophilic *E. coli*, and *Saccharomyces cerevisiae* PPases under the presence of 0.1 to 4.0 M NaCl were estimated. Unexpectedly, constant enthalpy changes were observed across broad ranges of NaCl concentration even though 4.0 M NaCl corresponds to a water activity value of 0.84 (Wakai et al. 2013b). Thus, the PPi hydrolysis is not loaded with any enthalpic penalty at high NaCl concentrations in the *H. japonica* cells. Currently, the influence of high salt concentration on the intracellular energy conversion is recognized, whereas this study was the first report demonstrating the maintenance of energy conversion of high-energy phosphate compounds in extremely halophilic cells.

In contrast, under high magnesium conditions (~200 mM Mg^{2+}) the enthalpy change values for PPi hydrolysis by *H. japonica* are smaller than those under low Mg^{2+} concentration (~10 mM) (unpublished data). This implies that Mg^{2+} strongly influences the thermodynamic parameter of PPi compared with Na^{+}, and under extremely high concentrations of Mg^{2+} the *H. japonica* enzyme may be unable to maintain the enthalpy change value for PPi hydrolysis. The observed limit of life in the high Mg^{2+} environment is approximately 2.3 M (Table 17.1), which strongly indicates that Mg^{2+} is a harsh cation for living organisms.

17.6 Summary and Perspective

Throughout this chapter, we summarized the biochemistry and thermodynamics of extremophilic enzymes related to energy conversion. With respect to biochemical aspects, thermophilic and piezophilic enzymes adapt to each harsh condition by forming a tolerant protein structure, and the progression of enzymatic reactions by these proteins depends on their structural stability. In contrast, halophilic enzymes adapt to high salt concentration by utilizing ions in order to acquire solubility. Regarding thermodynamic aspects, the progression of reactions in high temperatures is thermodynamically advantageous, whereas reaction progression in the presence of high salt concentration is thermodynamically disadvantageous because these are influenced by ions that alter the energy state of substrates and products. Under high salt conditions, the reaction can progress only in the range of salt concentrations that is thermodynamically reasonable. Therefore, the role of enzymes is to establish thermodynamically acceptable reaction fields, with extremophilic enzymes being evolved and adapted to form such reaction fields under extreme conditions.

In addition to the field of basic science as described above, studies on energy conversion and enzymes in extremophilic microorganisms are important in the field of applied science. Thermophilic, halophilic, psychrophilic, and piezophilic enzymes are already used in various industrial fields such as chemical, medical, and food industries. Recently, research has focused on synthetic biology, which strategically creates microorganisms possessing artificially designed metabolic pathways (Hara et al. 2014; Wakai et al. 2017b). In this area, specific attention has been paid to extremophilic enzymes, because such enzymes have higher thermal stability and tolerance for high salt and organic solvent conditions. Most recently, the design of de novo metabolic pathways using extremophilic enzymes and microorganisms has progressed through collaborations with computer science. In contrast, little progress has been made with respect to de novo enzyme design schemes. A fundamental understanding regarding the mechanisms of protein machineries in extremophilic enzymes is expected to contribute to progress in technologic fields for the creation of enzymes possessing artificially designed properties.

17.7 Conclusions

The aim of this chapter was to discern fundamentals of the mechanisms of protein machineries through achieving deep insight into extremophilic enzymes related to energy conversion. The role of an enzyme is to facilitate a specific chemical reaction under a given condition whether mild or extreme, for which it is essential to create a biochemically and thermodynamically acceptable reaction field in the enzyme. If unknown living organisms exist beyond the known edges of the limits of life, they must contain enzymes that are able to execute thermodynamically acceptable energy conversion under more harsh conditions. In such enzymes, various adaptation strategies as summarized in this chapter would likely be utilized corresponding to individual physicochemical parameters. However, the fundamental principle behind forming a thermodynamically acceptable reaction field has not yet been revealed. Therefore, the accumulation of findings from combinatorial studies of biochemistry, structural biology, and thermodynamics is essential in order to fundamentally understand the mechanisms of protein machineries attributed to energy conversion in living cells.

References

Adams MW, Perler FB, Kelly RM (1995) Extremozymes: expanding the limits of biocatalysis. Nat Biotechnol 13(7):662–668

Amend JP, Plyasunov AV (2001) Carbohydrates in thermophile metabolism: calculation of the standard molal thermodynamic properties of aqueous pentoses and hexoses at elevated temperatures and pressures. Geochim Cosmochima Acta 65:3901–3917

Amend JP, Shock EL (2001) Energetics of overall metabolic reactions of thermophilic and hyperthermophilic Archaea and Bacteria. FEMS Microbiol Rev 25:175–243

Boujelben I, Gomariz M, Martínez-García M, Santos F, Peña A, López C, Antón J, Maalej S (2012) Spatial and seasonal prokaryotic community dynamics in ponds of increasing salinity of Sfax solar saltern in Tunisia, vol 101. Antonie Van Leeuwenhoek, pp 845–857

Bianconi ML (2003) Calorimetric determination of thermodynamic parameters of reaction reveals different enthalpic compensations of the yeast hexokinase isozymes. J Biol Chem 278:18709–18713

Bischoff JL, Rosenbauer JR (1988) Liquid-vapor relations in the critical region of the system NaCl-H_2O from 380 to 415 °C: a refined determination of the critical point and two-phase boundary of seawater. Geochim Cosmochim Acta 52:2121–2126

Boonyaratanakornkit BB, Park CB, Clark DS (2002) Pressure effects on intra- and intermolecular interactions within proteins. Biochim Biophys Acta 1595:235–249

Bozal N, Montes MJ, Tudela E, Jiménez F, Guinea J (2002) *Shewanella frigidimarina* and *Shewanella livingstonensis* sp. nov. isolated from Antarctic coastal areas. Int J Syst Evol Microbiol 52:195–205

Cacciapuoti G, Porcelli M, Bertoldo C, De Rosa M, Zappia V (1994) Purification and characterization of extremely thermophilic and thermostable 5'-methylthioadenosine phosphorylase from the archaeon *Sulfolobus solfataricus*. Purine nucleoside phosphorylase activity and evidence for intersubunit disulfide bonds. J Biol Chem 269:24762–24769

Cacciapuoti G, Fuccio F, Petraccone L, Del Vecchio P, Porcelli M (2012) Role of disulfide bonds in conformational stability and folding of 5'-deoxy-5'-methylthioadenosine phosphorylase II from the hyperthermophilic archaeon Sulfolobus solfataricus. Biochim Biophys Acta 1824:1136–1143

Chan CH, Yu TH, Wong KB (2011) Stabilizing Salt-bridge enhances protein thermostability by reducing the heat capacity change of unfolding. PLoS ONE 6(6):e21624

Christian JH, Waltho JA (1962) Solute concentrations within cells of halophilic and non-halophilic bacteria. Biochim Biophys Acta 17:506–508

Consonni R, Santomo L, Fusi P, Tortora P, Zetta L (1999) A single-point mutation in the extreme heat- and pressure-resistant sso7d protein from *Sulfolobus solfataricus* leads to a major rearrangement of the hydrophobic core. Biochemistry 38:12709–12717

Czop M, Motyka J, Sracek O, Szuwarzyński M (2011) Geochemistry of the hyperalkaline Gorka pit lake (pH > 13) in the Chrzanow region, southern Poland. Water Air Soil Pollution 214:423–434

D'Amico S, Sohier JS, Feller G (2006) Kinetics and energetics of ligand binding determined by microcalorimetry: insights into active site mobility in a psychrophilic alpha-amylase. J Mol Biol 358:1296–1304

de Meis L (1989) Role of water in the energy of hydrolysis of phosphate compounds—energy transduction in biological membranes. Biochim Biophys Acta 973:333–349

Demirjian DC, Morís-Varas F, Cassidy CS (2001) Enzymes from extremophiles. Curr Opin Chem Biol 5:144–151

Di Giulio M (2005) A comparison of proteins from *Pyrococcus furiosus* and *Pyrococcus abyssi*: barophily in the physicochemical properties of amino acids and in the genetic code. Gene 346:1–6

Dubnovitsky AP, Kapetaniou EG, Papageorgiou AC (2005) Enzyme adaptation to alkaline pH: atomic resolution (1.08 Å) structure of phosphoserine aminotransferase from *Bacillus alcalophilus*. Protein Sci 14:97–110

Evilia C, Hou YM (2006) Acquisition of an insertion peptide for efficient aminoacylation by a halophile tRNA synthetase. Biochemistry 45:6835–6845

Feller G (2007) Life at low temperatures: is disorder the driving force? Extremophiles 11:211–216

Feller G (2010) Protein stability and enzyme activity at extreme biological temperatures. J Phys Condens Matter 22:323101

Feller G, Gerday C (2003) Psychrophilic enzymes: hot topics in cold adaptation. Nat Rev Microbiol 1:200–208

Frolow F, Harel M, Sussman JL, Mevarech M, Shoham M (1996) Insights into protein adaptation to a saturated salt environment from the crystal structure of a halophilic 2Fe-2S ferredoxin. Nat Struct Biol 3:452–458

Fujii S, Oki H, Kawahara K, Yamane D, Yamanaka M, Maruno T, Kobayashi Y, Masanari M, Wakai S, Nishihara H, Ohkubo T, Sambongi Y (2017) Structural and functional insights into thermally stable cytochrome *c*' from a thermophile. Protein Sci 26:737–748

Fukuchi S, Nishikawa K (2001) Protein surface amino acid compositions distinctively differ between thermophilic and mesophilic bacteria. J Mol Biol 309:835–843

George P, Witonsky RJ, Trachtman M, Wu C, Dorwart W, Richman L, Richman W, Shurayh F, Lentz B (1970) "Squiggle-H_2O". An enquiry into the importance of solvation effects in phosphate ester and anhydride reactions. Biochim Biophys Acta 223:1–15

Goto E, Kodama T, Minoda Y (1978) Growth and taxonomy of thermophilic hydrogen bacteria. Agric Biol Chem 42:1305–1308

Graziano G, Merlino A (2014) Molecular bases of protein halotolerance. Biochim Biophys Acta 1844:850–858

Hakamada S, Sonoyama T, Ichiki S, Nakamura S, Uchiyama S, Kobayashi Y, Sambongi Y (2008) Stabilization mechanism of cytochrome c_{552} from a moderately thermophilic bacterium, *Hydrogenophilus thermoluteolus*. Biosci Biotechnol Biochem 72:2103–2109

Hallsworth JE, Yakimov MM, Golyshin PN, Gillion JL, D'Auria G, de Lima Alves F, La Cono V, Genovese M, McKew BA, Hayes SL, Harris G, Giuliano L, Timmis KN, McGenity TJ (2007) Limits of life in $MgCl_2$-containing environments: chaotropicity defines the window. Environ Microbiol 9:801–813

Hara KY, Araki M, Okai N, Wakai S, Hasunuma T, Kondo A (2014) Development of bio-based fine chemical production through synthetic bioengineering. Microb Cell Fact 13:173

Harrison JP, Gheeraert N, Tsigelnitskiy D, Cockell CS (2013) The limits for life under multiple extremes. Trends Microbiol 21:204–212

Hasegawa J, Uchiyama S, Tanimoto Y, Mizutani M, Kobayashi Y, Sambongi Y, Igarashi Y (2000) Selected mutations in a mesophilic cytochrome *c* confer the stability of a thermophilic counterpart. J Biol Chem 275:37824–37828

Hau HH, Gralnick JA (2007) Ecology and biotechnology of the genus *Shewanella*. Annu Rev Microbiol 61:237–258

Hoehler TM (2007) An energy balance concept for habitability. Astrobiology 7:824–838

Hong J, Yoshida N, Chong SH, Lee C, Ham S, Hirata F (2012) Elucidating the molecular origin of hydrolysis energy of pyrophosphate in water. J Chem Theory Comput 8:2239–2246

Horikoshi K (1999) Alkaliphiles: some applications of their products for biotechnology. Microbiol Mol Biol Rev 63:735–750

Huang Y, Krauss G, Cottaz S, Driguez H, Lipps G (2005) A highly acid-stable and thermostable endo-beta-glucanase from the thermoacidophilic archaeon *Sulfolobus solfataricus*. Biochem J 385:581–588

Jetten MS, Stams AJ, Zehnder AJ (1989) Isolation and characterization of acetyl-coenzyme A synthetase from *Methanothrix soehngenii*. J Bacteriol 171:5430–5435

Kanao T, Matsumoto C, Shiraga K, Yoshida K, Takada J, Kamimura K (2010) Recombinant tetrathionate hydrolase from *Acidithiobacillus ferrooxidans* requires exposure to acidic conditions for proper folding. FEMS Microbiol Lett 309:43–47

Kang X, Carey J (1999) Role of heme in structural organization of cytochrome c probed by semisynthesis. Biochemistry 38:15944–15951

Kankare J, Salminen T, Lahti R, Cooperman BS, Baykov AA, Goldman A (1996) Structure of *Escherichia coli* inorganic pyrophosphatase at 2.2 A resolution. Acta Crystallogr D Biol Crystallogr 52:551–563

Karan R, Capes MD, DasSarma S (2012) Function and biotechnology of extremophilic enzymes in low water activity. Aquat Biosyst 8(1):4

Kastritis PL, Papandreou NC, Hamodrakas SJ (2007) Haloadaptation: insights from comparative modeling studies of halophilic archaeal DHFRs. Int J Biol Macromol 41:447–453

Kato Y, Fujii S, Kuribayashi TA, Masanari M, Sambongi Y (2015) Thermal stability of cytochrome *c*' from mesophilic *Shewanella amazonensis*. Biosci Biotechnol Biochem 79:1125–1129

Kato C, Li L, Nogi Y, Nakamura Y, Tamaoka J, Horikoshi K (1998) Extremely barophilic bacteria isolated from the Mariana Trench, Challenger Deep, at a depth of 11,000 meters. Appl Environ Microbiol 64:1510–1513

Kato C, Li L, Tamaoka J, Horikoshi K (1997) Molecular analyses of the sediment of the 11,000-m deep Mariana Trench. Extremophiles 1:117–123

Kato C, Nogi Y (2001) Correlation between phylogenetic structure and function: examples from deep-sea *Shewanella*. FEMS Microbiol Ecol 35:223–230

Kimura K, Morimatsu K, Inaoka T, Yamamoto K (2017) Injury and recovery of *Escherichia coli* ATCC25922 cells treated by high hydrostatic pressure at 400–600 MPa. J Biosci Bioeng 123:698–706

Kobayashi S, Fujii S, Koga A, Wakai S, Matubayasi N, Sambongi Y (2017) *Pseudomonas aeruginosa* cytochrome c_{551} denaturation by five systematic urea derivatives that differ in the alkyl chain length. Biosci Biotechnol Biochem 81:1274–1278

Koschinsky A, Garbe-Schönberg D, Sander S, Schmidt K, Gennerich HH, Strauss H (2008) Hydrothermal venting at pressure-temperature conditions above the critical point of seawater, 5 S on the Mid-Atlantic Ridge. Geology 36:615–618

Kumari S, Tishel R, Eisenbach M, Wolfe AJ (1995) Cloning, characterization, and functional expression of acs, the gene which encodes acetyl coenzyme A synthetase in *Escherichia coli.* J Bacteriol 177:2878–2886

Kuribayashi TA, Fujii S, Masanari M, Yamanaka M, Wakai S, Sambongi Y (2017) Difference in NaCl tolerance of membrane-bound 5'-nucleotidases purified from deep-sea and brackish water *Shewanella* species. Extremophiles 21:357–368

Kushner DJ (1993) Growth and nutrition of halophilic bacteria. In: Vreeland RH, Hochstein L (eds) The biology of halophilic bacteria. CRC Press, Boca Raton, FL, pp 87–103

Kusube M, Kyaw TS, Tanikawa K, Chastain RA, Hardy KM, Cameron J, Bartlett DH (2017) *Colwellia marinimaniae* sp. nov., a hyperpiezophilic species isolated from an amphipod within the Challenger Deep, Mariana Trench. Int J Syst Evol Microbiol 67:824–831

Lavire C, Normand P, Alekhina I, Bulat S, Prieur D, Birrien JL, Fournier P, Hänni C, Petit JR (2006) Presence of *Hydrogenophilus thermoluteolus* DNA in accretion ice in the subglacial Lake Vostok, Antarctica, assessed using rrs, cbb and hox. Environ Microbiol 8:2106–2114

Lee CF, Makhatadze GI, Wong KB (2005) Effects of charge-to-alanine substitutions on the stability of ribosomal protein L30e from *Thermococcus celer*. Biochemistry 44:16817–16825

Leppänen VM, Nummelin H, Hansen T, Lahti R, Schäfer G, Goldman A (1999) *Sulfolobus acidocaldarius* inorganic pyrophosphatase: structure, thermostability, and effect of metal ion in an archael pyrophosphatase. Protein Sci 8:1218–1231

Liu B, Bartlam M, Gao R, Zhou W, Pang H, Liu Y, Feng Y, Rao Z (2004) Crystal structure of the hyperthermophilic inorganic pyrophosphatase from the archaeon *Pyrococcus horikoshii.* Biophys J 86:420–427

Madern D, Ebel C (2007) Influence of an anion-binding site in the stabilization of halophilic malate dehydrogenase from *Haloarcula marismortui*. Biochimie 89:981–987

Madern D, Ebel C, Zaccai G (2000) Halophilic adaptation of enzymes. Extremophiles 4:91–98

Mancinelli R, Botti A, Bruni F, Ricci MA, Soper AK (2007) Hydration of sodium, potassium, and chloride ions in solution and the concept of structure maker/breaker. J Phys Chem B 111:13570–13577

Marg BL, Schweimer K, Sticht H, Oesterhelt D (2005) A two-alpha-helix extra domain mediates the halophilic character of a plant-type ferredoxin from halophilic archaea. Biochemistry 44:29–39

Masanari M, Fujii S, Kawahara K, Oki H, Tsujino H, Maruno T, Kobayashi Y, Ohkubo T, Wakai S, Sambongi Y (2016) Comparative study on stabilization mechanism of monomeric cytochrome c_5 from deep-sea piezophilic *Shewanella violacea*. Biosci Biotechnol Biochem 80:2365–2370

Masanari M, Wakai S, Ishida M, Kato C, Sambongi Y (2014) Correlation between the optimal growth pressures of four *Shewanella* species and the stabilities of their cytochromes c_5. Extremophiles 18:617–627

Masanari M, Wakai S, Tamegai H, Kurihara T, Kato C, Sambongi Y (2011) Thermal stability of cytochrome c_5 of pressure-sensitive *Shewanella livingstonensis*. Biosci Biotechnol Biochem 75:1859–1861

Mayer F, Küper U, Meyer C, Daxer S, Müller V, Rachel R, Huber H (2012) AMP-forming acetyl coenzyme A synthetase in the outermost membrane of the hyperthermophilic crenarchaeon *Ignicoccus hospitalis*. J Bacteriol 194:1572–1581

McMillan LJ, Hepowit NL, Maupin-Furlow JA (2015) Archaeal inorganic pyrophosphatase displays robust activity under high-salt conditions and in organic solvents. Appl Environ Microbiol 82:538–548

Médicis ED, Paquette J, Gauthier JJ, Shapcott D (1986) Magnesium and manganese content of halophilic bacteria. Appl Environ Microbiol 52:567–573

Mevarech M, Frolow F, Gloss LM (2000) Halophilic enzymes: proteins with a grain of salt. Biophys Chem 86:155–164

Miyashita Y, Ohmae E, Nakasone K, Katayanagi K (2015) Effects of salt on the structure, stability, and function of a halophilic dihydrofolate reductase from a hyperhalophilic archaeon, *Haloarcula japonica* strain TR-1. Extremophiles 19:479–493

Morin PE, Freire E (1991) Direct calorimetric analysis of the enzymatic activity of yeast cytochrome *c* oxidase. Biochemistry 30:8494–8500

Morozkina EV, Slutskaya ES, Fedorova TV, Tugay TI, Golubeva LI, Koroleva OV (2010) Extremophilic microorganisms: biochemical adaptation and biotechnological application. Appl Biochem Microbiol 46:1–14

Mykytczuk NC, Foote SJ, Omelon CR, Southam G, Greer CW, Whyte LG (2013) Bacterial growth at −15 °C; molecular insights from the permafrost bacterium *Planococcus halocryophilus* Or1. ISME J 7:1211–1226

Nakamura S, Ichiki S, Takashima H, Uchiyama S, Hasegawa J, Kobayashi Y, Sambongi Y, Ohkubo T (2006) Structure of cytochrome c_{552} from a moderate thermophilic bacterium, *Hydrogenophilus thermoluteolus*: comparative study on the thermostability of cytochrome *c*. Biochemistry 45:6115–6123

Nealson KH, Rowe AR (2016) Electromicrobiology: realities, grand challenges, goals and predictions. Microb Biotechnol 9:595–600

Nogi Y, Kato C, Horikoshi K (1998) Taxonomic studies of deep-sea barophilic *Shewanella* strains and description of *Shewanella violacea* sp. nov. Arch Microbiol 170:331–338

Obuchi M, Kawahara K, Motooka D, Nakamura S, Yamanaka M, Takeda T, Uchiyama S, Kobayashi Y, Ohkubo T, Sambongi Y (2009) Hyperstability and crystal structure of cytochrome c_{555} from hyperthermophilic *Aquifex aeolicus*. Acta Crystallogr D Biol Crystallogr 65:804–813

Oda K, Kodama R, Yoshidome T, Yamanaka M, Sambongi Y, Kinoshita M (2011) Effects of heme on the thermal stability of mesophilic and thermophilic cytochromes *c*: comparison between experimental and theoretical results. J Chem Phys 134:025101

Ogawa K, Sonoyama T, Takeda T, Ichiki S, Nakamura S, Kobayashi Y, Uchiyama S, Nakasone K, Takayama SJ, Mita H, Yamamoto Y, Sambongi Y (2007) Roles of a short connecting disulfide bond in the stability and function of psychrophilic *Shewanella violacea* cytochrome c_5. Extremophiles 11:797–807

Ohshida T, Hayashi J, Satomura T, Kawakami R, Ohshima T, Sakuraba H (2016) First characterization of extremely halophilic 2-deoxy-D-ribose-5-phosphate aldolase. Protein Expr Purif 126:62–68

Oikawa K, Nakamura S, Sonoyama T, Ohshima A, Kobayashi Y, Takayama SJ, Yamamoto Y, Uchiyama S, Hasegawa J, Sambongi Y (2005) Five amino acid residues responsible for the high stability of *Hydrogenobacter thermophilus* cytochrome c_{552}: reciprocal mutation analysis. J Biol Chem 280:5527–5532

Onodera M, Yatsunami R, Tsukimura W, Fukui T, Nakasone K, Takashina T, Nakamura S (2013) Gene analysis, expression, and characterization of an intracellular α-amylase from the extremely halophilic archaeon *Haloarcula japonica*. Biosci Biotechnol Biochem 77:281–288

Oren A (1983) *Halobacterium sodomense* sp. nov., a dead sea halobacterium with an extremely high magnesium requirement. Int J Syst Evol Microbiol 33:381–386

Oren A (2013) Life at high salt concentrations, intracellular KCl concentrations, and acidic proteomes. Front Microbiol 4:315

Ozawa K, Harashina T, Yatsunami R, Nakamura S (2005) Gene cloning, expression and partial characterization of cell division protein FtsZ1 from extremely halophilic archaeon *Haloarcula japonica* strain TR-1. Extremophiles 9:281–288

Popinako A, Antonov M, Tikhonov A, Tikhonova T, Popov V (2017) Structural adaptations of octaheme nitrite reductases from haloalkaliphilic *Thioalkalivibrio* bacteria to alkaline pH and high salinity. PLoS ONE 12:e0177392

Romero PJ, de Meis L (1989) Role of water in the energy of hydrolysis of phosphoanhydride and phosphoester bonds. J Biol Chem 264:7869–7873

Rosenbaum E, Gabel F, Durá MA, Finet S, Cléry-Barraud C, Masson P, Franzetti B (2012) Effects of hydrostatic pressure on the quaternary structure and enzymatic activity of a large peptidase complex from *Pyrococcus horikoshii*. Arch Biochem Biophys 517:104–110

Saint-Martin H, Ortega-Blake I, Leś A, Adamowicz L (1994) The role of hydration in the hydrolysis of pyrophosphate. A Monte Carlo simulation with polarizable-type interaction potentials. Biochim Biophys Acta 1207:12–23

Sarethy IP, Saxena Y, Kapoor A, Sharma M, Sharma SK, Gupta V, Gupta S (2011) Alkaliphilic bacteria: applications in industrial biotechnology. J Ind Microbiol Biotechnol 38:769–790

Siddiqui KS, Cavicchioli R (2006) Cold-adapted enzymes. Annu Rev Biochem 75:403–433

Stetter KO (1999) Extremophiles and their adaptation to hot environments. FEBS Lett 452:22–25

Schink B (1997) Energetics of syntrophic cooperation in methanogenic degradation. Microbiol Mol Biol Rev 61:262–280

Schleper C, Pühler G, Kühlmorgen B, Zillig W (1995a) Life at extremely low pH. Nature 375:741–742

Schleper C, Puehler G, Holz I, Gambacorta A, Janekovic D, Santarius U, Klenk HP, Zillig W (1995b) *Picrophilus* gen. nov., fam. nov.: a novel aerobic, heterotrophic, thermoacidophilic genus and family comprising archaea capable of growth around pH 0. J Bacteriol 177:7050–7059

Sharma A, Kawarabayasi Y, Satyanarayana T (2012) Acidophilic bacteria and archaea: acid stable biocatalysts and their potential applications. Extremophiles 16:1–19

Siglioccolo A, Paiardini A, Piscitelli M, Pascarella S (2011) Structural adaptation of extreme halophilic proteins through decrease of conserved hydrophobic contact surface. BMC Struct Biol 11:50

Siliakus MF, van der Oost J, Kengen SWM (2017) Adaptations of archaeal and bacterial membranes to variations in temperature, pH and pressure. Extremophiles 21:651–670

Soppa J (2006) From genomes to function: haloarchaea as model organisms. Microbiology 152:585–590

Sun MM, Caillot R, Mak G, Robb FT, Clark DS (2001) Mechanism of pressure-induced thermostabilization of proteins: studies of glutamate dehydrogenases from the hyperthermophile *Thermococcus litoralis*. Protein Sci 10:1750–1757

Tadeo X, López-Méndez B, Trigueros T, Laín A, Castaño D, Millet O (2009) Structural basis for the aminoacid composition of proteins from halophilic archea. PLoS Biol 7:e1000257

Takai K, Gamo T, Tsunogai U, Nakayama N, Hirayama H, Nealson KH, Horikoshi K (2004) Geochemical and microbiological evidence for a hydrogen-based, hyperthermophilic subsurface lithoautotrophic microbial ecosystem (HyperSLiME) beneath an active deep-sea hydrothermal field. Extremophiles 8:269–282

Takai K, Moser DP, Onstott TC, Spoelstra N, Pfiffner SM, Dohnalkova A, Fredrickson JK (2001) *Alkaliphilus transvaalensis* gen. nov., sp. nov., an extremely alkaliphilic bacterium isolated from a deep South African gold mine. Int J Syst Evol Microbiol 51:1245–1256

Takai K, Nakamura K, Toki T, Tsunogai U, Miyazaki M, Miyazaki J, Hirayama H, Nakagawa S, Nunoura T, Horikoshi K (2008) Cell proliferation at 122 °C and isotopically heavy CH_4

production by a hyperthermophilic methanogen under high-pressure cultivation. Proc Natl Acad Sci U S A 105:10949–10954
Takeda T, Sonoyama T, Takayama SJ, Mita H, Yamamoto Y, Sambongi Y (2009) Correlation between the stability and redox potential of three homologous cytochromes *c* from two thermophiles and one mesophile. Biosci Biotechnol Biochem 73:366–371
Takenaka S, Wakai S, Tamegai H, Uchiyama S, Sambongi Y (2010) Comparative analysis of highly homologous *Shewanella* cytochromes c_5 for stability and function. Biosci Biotechnol Biochem 74:1079–1083
Taupin CM, Härtlein M, Leberman R (1997) Seryl-tRNA synthetase from the extreme halophile *Haloarcula marismortui*–isolation, characterization and sequencing of the gene and its expression in *Escherichia coli*. Eur J Biochem 243:141–150
Teplyakov A, Obmolova G, Wilson KS, Ishii K, Kaji H, Samejima T, Kuranova I (1994) Crystal structure of inorganic pyrophosphatase from *Thermus thermophilus*. Protein Sci 3:1098–1107
Todd MJ, Gomez J (2001) Enzyme kinetics determined using calorimetry: a general assay for enzyme activity? Anal Biochem 296:179–187
van der Wielen PW, Bolhuis H, Borin S, Daffonchio D, Corselli C, Giuliano L, D'Auria G, de Lange GJ, Huebner A, Varnavas SP, Thomson J, Tamburini C, Marty D, McGenity TJ, Timmis KN, Party BioDeep Scientific (2005) The enigma of prokaryotic life in deep hypersaline anoxic basins. Science 307:121–123
Venkateswaran K, Dollhopf ME, Aller R, Stackebrandt E, Nealson KH (1998) *Shewanella amazonensis* sp. nov., a novel metal-reducing facultative anaerobe from Amazonian shelf muds. Int J Syst Bacteriol 48:965–972
Vieille C, Zeikus GJ (2001) Hyperthermophilic enzymes: sources, uses, and molecular mechanisms for thermostability. Microbiol Mol Biol Rev 65:1–43
Wakai S, Abe A, Fujii S, Nakasone K, Sambongi Y (2017a) Pyrophosphate hydrolysis in the extremely halophilic archaeon *Haloarcula japonica* is catalyzed by a single enzyme with a broad ionic strength range. Extremophiles 21:471–477
Wakai S, Arazoe T, Ogino C, Kondo A (2017b) Future insights in fungal metabolic engineering. Bioresour Technol 245(Pt B):1314–1326
Wakai S, Masanari M, Ikeda T, Yamaguchi N, Ueshima S, Watanabe K, Nishihara H, Sambongi Y (2013a) Oxidative phosphorylation in a thermophilic, facultative chemoautotroph, *Hydrogenophilus thermoluteolus*, living prevalently in geothermal niches. Environ Microbiol Rep 5:235–242
Wakai S, Kidokoro S, Masaki K, Nakasone K, Sambongi Y (2013b) Constant enthalpy change value during pyrophosphate hydrolysis within the physiological limits of NaCl. J Biol Chem 288:29247–29251
Whitman WB, Coleman DC, Wiebe WJ (1998) Prokaryotes: the unseen majority. Proc Natl Acad Sci USA 95:6578–6583
Wolfenden R (2006) Degrees of difficulty of water-consuming reactions in the absence of enzymes. Chem Rev 106:3379–3396
Yamanaka M, Masanari M, Sambongi Y (2011) Conferment of folding ability to a naturally unfolded apocytochrome *c* through introduction of hydrophobic amino acid residues. Biochemistry 50:2313–2320
Yamanaka M, Mita H, Yamamoto Y, Sambongi Y (2009) Heme is not required for *Aquifex aeolicus* cytochrome c_{555} polypeptide folding. Biosci Biotechnol Biochem 73:2022–2025
Zhang G, Ge H (2013) Protein hypersaline adaptation: insight from amino acids with machine learning algorithms. Protein J 32:239–245

Chapter 18
Functioning Mechanism of ATP-Driven Proteins Inferred on the Basis of Water-Entropy Effect

Masahiro Kinoshita

Abstract There is a class of proteins called "motor proteins" or "protein machineries" which utilizes the ATP hydrolysis cycle: binding of ATP to a protein, hydrolysis of ATP, and dissociation of ADP and Pi from the protein. In the literature, the functioning mechanism of an ATP-driven protein has been discussed using the concept that it does work by utilizing the chemical energy stored in an ATP molecule or the free energy of ATP hydrolysis. In this chapter, we present a completely different view and argue that a protein is involved in an irreversible chemical reaction accompanying a decrease in system free energy, ATP hydrolysis, and during each hydrolysis cycle the protein undergoes a series of structural changes, leading to the exhibition of high function. The force which makes myosin move along F-actin, for example, is generated not by ATP but by water. In particular, the entropic force originating from the translational displacement of water molecules in the system plays a pivotal role. The concept of chemical–mechanical energy conversion is physically irrelevant.

Keywords ATP-driven protein • ATP hydrolysis cycle • Actomyosin Unidirectional movement • Water • Entropic force

18.1 Introduction

How do the ATP-driven proteins function? Why does the rotation of the central subunit in F_1-ATPase (Adachi et al. 2007; Okuno et al. 2008; Masaike et al. 2008) or the movement of myosin along F-actin (Yanagida et al. 2000; Okada et al. 2007) occur unidirectionally? The aim of this chapter is to give an answer to this question. In protein folding, for example, the initial and final states of the protein are unfolded and folded, respectively, and apparently different. Therefore, it can readily be understood that protein folding spontaneously occurs as an *irreversible* process

M. Kinoshita (✉)
Institute of Advanced Energy, Kyoto University, Uji, Kyoto 611-0011, Japan
e-mail: kinoshit@iae.kyoto-u.ac.jp

© Springer Nature Singapore Pte Ltd. 2018

M. Suzuki (ed.), *The Role of Water in ATP Hydrolysis Energy Transduction by Protein Machinery*, https://doi.org/10.1007/978-981-10-8459-1_18

accompanying a decrease in system free energy. By contrast, the conformations of F_1-ATPase before and after a 120° rotation of the central subunit in each ATP hydrolysis cycle are the same. The conformations of actomyosin before and after myosin performs a move along F-actin are essentially the same. In these cases, it appears that an equilibrium state is first broken and then recovered. However, once an equilibrium state is reached, it remains stable and cannot be broken unless work or energy is applied to it. Therefore, one takes the view that the chemical energy stored in an ATP molecule or the free energy of ATP hydrolysis is converted to the mechanical energy required for breaking the equilibrium state and performing the directed rotation or movement. In this chapter, we question this view and suggest a new one.

18.2 Crucial Importance of Water-Entropy Effect in Biological Systems

In a biological, microscopic self-assembly process, contact of oppositely (positively and negatively) charged groups of a biomolecule or biomolecules (see Fig. 18.1) leads to a gain of electrostatic attractive interaction between these two groups. However, the contact occurs not in vacuum but in aqueous solution. An oxygen atom (O) in a water molecule carries a negative partial charge, and a hydrogen atom (H) in it carries a positive one. Therefore, the decrease in energy due to a gain of electrostatic attractive interaction mentioned above is accompanied by the increase in energy due to a loss of electrostatic attractive interactions between the positively charged group and water oxygen atoms and between the negatively charged group and water hydrogen atoms. Roughly half of the increase is cancelled out by the decrease arising from structure reorganization of water including the water released to the bulk upon the self-assembly process: Still, the remaining increase (we refer to this as electrostatic "energetic dehydration") is almost as large as the decrease mentioned above. In other words, when the energetic dehydration occurs during the self-assembly process, a gain of electrostatic attractive interaction between

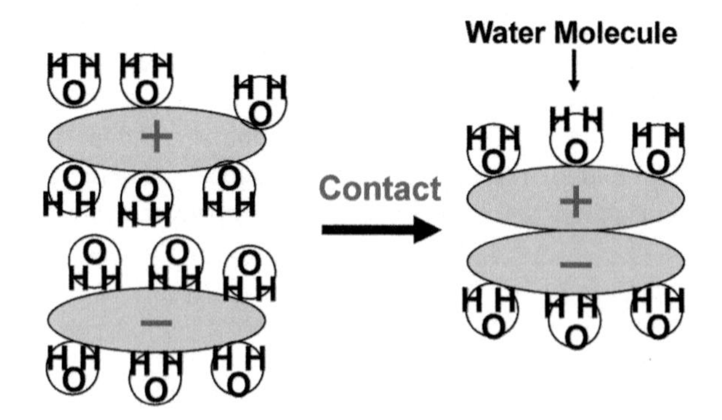

Fig. 18.1 Contact of oppositely (positively and negatively) charged groups of a biomolecule or biomolecules in water

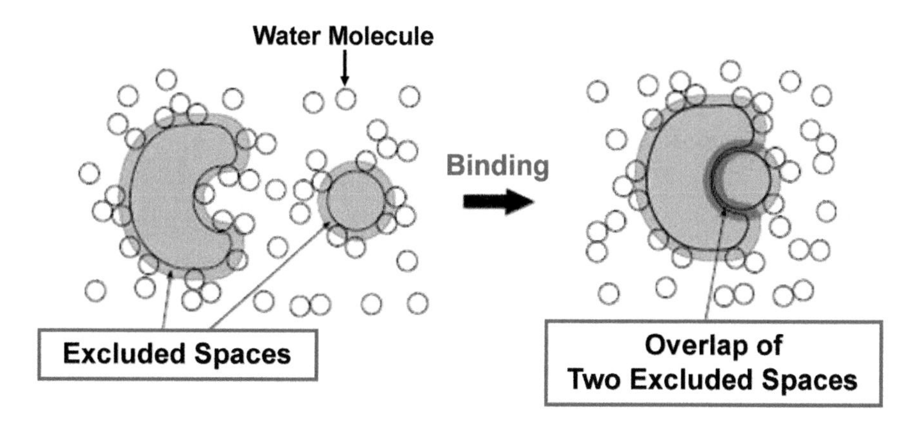

Fig. 18.2 Binding of two biomolecules in water

oppositely charged groups is essential for compensating the energetic dehydration. However, the gain itself does not drive the self-assembly process (Hayashi et al. 2014, 2015; Oshima et al. 2016). We note that a gain of van der Waals (vdW) attractive interaction between groups of a biomolecule or biomolecules upon a self-assembly process is also accompanied by vdW energetic dehydration.

Upon contact of two groups of a biomolecule or binding of two biomolecules (see Fig. 18.2), the two excluded spaces (an excluded space (ES) is a space which the centers of water molecules cannot enter) overlap, and the total ES decreases by the overlapped space marked in pink. This leads to an increase in the total volume available to the translational displacement of water molecules in the entire system, which is followed by a large gain of water entropy (Hayashi et al. 2014, 2015; Oshima et al. 2016). As we have shown for a number of examples, this water-entropy gain can predominate over the conformational entropy loss of a biomolecule or two biomolecules and crucially drive the contact or binding (Hayashi et al. 2014, 2015; Oshima et al. 2016; Yoshidome et al. 2008).

The water-entropy gain upon a biological self-assembly process is remarkably large. For instance, folding of apoplastocyanin with 99 residues is accompanied by a water-entropy gain of ~670 k_B (k_B is the Boltzmann constant) (Yoshidome et al. 2008). This gain is ascribed to a large decrease in the total excluded volume (EV) generated by the protein upon folding where the backbone and side chains are closely packed. Only by a theory emphasizing this entropic EV effect, all of protein folding and cold (Yoshidome and Kinoshita 2012), pressure (Harano et al. 2008), and thermal (Oda et al. 2011) denaturating can be elucidated in a unified manner within the same theoretical framework (Kinoshita 2013; Oshima and Kinoshita 2015).

We refer to the driving force by the water-entropy effect as “entropic force.” The entropic force is much weaker than the electrostatic force in vacuum but significantly stronger than the electrostatic force screened by water and ions (Na^+ and Cl^-) in aqueous solution. In this chapter, we discuss the functioning mechanism of ATP-driven proteins with the emphasis on the entropic force.

18.3 Outline of Functioning Mechanism of ATP-Driven Proteins

18.3.1 Chemical Reaction: Hydrolysis or Synthesis of ATP

It is well known that the standard free-energy change $\Delta G°$ of the chemical reaction,

$$ATP + H_2O \rightarrow ADP + Pi, \quad (18.1)$$

is negative (Kodama 1985) (it is more strict to write "ATP^{4-}+H_2O → ADP^{3-}+ Pi^{2-}+H^+"). The free-energy change upon the chemical reaction ΔG is related to ΔG ° through

$$\Delta G = \Delta G° + RT \ln(Z) \quad (18.2)$$

where R is the gas constant, T is the absolute temperature, and Z is dependent on concentrations of ATP, ADP, and Pi. In the equilibrium state, $\Delta G = 0$ and

$$\Delta G° = -RT \ln\ (K) \quad (18.3)$$

where K is the value of Z in the equilibrium state referred to as "equilibrium constant."

When the concentrations of ATP and ADP are sufficiently high and low, respectively (and the concentration of Pi is low enough), ln(Z) is negative and ΔG takes a significantly large, negative value: The chemical reaction proceeds in the forward (right) direction, and the ATP hydrolysis occurs. On the other hand, if the concentration of ATP is almost zero but that of ADP is very high (and that of Pi is high enough), ln(Z) is positive and quite large, leading to positive ΔG: The chemical reaction proceeds in the backward (left) direction, and the ATP synthesis occurs. The chemical reaction is an irreversible process except when it is in the equilibrium state.

However, the chemical reaction can hardly occur in bulk aqueous solution without a catalyst. An ATP-driven protein acts as the catalyst. In other words, the protein is forced to be coupled with the chemical reaction. An ATP-driven protein functions when its affinity with ATP is higher than that with ADP and under the solution condition that the ATP hydrolysis occurs.

18.3.2 A Protein Coupled with Irreversible Chemical Reaction

As stated in "Introduction," the conformations of an ATP-driven protein or protein complex before and after it exhibits a function are the same. In our view, the protein

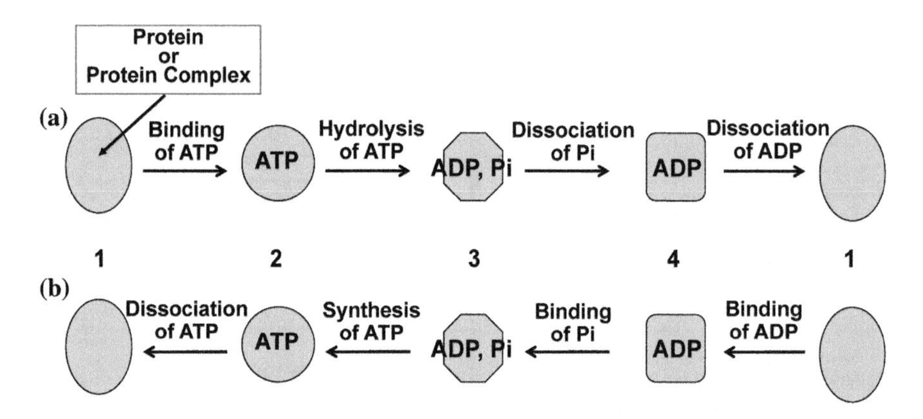

Fig. 18.3 **a** One cycle of ATP hydrolysis. The concentrations of ATP and ADP are sufficiently high and low, respectively. The protein or protein complex changes its structure as "$1 \rightarrow 2 \rightarrow 3 \rightarrow 4 \rightarrow 1$". A different shape implies a different structure. The change from a structure to the next one occurs *continuously*. **b** One cycle of ATP synthesis. The ATP concentration is almost zero, and the ADP one is very high. The protein or protein complex changes its structure as "$1 \rightarrow 4 \rightarrow 3 \rightarrow 2 \rightarrow 1$"

or protein complex is coupled with the ATP hydrolysis, an irreversible chemical reaction, with the result that the unidirectional rotation or movement occurs as an irreversible process accompanying a decrease in system free energy, just as in the case of protein folding.

When the concentrations of ATP and ADP are sufficiently high and low, respectively, the ATP hydrolysis occurs (see Sect. 18.3.1). As illustrated in Fig. 18.3a, one cycle comprises the following events: binding of ATP to the protein (or protein complex), hydrolysis of ATP into ADP and Pi where the protein acts as the catalyst, and dissociation of ADP and Pi from the protein (in the figure, it is assumed that dissociation of Pi occurs first). The events in the right direction preferentially occur. The system free energy decreases in each step indicated by the arrow. The protein takes the most stable structure so that the system free energy can be minimized: There are *at least* four different structures. If the protein was not coupled with the ATP hydrolysis, the protein structure would not change from structure 1. Due to the coupling, the protein sequentially changes its structure and exhibits high function accordingly. In one cycle, the protein structure returns to structure 1, and the system free energy decreases by ΔG given by Eq. (18.2). The change in system free energy during one cycle is illustrated in Fig. 18.4.

As Kodama (1985) showed for the actomyosin system, the free-energy change upon the dissociation of ADP from myosin can become positive, depending on the ATP and ADP concentrations. Even in such a case, when the free-energy change is as small as $\sim k_B T$ (Kodama 1985), ADP can readily dissociate and the net change in system free energy is still equal to ΔG given by Eq. (18.2).

If the ATP concentration is almost zero and the ADP one is very high, one cycle can be illustrated in Fig. 18.3b. The events in the left direction preferentially occur.

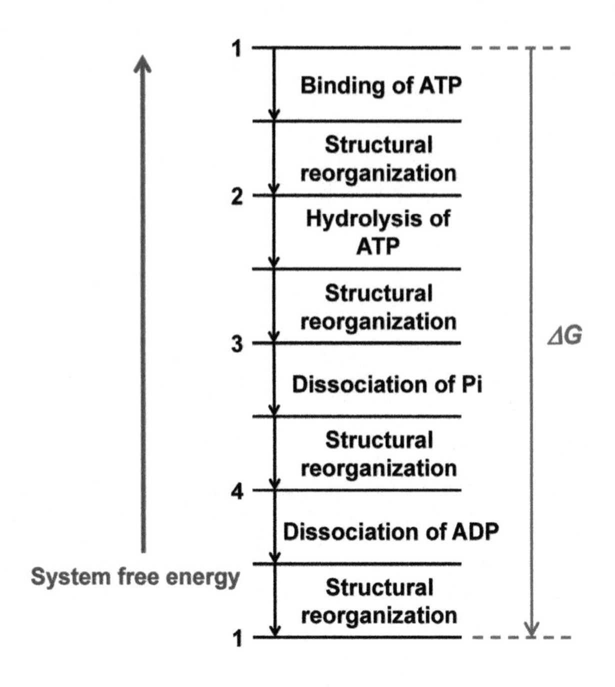

Fig. 18.4 Change in system free energy during one cycle of ATP hydrolysis. The concentrations of ATP and ADP are sufficiently high and low, respectively. The protein or protein complex changes its structure as "$1 \rightarrow 2 \rightarrow 3 \rightarrow 4 \rightarrow 1$" illustrated in Fig. 18.3a. "Structural reorganization" implies each change of the protein or protein-complex structure. The net change in system free energy is ΔG given by Eq. (18.2). Also see the third paragraph in Sect. 18.3.2

What if the chemical reaction is in the equilibrium state? Even in the equilibrium state, the reaction is never stopped: The reaction rates in the right and left directions are equal. The events in the right direction and those in the left one occur with equal frequency. For instance, even after ATP binds to the protein, it can dissociate from the protein before hydrolysis begins; and even after hydrolysis of ATP into ADP and Pi occurs, synthesis of ATP can possibly occur before ADP or Pi dissociates.

18.4 Entropic Force and Entropic Potential

18.4.1 Entropic Force and Entropic Potential Induced Between a Planar Wall and a Large Sphere

We consider a planar hard wall and a large hard sphere immersed in small hard spheres. The wall–sphere and sphere–sphere potentials include no attractive tails. In such a system, all the accessible system configurations share the same energy and the system behavior becomes purely entropic in origin. Therefore, the potential of mean force and the mean force can be referred to as "entropic potential" and "entropic force," respectively. Due to the effect of the translational motion of small spheres, a force is induced between the wall and the large sphere. Figure 18.5 shows an example force F_{wall} induced between a planar wall and a large sphere. The dimensionless force is plotted against h (h is the surface separation) divided by the

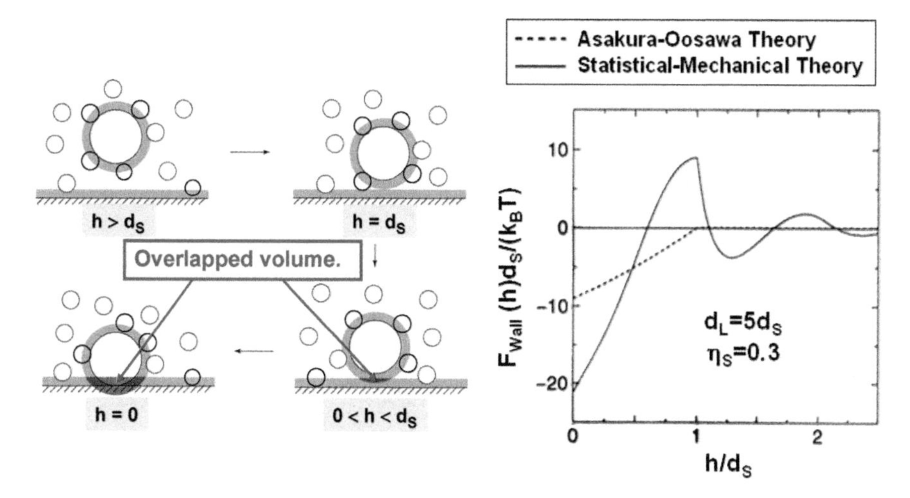

Fig. 18.5 Example of entropic force F_{Wall} induced between a large sphere and a planar wall immersed in small spheres. T, k_B, h, d_L, and d_S denote the absolute temperature, Boltzmann constant, surface separation, diameter of the large sphere, and diameter of the small spheres, respectively. η_S is the packing fraction of the small spheres ($\eta_S = 0.383$ for bulk water at ambient temperature and pressure). As η_S increases, the difference between the broken and solid curves becomes larger

diameter of small spheres d_S. The Asakura–Oosawa (AO) theory (Asakura and Oosawa 1954, 1958), which looks at only the overlapped volume marked in black, gives the broken line. The AO force is zero as long as there is no overlap of the two EVs ($h > d_S$). An attractive force is induced for $h < d_S$, and it becomes monotonically stronger with decreasing h. On the other hand, the exact force represented by the solid line is longer ranged and oscillatory with a periodicity of d_S. The most significant difference between the forces obtained via the two routes is in the high, repulsive peak of the exact one occurring at $h \sim d_S$. To reproduce the solid line, an elaborate statistical–mechanical theory such as an integral equation theory (Hansen and McDonald 2006; Kinoshita 2002, 2006) is required. The potential $\Phi_{\mathrm{wall}}(h)$, which is related to $F_{\mathrm{wall}}(h)$ by $F_{\mathrm{wall}}(h) = -d\Phi_{\mathrm{wall}}(h)/dh$, is shown in Fig. 18.6.

In what follows, the physical origin of the exact force is briefly discussed (Kinoshita 2002, 2006). The induced force $F_{\mathrm{wall}}(h)$ is related to the density of small spheres at contact around the large sphere. The wall and the large sphere generate EVs for small spheres. Further, each small sphere also generates an EV for *the other small spheres* and is promoted to contact the wall and the large sphere, with the result that the contact density at the wall surface and the large sphere is considerably in excess of the bulk density ρ_S. When the large sphere is near contact with the wall, an additional, important factor arises: The contact density is further enhanced in the channel confined between the two surfaces for $h \sim nd_s$ ($n = 1, 2, \ldots$). This is particularly true for $h \sim d_S$ where small sphere can contact two surfaces with the result of a more pronounced decrease in the EVs for the other small

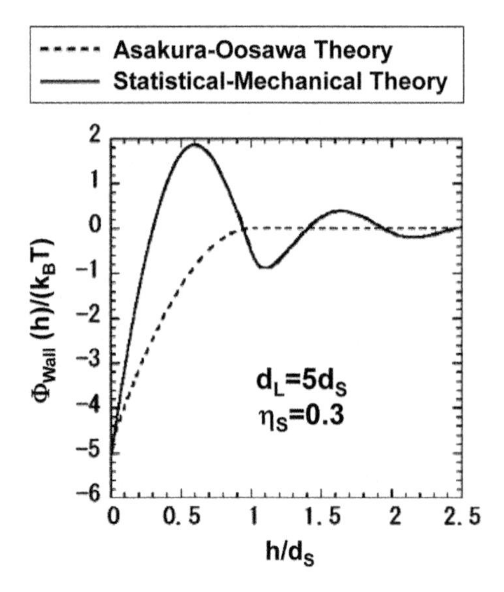

Fig. 18.6 Example of entropic potential Φ_{Wall} induced between a large sphere and a planar wall immersed in small spheres. Φ_{Wall} is related to $F_{Wall}(h)$ shown in Fig. 18.5 through $F_{Wall}(h) = -d\Phi_{Wall}(h)/dh$. T, k_B, h, d_L, and d_S denote the absolute temperature, Boltzmann constant, surface separation, diameter of the large sphere, and diameter of the small spheres, respectively. η_S is the packing fraction of the small spheres ($\eta_S = 0.383$ for bulk water at ambient temperature and pressure). The entropic potential is the potential of mean force (the free energy of small spheres for "h = finite" *relative to* that for "$h \to \infty$") in the presence of no energetic factors. It should be noted that the Asakura–Oosawa theory gives a fairly accurate value of the entropic potential *at contact* due to fortuitous cancellation of errors (Kinoshita 2002, 2006)

spheres. Choosing the large sphere center as the origin and using polar coordinates, we express the contact density as $\rho_C(r = (d_L + d_S)/2, \theta)$ (θ is measured from the positive axis normal to the flat surface; see Fig. 18.7). The force is induced by collisions of small spheres with the large sphere. The force thus acting on the large sphere surface becomes stronger as the collision frequency increases. The collision frequency becomes higher as ρ_C increases. The forces originating from ρ_C for $0 \leq \theta < \pi/2$ and from ρ_C for $\pi/2 < \theta \leq \pi$ (ρ_C is symmetrical about $\theta = \pi$), respectively, constitute attractive and repulsive components of the net force. The two components are both very strong and comparable in magnitude, and the net force (i.e., the sum of the two components) is much weaker. See Fig. 18.7. At $h = d_S$, for example, ρ_C at $\theta \sim \pi$ is considerably high with the result that the repulsive component dominates, thus making the net force repulsive. At $h = 0$, vanishing of ρ_C for $\theta_{max} < \theta \leq \pi$ where $\cos\theta_{max} = -(d_L - d_S)/(d_L + d_S)$ dominates, leading to an attractive net force. In summary, the exact force stems from the microscopic packing effects of small spheres near the large sphere and the wall. The AO theory neglects all of these effects.

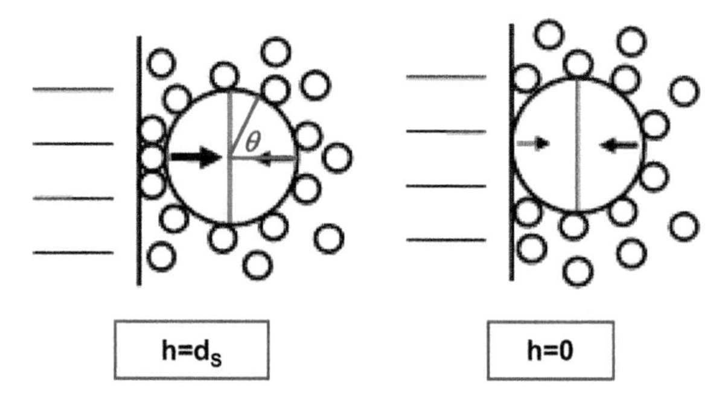

Fig. 18.7 Attractive (right arrow) and repulsive (left arrow) components of the net force acting on the large sphere. The contact density of the small spheres is expressed as $\rho_C(r = (d_L + d_S)/2, \theta)$ where d_L and d_S denote the diameter of the large sphere and that of the small spheres, respectively, and polar coordinates (r, θ) are used by choosing the center of the large sphere as the origin

The small spheres form a solvent, and the wall and the large sphere correspond to solutes. In biological systems, the solvent is water characterized by hydrogen bonds. However, in the entropic gain upon the solute contact and related processes, the translational entropy predominates over the rotational entropy (Yoshidome et al. 2008; Harano et al. 2008; Kinoshita 2008). The basic physics of the entropic effect considered in this chapter can be captured by modeling water as hard spheres in many cases where solute molecules consist of nonpolar, polar, and positively or negatively charged groups, *as long as the diameter and number density are set at the values of water*. The entropic EV effect by a solvent becomes larger as the solvent density increases and the solvent diameter decreases. Thanks to the hydrogen bonding, water can exist as a dense liquid despite its quite a small molecular size, leading to an exceptionally large entropic effect (Soda 1993). Hereafter, the small spheres are referred to as water molecules.

18.4.2 Entropic Force and Entropic Potential Acting on a Large Sphere Along a Large Body

Suppose that a large sphere is sitting on a large body. For the large sphere, geometric features of the large body induce the entropic force (or potential) acting on the large sphere in a specific direction *along* the large body (Kinoshita 2002, 2006; Dinsmore et al. 1996), as illustrated in Fig. 18.8. The large sphere and the large body generate the EVs for water molecules. The large sphere on the large body tends to be moved so that the overlapped volume marked in black, total volume available to the translational displacement of water molecules, and translational entropy of water can increase. For example, the large sphere is locally repelled from a step edge, attracted to a corner, and moved from a convex surface to a concave one.

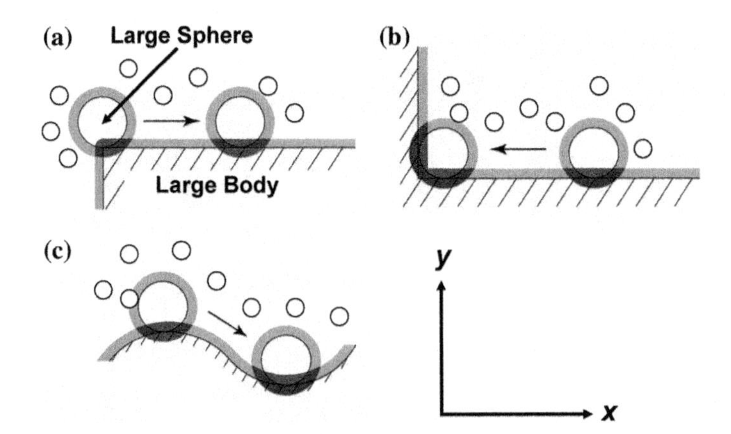

Fig. 18.8 Entropic force acting on a large sphere sitting on a large body. They are immersed in small spheres. The force induced *along* the large body, which is the entropic force between the large sphere and the large body along a specific trajectory, is shown

From the explanations made in Sect. 18.4.1, the large sphere feels an entropic potential field spatially formed. The field is three-dimensional (Kinoshita 2006), but we are especially interested in the potentials along the x- and y-axes (see Fig. 18.8). The y-axis is normal to the large body surface. Since the field is influenced by the microstructure of water within the confined space, an elaborate statistical–mechanical theory (not the AO theory (Asakura and Oosawa 1954, 1958)) is required for calculating the field.

18.4.3 Control of a Movement of a Large Solute Along a Large Body

In Sect. 18.4.2, the large sphere corresponds to a solute. An important point is that the overlapped volume (the volume marked in black in Fig. 18.8) and the entropic potential field felt by the solute are strongly dependent on the solute shape as well as on the geometric features of the large body surface ("geometry" is the most important factor in discussing the entropic force or potential). In other words, the field and the movement of the solute can be controlled by adjusting the solute geometry (i.e., overall shape and polyatomic structure). We believe that this mechanism is utilized in the unidirectional movement of myosin along F-actin. The adjustment is achieved using the ATP hydrolysis cycle.

18.5 Mechanism of Unidirectional Movement of Myosin Head (S1) Along F-Actin

18.5.1 Physical Basis

It was experimentally shown that the binding of myosin to F-actin is entropically driven (Katoh and Morita 1996). Also, the binding is much stronger for myosin to which ADP or nothing is bound than for myosin to which ATP is bound (Kodama 1985). Our recent analysis based on statistical thermodynamics (Oshima et al. 2016) has also shown that the binding of myosin to F-actin is accompanied by positive changes in entropy and in enthalpy (i.e., entropy-driven and enthalpy opposed).

According to experimental results (Coureux et al. 2003, 2004), myosin with ATP or ADP + Pi bound to it has a cleft near the myosin/F-actin interface and this cleft is closed in myosin with ADP or nothing bound to it. Due to the cleft, the decrease in the total EV for water molecules upon the contact of myosin to F-actin becomes smaller, leading to weaker entropic force. F-actin and myosin possess the polyatomic structures. When myosin is strongly bound to a rigor binding site of F-actin, their interface is tightly packed. When ATP is bound to myosin, the structure of myosin undergoes a large change. As a result, the tight packing of the interfaces is lost, forming the cleft. When the tight packing of myosin and ATP occurs, the packing of myosin and F-actin is inevitably loosened. The former should be more important for increasing the water entropy.

A significant amount of information has been obtained by experimental studies (Yanagida et al. 2000; Okada et al. 2007; Howard 2001; Iwaki et al. 2013). It is generally believed that the two-headed structure of myosin is essential in the unidirectional movement along F-actin. On the other hand, Kitamura et al. (1999, 2005) performed an interesting experiment using a single head of myosin, S1. They prevented the detachment of S1 (i.e., S1 remained attached to F-actin) using a novel experimental technique and confined its movement within a one-dimensional space. They found that S1 realizes the unidirectional movement. We performed a theoretical analysis based on statistical mechanics for this behavior of S1 (Amano et al. 2010).

In what follows, we consider S1 and describe the mechanism of its unidirectional movement on the basis of the experimental and theoretical results mentioned above, which should provide a clue to more complex issues such as the unidirectional walk of myosin V with two conjoined legs along F-actin.

18.5.2 *Physical Picture of Unidirectional Movement Coupled with ATP Hydrolysis Cycle*

We consider the solution condition under which the concentrations of ATP and ADP are sufficiently high and low, respectively, and the ATP hydrolysis occurs. We consider two cases: In case (1), S1 can get detached from F-actin (it is not forcibly attached to F-actin); and in case (2), as in the experiment by Kitamura et al. (1999, 2005), S1 is not allowed to get detached from F-actin (it remains attached to F-actin).

18.5.2.1 Case (1)

On the basis of the experimental information available for myosin (Yanagida et al. 2000; Okada et al. 2007; Howard 2001; Iwaki et al. 2013), we infer the behavior of S1 exhibited when it is allowed to get detached from F-actin. As illustrated in Fig. 18.9, there can be five different geometries (structures) of S1 and any change from a geometry to the next one occurs continuously. In our discussion, however, we employ the following simplifications (see Fig. 18.9) and examine to what extent the experimental result can be explained and what problems remain to be solved in Sect. 18.5.3. When only ADP or nothing is bound to S1, S1 takes geometry I: S1

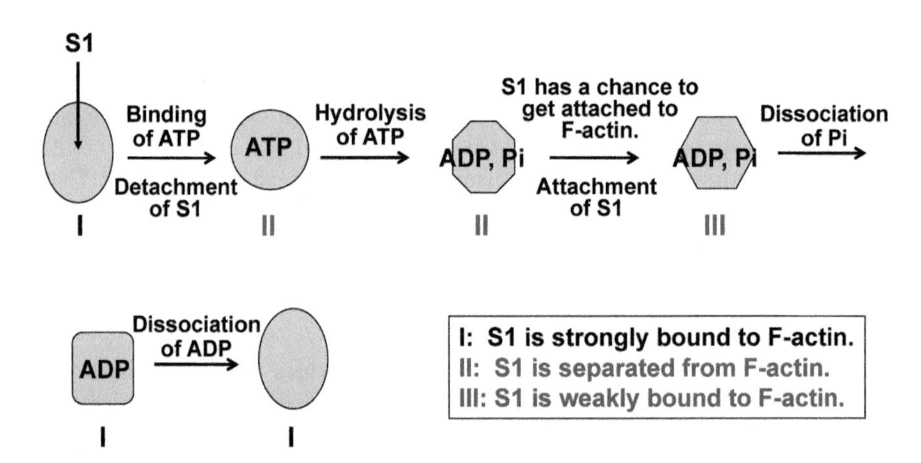

Fig. 18.9 Three geometries (I, II, and III) of a single head of myosin (S1) during one cycle of ATP hydrolysis in case (1). When ADP or nothing is bound to S1, S1 takes geometry I: S1 is strongly bound to F-actin. When ATP is bound to S1, S1 takes geometry II: S1 can readily get detached from F-actin. When ADP and Pi are bound to S1 and S1 is separated from F-actin, S1 takes geometry II: There is a chance for S1 to get attached to F-actin. When ADP and Pi are bound to S1, S1 takes geometry III: S1 is weakly bound to F-actin. Not only geometry II but also geometry III has a cleft near the myosin/F-actin interface, but the cleft in geometry III is less open. It is assumed that any change from one of the three geometries to the next geometry occurs in a stepwise manner

strongly binds to F-actin and can hardly move. When ATP binds to S1, S1 changes its geometry to geometry II, gets detached from F-actin, and makes a biased Brownian motion (i.e., a Brownian motion in the presence of a potential field). During the Brownian motion of S1, ATP is hydrolyzed into ADP and Pi, but Pi does not dissociate from S1 as long as S1 is separated from F-actin. S1 either has a chance to get attached to F-actin or becomes infinitely separated from F-actin. (In myosin V, the infinite separation is prohibited for the following reason: While one of the two heads is separated from F-actin, the other is bound to F-actin. See the third paragraph in "18.6 Roles of ATP Hydrolysis Cycle and Water" for more details.) If S1 has a chance to get attached to F-actin, S1 changes its geometry to geometry III and Pi dissociates from S1. As a result, S1 changes its geometry to geometry I and S1 strongly binds to F-actin. During one cycle of ATP hydrolysis, S1 changes its geometry as "I → II → III → I". Dissociation of Pi occurs *rapidly* only after S1 weakly binds to F-actin. Therefore, the time during which S1 is in geometry III is quite short.

As described above, the spatial distribution of the entropic potential field felt by S1 is strongly dependent on the geometry (more accurately, polyatomic structure) of S1. When S1 takes geometry I, it is difficult for S1 to get detached from F-actin because the potential along the y-axis (see Fig. 18.10) qualitatively looks like the solid curve in Fig. 18.6 possessing a high free-energy barrier well exceeding k_BT. When S1 takes geometry II with the cleft mentioned above, the potential along the y-axis is still oscillatory as the solid curve in Fig. 18.6, but it is significantly damped (i.e., its amplitudes are much smaller) and the barrier is much lower ($\sim k_BT$): It is quite easy for S1 to overcome the barrier and get detached from F-actin. Even after S1 gets detached from F-actin, it feels the potential field and makes a biased Brownian motion. S1 may have a chance to get attached to F-actin due to the potential field. If S1 becomes sufficiently separated from F-actin, it feels essentially no potential field: S1 either has a chance to get attached to F-actin or becomes infinitely separated from F-actin (presumably, the probability of the latter is higher). When S1 gets attached to F-actin, the time during which S1 is in geometry II should be significantly long (the longest). This is because there is only a weak or essentially no potential field felt by S1, leading to the almost random

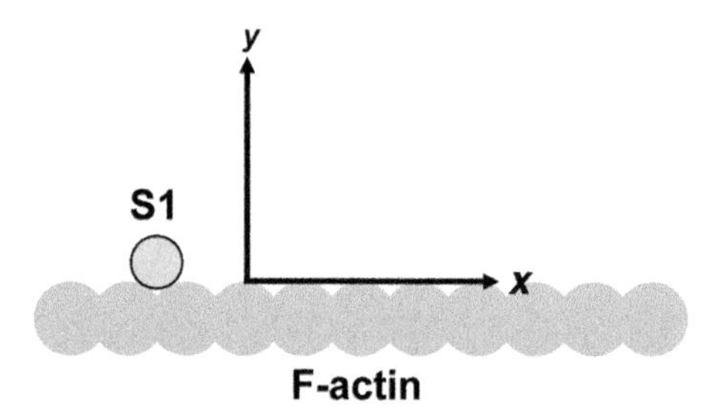

Fig. 18.10 Definition of x- and y-axes. The geometry of F-actin is like a double helical structure formed by two sets of connected spheres, but it is drawn as a set of one-dimensionally connected spheres for simplicity. A sphere represents a G-actin molecule. The water molecules are not shown

nature of the motion. When S1 takes geometry III, the amplitudes of the potential along the *y*-axis are smaller than in the case of geometry I but somewhat larger than in the case of geometry II: S1 remains attached to F-actin.

18.5.2.2 Case (2)

In this case, S1 remains attached to F-actin (i.e., S1 is always sitting on F-actin irrespective of the potential along the *y*-axis). It moves only in accordance with the potential along the *x*-axis. The change in geometric characteristics can be illustrated in Fig. 18.11. Pi dissociates as soon as the hydrolysis of ATP is finished because S1 is not separated from F-actin but weakly bound to it (the time during which S1 is in geometry III is still quite short), and the time during which S1 is in geometry II is far shorter than in case (1). The time during which S1 is in geometry I is the longest. The potential along the *x*-axis can be characterized as follows. When S1 is in geometry I, the potential felt by S1 looks like the black curve shown in Fig. 18.12. The potential felt by S1 with geometries II and III looks like the red curve shown in Fig. 18.12. The black and red curves are drawn on the basis of our statistical–mechanical analysis using a simple model of the S1 and F-actin system (Amano et al. 2010). (In case (1), the potential felt by S1 with geometry I looks like the black curve, and that felt by S1 with geometry III looks like the red curve.)

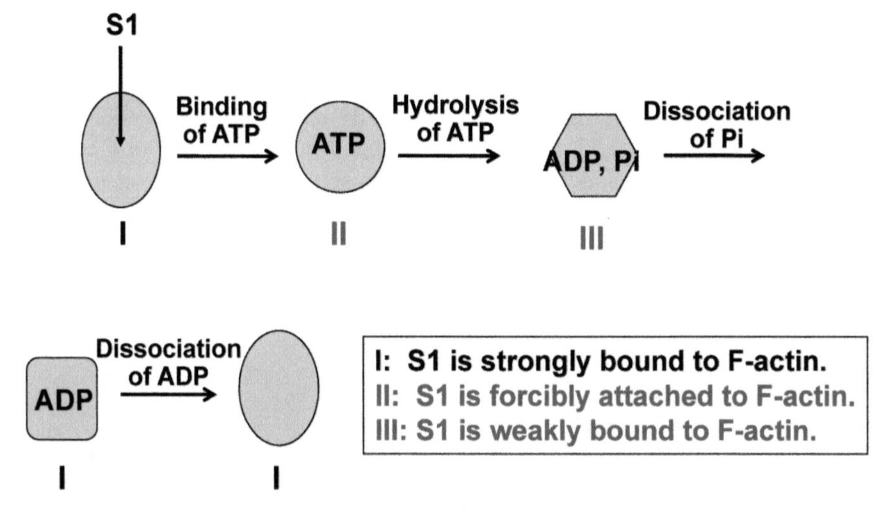

Fig. 18.11 Three geometries (I, II, and III) of a single head of myosin (S1) during one cycle of ATP hydrolysis in case (2). When ADP or nothing is bound to S1, S1 takes geometry I: S1 is strongly bound to F-actin. When ATP is bound to S1, S1 takes geometry II: However, S1 is not allowed to get detached from F-actin. When ADP and Pi are bound to S1, S1 takes geometry III: S1 is weakly bound to F-actin. Not only geometry II but also geometry III has a cleft near the myosin/F-actin interface, but the cleft in geometry III is less open. It is assumed that any change from one of the three geometries to the next geometry occurs in a stepwise manner

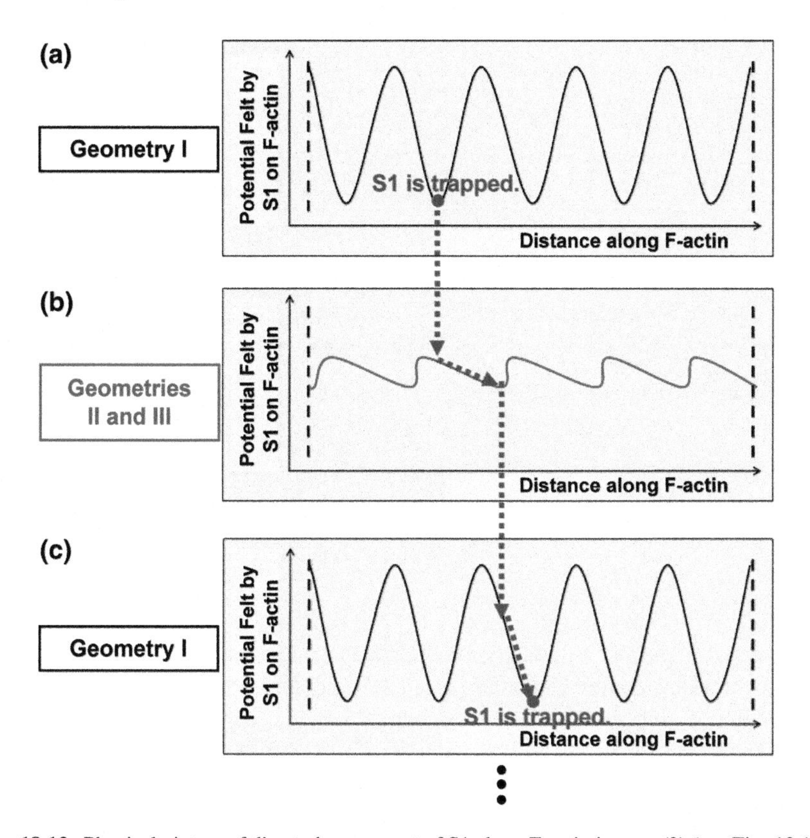

Fig. 18.12 Physical picture of directed movement of S1 along F-actin in case (2) (see Fig. 18.11). **a**, **c** Potential field felt by S1 with geometry I on F-actin. **b** Potential field felt by S1 with geometries II and III on F-actin. The movement of S1 is controlled by adjusting the S1 geometry (a change in the S1 geometry leads to that in the potential field felt by S1): The adjustment is performed by the ATP hydrolysis cycle

Our theoretical study (Amano et al. 2010) suggested that compared to the black curve, the red curve is more asymmetrical about the *y*-axis and the amplitudes are much smaller. The unidirectional movement of S1 is realized by repeated switches from one of the curves to the other. In the presence of the red-curve potential, a biased Brownian motion of S1 occurs: Due to the sufficiently small amplitudes, S1 sometimes moves by more than a single step and even backward, which is in accord with the experimental observation (Kitamura et al. 1999, 2005). The ATP hydrolysis cycle is utilized for controlling the movement of S1 through the adjustment of S1 geometry. While S1 feels the potential of the black curve, it is trapped on the so-called rigor binding site. S1 trapped on a rigor binding site is illustrated in Fig. 18.13a. It becomes apparent from a simple geometric consideration that the overlap of the EVs generated by S1 and two G-actin molecules becomes largest and the water entropy is maximized when S1 is on a rigor binding site (see Fig. 18.13b). The distance between two adjacent rigor binding sites is

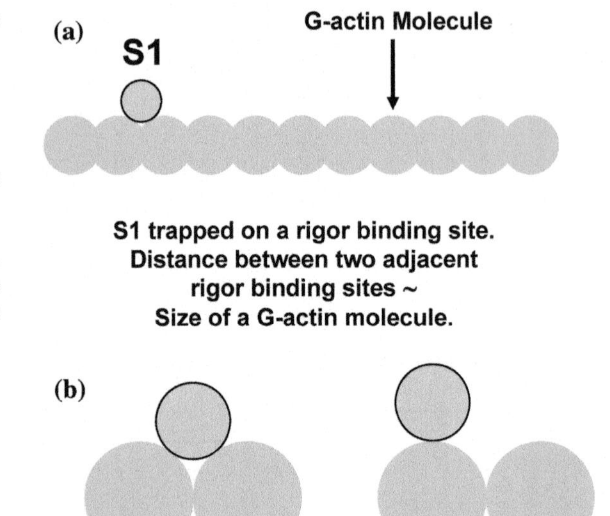

Fig. 18.13 **a** S1 trapped on a rigor binding site. On the rigor binding site, S1 is in contact with two G-actin molecules. **b** The overlap of excluded volumes occurs for S1 and two G-actin molecules in the left, whereas it occurs for S1 and only a single G-actin molecule in the right. The overlapped volume in the left is significantly larger than that in the right. The water molecules are not shown

almost equal to the size of a G-actin molecule. This result is also in accord with the experimental observation (Kitamura et al. 1999, 2005).

The time during which S1 is in geometry I is that during which S1 is on the black curve. The time during which S1 is in geometries II and III is that during which S1 is on the red curve. There is one important point to be noted: It is required that S1 be on the red curve only for a sufficiently short length of time. If S1 was on the red curve for too long a time, the unidirectional behavior would be lost. This is much more essential than the asymmetry of the red curve mentioned above. According to the experimental data of Kitamura et al. (1999, 2005) shown in Fig. 18.14, this requirement is certainly met: S1 spends a much shorter time on the red curve than on the black curve.

18.5.3 Problems to Be Solved

While S1 is on the black curve, S1 cannot move. While it is on the red curve, it moves in the right direction. Repeated switches from one of the black and red curves to the other lead to the unidirectional movement of S1. What will happen under the condition that the ATP concentration is almost zero, the ADP one is very high, and the ATP synthesis occurs (i.e., under the solution condition of ATP synthesis)? In the present picture, S1 also moves in the right direction. What will happen under the condition that the chemical reaction is in equilibrium (i.e., under the solution condition of chemical equilibrium)? Even in the equilibrium state, the reaction is never stopped: The reaction rates in the right and left directions are

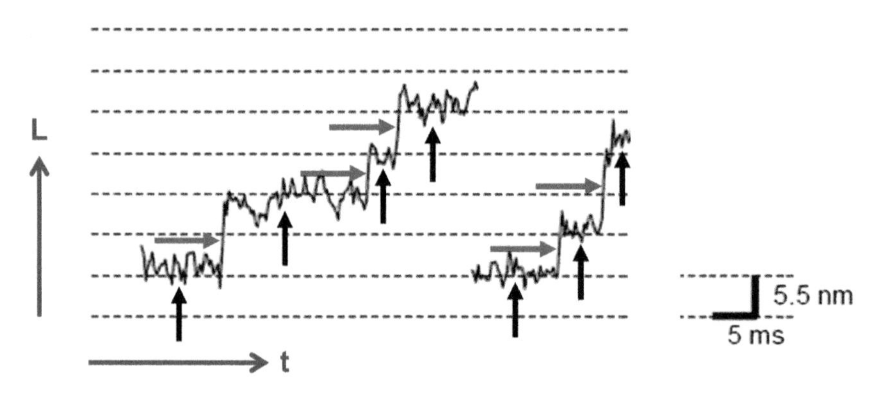

Fig. 18.14 Distance by which S1 moves plotted against time. Two sets of experimental data reported by Kitamura et al. (1999, 2005) are shown. Also see Fig. 18.12

equal. In the present picture, it appears that the repeated switches sill occur and S1 moves in the right direction, which conflicts with the second law of thermodynamics. In reality, S1 can achieve nothing but a random walk.

Under the solution condition of chemical equilibrium, we can present two propositions to overcome the problem mentioned above. First, the time during which S1 is on the red curve becomes substantially longer with the result that the unidirectional behavior is lost. Second, the physical picture needs to be modified so that S1 can move in the *left* direction under the solution condition of ATP synthesis. As mentioned above, there can be five different geometries of S1 and any geometrical change occurs not in a stepwise manner but *continuously*. If this continuous aspect was taken into account, for example, the red curve under the solution condition of ATP hydrolysis in Fig. 18.12 would become downward to the right. The red curve under the solution condition of ATP synthesis would become downward to the left. As a consequence, S1 moves in the right and left directions under the solution conditions of ATP hydrolysis and of ATP synthesis, respectively: Under the solution condition of chemical equilibrium, S1 moves in the right and left directions with the same frequency; i.e., it makes a random walk.

18.5.4 Drawbacks of Conventional View

According to the prevailing view, an ATP-driven protein functions by converting the chemical energy stored in an ATP molecule or the free energy of ATP hydrolysis to work: In the case of the unidirectional movement of myosin, for example, the work is necessary for myosin to move against the viscous resistance force by water. We disagree with this view. It sounds like the following: Work is necessary for a protein to fold against viscous resistance force by water. If it was valid, the distance by which S1 studied by Kitamura et al. (1999, 2005) moves would be the same for all of the hydrolysis cycles. Actually, it sometimes moves by more than a single step and even backward per hydrolysis cycle (Kitamura et al. 1999, 2005). It should be emphasized that the system comprises not only myosin but also the surrounding aqueous solution: The system does not do any work. Heat is generated but it is absorbed by the external bath. It is interesting that the movement is hindered by water in the prevailing view, whereas it is promoted by water in our view. In what follows, "prevailing" is replaced by "conventional."

Iwaki et al. experimentally studied the movement of myosin along F-actin by dissolving sucrose in aqueous solution (Iwaki unpublished results) to a concentration of 1 mol/l. By the sucrose addition, the viscosity of aqueous solution becomes about six times higher, giving rise to six times stronger viscous resistance force in the conventional view. Nevertheless, the sucrose dissolution had essentially no effects on the myosin movement. When the sucrose concentration reaches 2 mol/l, the myosin movement is stopped but this is because the myosin binding to F-actin becomes substantially stronger and even myosin with ATP bound to it is not detached from F-actin. We showed that the dissolution of sucrose in water enhances the water-entropy effect and, for example, the thermostability of a protein becomes significantly higher (Oshima and Kinoshita 2013). In the myosin case, the amplitudes of the entropic potential along the y-axis are enlarged by the sucrose dissolution, and it becomes difficult even for S1 with ATP bound to overcome the free-energy barrier and get detached from F-actin. The experimental result of Iwaki et al. (Kitamura et al. 2005) mentioned above is consistent not with the conventional view but with ours.

What we claim is that the ATP role is essential for the following reason (this is different from the reason widely accepted). With no ATP hydrolysis cycle, the unidirectional movement of S1 is not realized: S1 on a rigor binding site will not move because the system free energy becomes lowest when S1 is on the rigor binding site. In the presence of the cycle, the movement of S1 spontaneously occurs, because thanks to the cycle the system free energy continues to decrease.

Under the solution condition of ATP hydrolysis, myosin moves in the right direction. Under that of ATP synthesis, it moves in the right or left direction (hopefully, in the left direction as discussed in Sect. 18.5.3). Under that of chemical equilibrium, the myosin movement must be random. Under the latter two conditions, detailed experimental studies are strongly desired. The new experimental techniques recently developed by Iwaki et al. (2015, 2016) should be promising for this end. (Under the second condition, however, the chemical equilibrium is reached so fast that no experimental measurements would be feasible.)

18.6 Roles of ATP Hydrolysis Cycle and Water

An ATP-driven protein or protein complex is involved in an irreversible process, ATP hydrolysis. ATP binds to the protein, hydrolysis occurs, and ADP and Pi dissociate. During this cycle, the protein exhibits a series of structural changes, leading to its functioning. The cycle spontaneously occurs and accompanies a decrease in system free energy just as in the case of protein folding. Water plays the leading role. For example, water never hinders the myosin movement through the viscous resistance force: Water drives it primarily by the entropic force.

The force which makes myosin move is generated not by ATP but by water. More specifically, it is generated primarily by the entropic effect originating from the translational displacement of water molecules in the entire system. The role of ATP is to induce the structural change of the protein through its hydrolysis cycle and adjust the spatial distribution of entropic force (or potential) field acting on the protein (note that the distribution is strongly dependent on the protein structure).

Our view, the force which makes myosin move is generated not by ATP, is consistent with a recent suggestion based on an experimental study using high-speed atomic force microscopy (Ngo et al. 2016): The ATP hydrolysis is not required for the force generation. More strikingly, it has recently been found that even in the absence of ATP and ATP hydrolysis cycle, when the head of the hind leg (trailing head) of myosin V is *artificially* (forcibly) detached from F-actin using a novel experimental technique, myosin V makes a forward walk (Kodera unpublished results). This intriguing behavior can be applied to S1 as illustrated in Fig. 18.15. As described above, S1 (corresponding to the trailing head) must overcome a significantly high free-energy barrier to get detached from F-actin. It is achieved not by the structural change of S1 upon the ATP binding but by the artificial treatment. That is, the equilibrium state is broken by applying the artificial work. In the case of myosin V, it is likely that while the trailing head is moving, the head of the foreleg (leading head) is strongly bound to F-actin. However, the movement is restrained because the two legs are conjoined. Further, the most stable structure of myosin V must be retained. For example, the angle between the two

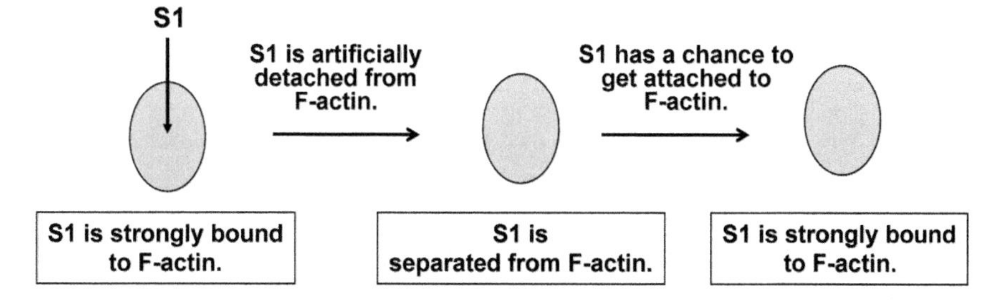

Fig. 18.15 How does Fig. 18.9 change when there is no ATP hydrolysis cycle?: S1 keeps being strongly bound to F-actin. What if S1 is artificially detached from F-actin?: S1 exhibits essentially no structural changes

legs is to be kept constant. When the trailing head is detached from F-actin and the conformation of the hind leg is perturbed, the angle is presumably made larger. As a consequence, a water-mediated force acts on the trailing head to return the angle to the most stable value *with the leading head strongly bound to F-actin*. The trailing head then moves forward. This discussion on the unidirectional walk of myosin V along F-actin is quite speculative but worth describing here.

In general, when a ligand binds to a receptor, the latter exhibits structural change so that the system free energy can be minimized. The binding of ATP to a protein is no exception: Upon the binding, the protein changes its structure, which is a matter of course. Strangely, only when the ligand is ATP, one claims that the chemical energy stored in ATP or the free energy of ATP hydrolysis changes the receptor structure. Take an enzyme catalyzing a hydrolysis of a ligand as an example. Is the structural change of the enzyme induced by the free energy of hydrolysis? The answer is No. In our opinion, the concept of chemical–mechanical energy conversion is physically irrelevant. The naming, "motor protein" or "protein machinery," is inappropriate in this sense ("ATP-driven protein" is appropriate).

Why was ATP chosen for the functioning of significantly many proteins in biological systems? The reason could be as follows: The protein structural change induced during the ATP hydrolysis cycle is relatively larger; nevertheless, the free energy of hydrolysis is not very large. If the latter was too large, the synthesis of ATP would become difficult (note that while ATP-driven proteins are functioning, to prevent a decrease in the ATP concentration, ATP synthesis must be performed somewhere).

A much more detailed discussion on the ATP-driven proteins is given in our recent review (Kinoshita 2016), but some of important points are updated in this review. We intend to construct a physical picture of the unidirectional walk of myosin V along F-actin in the near future. We thank M. Iwaki and M. Suzuki for a number of invaluable comments and suggestions.

References

Adachi K, Oiwa K, Nishizaka T, Furuike S, Noji H, Itoh H, Yoshida M, Kinosita M Jr (2007) Cell 130:309

Amano K, Yoshidome T, Iwaki M, Suzuki M, Kinoshita M (2010) J Chem Phys 133:045103

Asakura S, Oosawa F (1954) J Chem Phys 22:1255

Asakura S, Oosawa F (1958) J Polym Sci 33:183

Coureux P, Wells AL, Menetrey J, Yengo CM, Morris CA, Sweeney HL, Houdusse A (2003) Nature 425:419

Coureux P, Sweeney HL, Houdusse A (2004) EMBO J 23:4527

Dinsmore AD, Yodh AG, Pine DJ (1996) Nature 383:239

Hansen J-P, McDonald LR (2006) Theory of simple liquids, 3rd edn. Academic, London

Harano Y, Yoshidome T, Kinoshita M (2008) J Chem Phys 129:145103

Hayashi T, Oshima H, Mashima T, Nagata T, Katahira M, Kinoshita M (2014) Nucleic Acids Res 42:6861

Hayashi T, Oshima H, Yasuda S, Kinoshita M (2015) J Phys Chem B 119:14120

Howard J (2001) Mechanics of motor proteins and the cytoskeleton. Sinauer Associates Inc, Massachusetts
Iwaki M et al (unpublished results)
Iwaki M, Marcucci L, Togashi Y, Yanagida T (2013) Single molecule and collective dynamics of motor protein coupled with mechano-sensitive chemical reaction. In: Mikhailov AS (ed) Engineering of chemical complexity. World Scientific Publishing, Singapore
Iwaki M, Iwane AH, Ikezaki K, Yanagida T (2015) Nano Lett 15:2456
Iwaki M, Wickham SF, Ikezaki K, Yanagida T, Shih WM (2016) Nat Commun 7:13715
Katoh T, Morita F (1996) J Biochem 120:189
Kinoshita M (2002) J Chem Phys 116:3493
Kinoshita M (2006) Chem Eng Sci 61:2150
Kinoshita M (2008) J Chem Phys 128:024507
Kinoshita M (2013) Biophys Rev 5:283
Kinoshita M (2016) Mechanism of functional expression of the molecular machines. Springer briefs in molecular science. Springer, ISBN: 978-981-10-1484-0 (2016)
Kitamura K, Tokunaga M, Iwane AH, Yanagida T (1999) Nature 397:129
Kitamura K, Tokunaga M, Esaki S, Iwane AH, Yanagida T (2005) Biophysics 1:1
Kodama T (1985) Physiol Rev 65:467
Kodera N et al (unpublished results)
Masaike T, Koyama-Horibe F, Oiwa K, Yoshida M, Nishizaka T (2008) Nat Struct Mol Biol 15:1326
Ngo KX, Umeki N, Kijima ST, Kodera N, Ueno H, Furutani-Umezu N, Nakajima J, Noguchi TQP, Nagasaki A, Tokuraku K, Uyeda TQP (2016) Sci Rep 6:35449
Oda K, Kodama R, Yoshidome T, Yamanaka M, Sambongi Y, Kinoshita M (2011) J Chem Phys 134:025101
Okada T, Tanaka H, Iwane AH, Kitamura K, Ikebe M, Yanagida T (2007) Biochem Biophys Res Commun 354:379
Okuno D, Fujisawa R, Iino R, Hirono-Hara Y, Imamura H, Noji H (2008) Proc Natl Acad Sci USA 105:20722
Oshima H, Kinoshita M (2013) J Chem Phys 138:245101
Oshima H, Kinoshita M (2015) J Chem Phys 142:145103
Oshima H, Hayashi T, Kinoshita M (2016) Biophys J 110:2496
Soda K (1993) J Phys Soc Jpn 62:1782
Yanagida T, Esaki S, Iwane AH, Inoue Y, Ishijima A, Kitamura K, Tanaka H, Tokunaga M (2000) Phil Trans R Soc Lond B 355:441
Yoshidome T, Kinoshita M (2012) Phys Chem Chem Phys 14:14554
Yoshidome T, Kinoshita M, Hirota S, Baden N, Terazima M (2008) J Chem Phys 128:225104

Chapter 19
Controlling the Motility of ATP-Driven Molecular Motors Using High Hydrostatic Pressure

Masayoshi Nishiyama

Abstract High-pressure microscopy is a powerful technique for visualizing the effects of hydrostatic pressure on research targets. It can be used for monitoring the pressure-induced changes in the structure and function of molecular machines in vitro and in vivo. This chapter focuses on the use of high-pressure microscopy to measure the dynamic properties of molecular machines. We describe a high-pressure microscope that is optimized both for the best image formation and for stability under high hydrostatic pressure. The developed system allows us to visualize the motility of ATP-driven molecular motors under high pressure. The techniques described could be extended to study the detailed mechanism by which molecular machines work efficiently in collaboration with water molecules.

Keywords High-pressure microscopy • Molecular motors • Kinesin F_1-ATPase • Single-molecule measurement

19.1 Introduction

Movement is one of the central attributes of life. This biogenic function is carried out by various nanometer-sized molecular machines within cells. Molecular motors are representative examples of molecular machines in which the characteristic features of several proteins are integrated. These features include enzymatic activity, energy conversion, molecular recognition, and self-assembly. These biologically important reactions occur in association with water molecules that surround the proteins (Chaplin 2006). In the absence of water molecules, proteins have a rigid structure and lack activity.

The detailed mechanisms underlying the roles of water molecules in contributing to the structure and function of protein molecules have not been revealed.

M. Nishiyama (✉)
The HAKUBI Center for Advanced Research/Graduate School of Medicine, Kyoto University, Kyoto 606–8501, Japan
e-mail: mnishiyama@mbc.kuicr.kyoto-u.ac.jp

© Springer Nature Singapore Pte Ltd. 2018

M. Suzuki (ed.), *The Role of Water in ATP Hydrolysis Energy Transduction by Protein Machinery*, https://doi.org/10.1007/978-981-10-8459-1_19

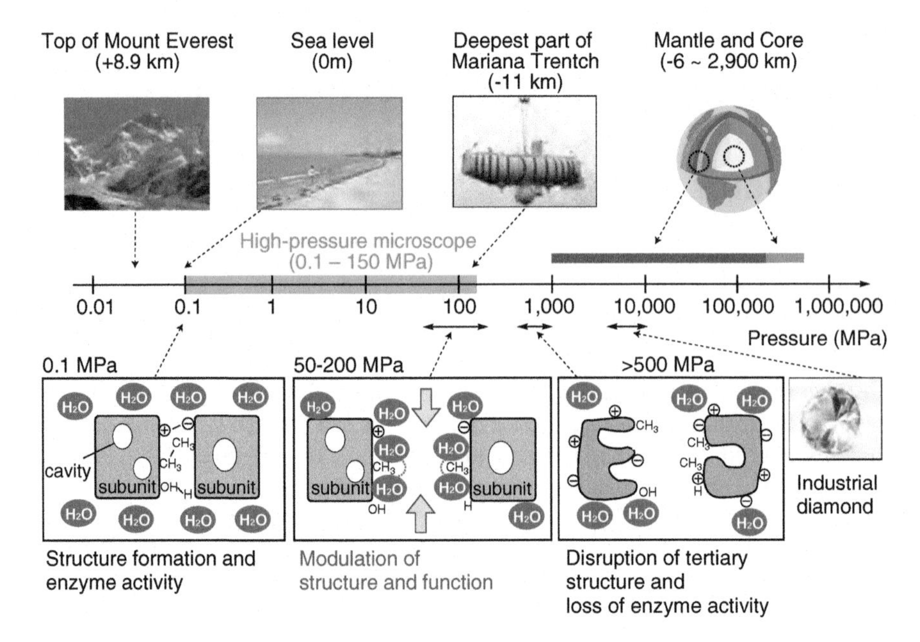

Fig. 19.1 Protein structure and function at high pressures. The application of high pressures induced the dissociation of oligomeric proteins and subunit denaturation. Water molecules at the >500 MPa range have been omitted for clarity. Modified from a reference with permission (Nishiyama and Sowa 2013)

Pressure application is a powerful method for modulating interactions between proteins and water molecules. Figure 19.1 illustrates the mechanisms that are believed to contribute to the pressure-induced changes in the structure and function of protein complexes. At ambient pressure (0.1 MPa), protein subunits interact via hydrogen bonding, hydrophobic, and electrostatic interactions. Water molecules are hardly able to access the cavities in the structures of the protein subunits. Under this condition, the protein complex maintains a stable structure and shows normal enzyme activities. At 50–200 MPa, the application of pressure promotes the dissociation of oligomeric proteins, electrostriction, and clathrate formation around hydrophobic residues (Boonyaratanakornkit et al. 2002). The applied pressure weakens protein–protein and protein–ligand interactions in solution (Akasaka 2006; Bartlett 2002; Luong et al. 2015; Mozhaev et al. 1996; Winter 2015). However, water molecules still have difficulty in penetrating the cavities in the structure of protein subunits. The structure and function of protein molecules may be partially altered at this pressure. At pressures >500 MPa, protein unfolding becomes irreversible and aggregation occurs because of an increased protein exposure to the solvent (Payne et al. 1997; Webb et al. 2001). Our research aim is to manipulate the structure and function of protein molecules without using chemical materials. We have been developing high-pressure microscopy for tracking the dynamic properties of molecular machines (Nishiyama 2017).

In this chapter, we describe the measurement of the dynamic properties of molecular machines using high-pressure microscopy. We start by describing the outline of our high-pressure microscope (Sect. 19.2). We describe the experimental results of motility assays of kinesin (Sect. 19.3) and F_1-ATPase (Sect. 19.4). Finally, we briefly comment on the future direction of research in this field (Sect. 19.5).

19.2 High-Pressure Microscope

High-pressure techniques have been used in various spectroscopy applications including conventional spectroscopy (Maeno and Akasaka 2015; Watanabe et al. 2013; Fujii et al. 2018), Fourier transform infrared spectroscopy (Dzwolak et al. 2002), nuclear magnetic resonance (Kitahara 2015; Roche et al. 2015), small-angle X-ray and neutron scattering (Fujisawa 2015; Winter 2002), X-ray crystallography (Fourme et al. 2012; Watanabe 2015), and time-resolved fluorescence anisotropy (Abe 2015). These methods are suitable for measuring the pressure-induced changes in the structure of protein molecules. In contrast to these spectroscopic techniques, optical microscopy is suitable for the observation of relatively large objects in the sub-micrometer size range or larger.

High-pressure microscope is composed of pressure apparatuses and a microscope. The most important component is a high-pressure chamber for optical microscopy that encloses the research targets and acquires the microscopic image under high pressure. Various high-pressure chambers have been reported (Vass et al. 2010). However, most of them have focused on the aspect of pressure resistance; therefore, the microscopy performance has generally been substandard. The reason for this is that the optical window of these chambers is frequently made of quartz ($n_e = 1.458$), sapphire ($n_e = 1.771$), or diamond ($n_e = 2.424$). Most commercially available objective lenses are not designed for the acquisition of images using these materials, and this incompatibility leads to a decrease in the resolving power of images. Additionally, in the case of sapphire, there is an optical problem of birefringence. To address these problems, we constructed a high-pressure chamber for optical microscopy with glass windows (BK7, $n_e = 1.519$). The chamber allows us to achieve the best image formation anticipated from commercially available objective lenses even at high pressures (Nishiyama 2017; Nishiyama and Sowa 2012).

Figure 19.2a–c displays photographs of the developed high-pressure microscope. The high-pressure chamber was connected to a separator, pressure gauge, and hand pump. The separator conferred the advantage of reducing the total dead volume of buffer solution in the pressure line (Nishiyama et al. 2009; Nishiyama and Kojima 2012) (Fig. 19.2d). Our apparatus can apply pressure up to 150 MPa, a limit that is imposed by the capabilities of the hand pump. This maximum pressure is 1.5-fold higher than that of the deepest part of the Mariana Trench (~11,000 m in depth), which is the highest pressure noted outside the Earth's crust. This ability to

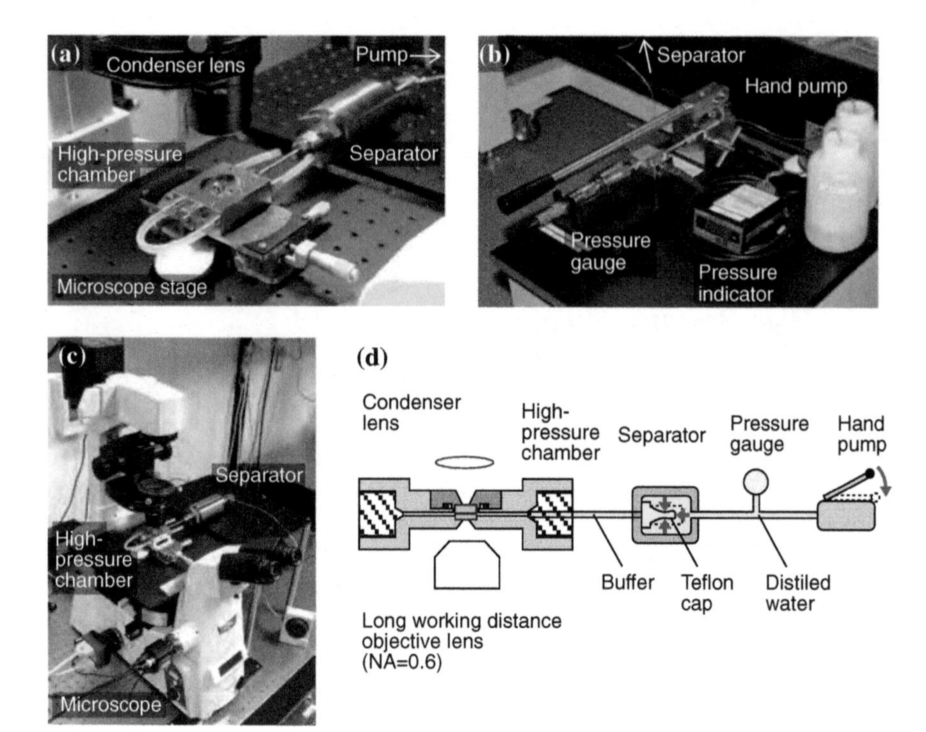

Fig. 19.2 High-pressure microscope. **a–c** Photographs of the experimental setup; **a** High-pressure chamber and separator; **b** Hand pump; **c** Main body. **d** Schematic diagram of the experimental setup (not to scale). Hydrostatic pressure was applied to the pressure line using a hand pump. The inside of the Teflon™ cap was filled with a buffer solution and connected to the chamber. The water pressure was transduced to the buffer solution by the deformation of a thin Teflon™ cap (~0.2 mm thickness) in the separator, and the pressure was then transmitted to the chamber. The hydrostatic pressure in the pressure line was measured using a gauge. The high-pressure chamber was mounted on a custom-made microscope stage, and the position could be manipulated with two manipulators in the X and Y directions. The mounting system succeeded in reducing the drift and vibration of the system regardless of the applied pressure. It allowed us to track the same research objects under various pressure conditions and study the effect of pressure (Nishiyama et al. 2013)

withstand such a high maximum pressure is sufficient to study biological activity at almost all pressures to which life is known to be subjected on Earth.

19.3 Sliding Movements of Kinesin Motors Along a Microtubule

Kinesin-1 is an ATP-driven molecular motor that moves stepwise in regular 8-nm steps along a microtubule (Kojima et al. 1997; Nishiyama et al. 2001; Svoboda et al. 1993). This movement is explained by a hand-over-hand model in which the two heads of kinesin work in a coordinated manner (Vale and Milligan 2000; Ishii et al. 2004; Block 2007; Cross 2016). One head remains bound to the microtubule, while the other steps from the αß-tubulin dimer behind the attached head to the dimer in front. The overall movement is 8 nm per ATPase cycle (Kojima et al. 1997; Schnitzer and Block 1997). The stepping motion is strongly dependent on physical and chemical conditions such as applied force (Svoboda and Block 1994; Nishiyama et al. 2003), temperature (Mazumdar and Cross 1998; Kawaguchi and Ishiwata 2000; Taniguchi et al. 2005), and anesthetic agents (Miyamoto et al. 2000). The application of pressure is also expected to change the stepping motion, but this has not been studied. We studied the pressure-induced effects on the motility of kinesin along microtubules by performing microtubule-sliding assays under high-pressure conditions (Nishiyama et al. 2009).

In advance, kinesin-1 molecules were attached to the surface of the observation window of the high-pressure chamber. Then, single microtubules were associated with multiple kinesin-1 molecules. In this experimental system, the motility of kinesin motors could be measured by the sliding motion of microtubules. When pressure was applied to the sample, the sliding velocity of microtubules immediately decreased from 790 ± 40 nm s^{-1} (mean ± standard deviation [SD]) at 0.1 MPa to 340 ± 30 nm s^{-1} at 130 MPa (Fig. 19.3a). Most of the microtubules moved smoothly and continuously, even under pressurized conditions. The sliding velocity was constant over time at each pressure. After the release of pressure, the sliding velocity immediately increased to 800 ± 40 nm s^{-1}. Thus, the application of pressure directly and reversibly inhibits the motility of kinesin along microtubules.

A systematic analysis of the sliding velocity of kinesin motors was then performed. Figure 19.3b shows the mean sliding velocities for each pressure and ATP concentration. The sliding velocity at each pressure followed normal Michaelis–Menten kinetics. When the pressure was increased from 0.1 to 130 MPa, the maximum sliding velocity, V_{max}, decreased from 900 to 320 nm s^{-1} and the substrate concentration where the sliding velocity was at half maximum, K_m, slightly decreased from 39 to 23 μM. Figure 19.3c shows the pressure–velocity relationships at ATP concentrations of 1 mM and 10 μM. The sliding velocity decreased monotonically with increases in pressure. Further analysis showed that the applied pressure mainly affected the stepping motion and not the ATP-binding reaction. Interestingly, the pressure–velocity relationship (Fig. 19.3c) was very close to the force–velocity relationship of single kinesin molecules (Nishiyama et al. 2002). The application of either ~100 MPa (= 1 pN $Å^{-2}$) of pressure or ~3 pN of force decreased the sliding velocity at 1 mM ATP by half.

Pressure is an isotropic mechanical action, which is different from directional force. Then, why do pressure and force similarly affect the stepping motion of kinesin motors? The nucleotide hydrolysis process causes a conformational change in the kinesin head, which includes the neck-linker region (Vale and Milligan 2000). This change decreases the distance between the kinesin head and tubulin (Kuffel and Zielkiewicz 2013), lowering the energy requirement for the floating head to proceed to the next binding site on the microtubule. Finally, the kinesin motors complete the stepping reaction. When a directional force is applied to kinesin motors, the force is

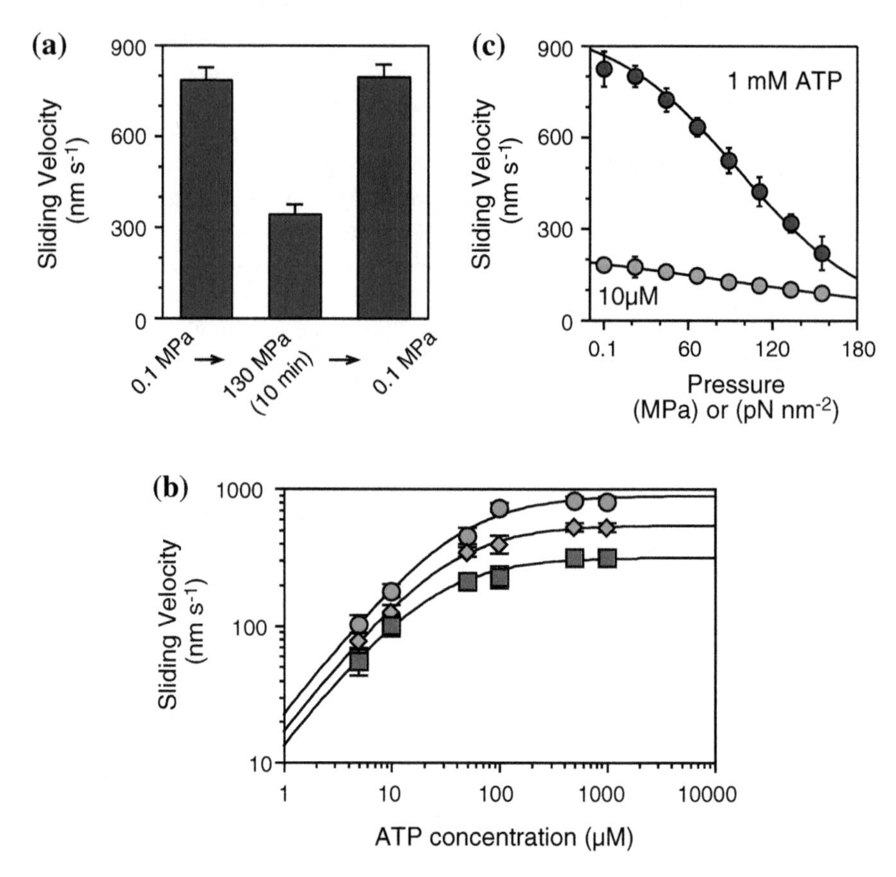

Fig. 19.3 (mean ± SD, n = 5–82). **a** Reversible inhibition of sliding movement of kinesin motors. [ATP] = 1 mM. **b** Sliding velocities as a function of ATP concentration at 0.1 (*circles*), 88 (*diamonds*), and 130 MPa (*squares*). Curves show the line of best fit with the Michaelis–Menten equation, $V = V_{\max} \cdot [\text{ATP}]/(K_m + [\text{ATP}])$; $V_{\max} = 900 \pm 50$, 550 ± 20, and 320 ± 10 nm s^{-1} and $K_m = 39 \pm 4$, 31 ± 2, and 23 ± 2 μM at 0.1, 88, and 130 MPa, respectively. **c** Pressure dependence of sliding velocity. [ATP] = 1 mM (*solid circles*) and 10 μM (*open circles*)

thought to inhibit the conformational change in the kinesin head during the stepping reaction and/or the kinesin–microtubule interaction.

In consideration of the similarity between the force-induced and pressure-induced effects on kinesin motility, the application of ~100 MPa of pressure generally enhances the structural fluctuation of protein molecules (Akasaka 2006; Bartlett 2002; Luong et al. 2015; Mozhaev et al. 1996). Thus, excessive fluctuation in the kinesin head and microtubule might prevent the ordered structural change and/or binding reaction of the floating head to the microtubule during the stepping reaction. Alternatively, pressure could enhance the association of water molecules with the exposed regions of the floating head and the microtubule. As a result, the mutual binding sites of the floating head and the microtubule would then be tightly covered by water molecules. Such a shielding effect of hydration might have prevented the interaction between the kinesin head and the microtubule.

19.4 Single-Molecule Analysis of the Rotation of F_1-ATPase

F_1-ATPase is the water-soluble part of ATP synthase and is an ATP-driven rotary molecular motor that rotates the rotary shaft (γ-subunit) against the surrounding stator ring ($\alpha_3\beta_3$-subunits), thereby hydrolyzing ATP (Noji et al. 1997, 2017). The stepping motion is strongly dependent on environmental conditions (Furuike et al. 2008; Toyabe et al. 2011). An elementary step of the mechanochemical reaction of F_1 is a 120° rotation that is coupled with a single turnover of ATP hydrolysis. The 120° rotation step can be divided into an 80° substep that is triggered after ATP binding and a 40° substep that is triggered after ATP hydrolysis. The angular positions before the 80° and 40° substeps are therefore termed the "binding angle" and the "catalytic angle," respectively. Although the mechanochemical coupling mechanism of F_1-ATPase has been well studied (Yasuda et al. 2001; Adachi et al. 2007; Adachi et al. 2012), the molecular details of individual reaction steps remain unclear. We conducted a single-molecule rotation assay of F_1 from thermophilic bacteria under various pressures (Okuno et al. 2013). This is the first experiment (to the best of our knowledge) to measure the enzyme activity of single molecular machines at high pressures.

Figure 19.4a illustrates a schematic drawing of the rotation measurements of single F_1 motors. A single bead was attached to the γ-subunit of a single F_1 motor for monitoring the rotational motion. Figure 19.4b displays the time courses and XY plots of the rotation of single F_1 molecules under an ATP-saturating concentration (2 mM). At ambient conditions (0.1 MPa and room temperature), F_1 showed smooth rotation without evident pauses in the presence of 2 mM ATP. The rotational velocity was 9.5 ± 3.5 rps (mean ± SD) at 0.1 MPa. When 140 MPa of pressure was applied to the system, the rotary motion drastically slowed down. F_1 showed discrete 120° steps, meaning that the applied pressure did not change the step size and directionality of the rotation. The rotational velocity was 3.0 ± 1.1 rps and remained constant over time. After the release of pressure, the rotational velocity immediately returned to 9.3 ± 4.3 rps, which was almost the same value as the initial rate.

Similar effects of high pressures were also observed under ATP-limiting conditions (200 nM). At 0.1 MPa, F_1 showed clear 120° steps pausing at three ATP-binding angles (Noji et al. 2017) (Fig. 19.4c). At 120 MPa, the dwell time was evidently lengthened. The pressure-induced decrease of the rotational velocity was also reversible. Thus, the application of pressure acts as an inhibitor that directly and reversibly alters the rotational motion of F_1. Under high pressures, F_1 showed distinctive pauses at three positions under both ATP-saturating and ATP-limiting conditions.

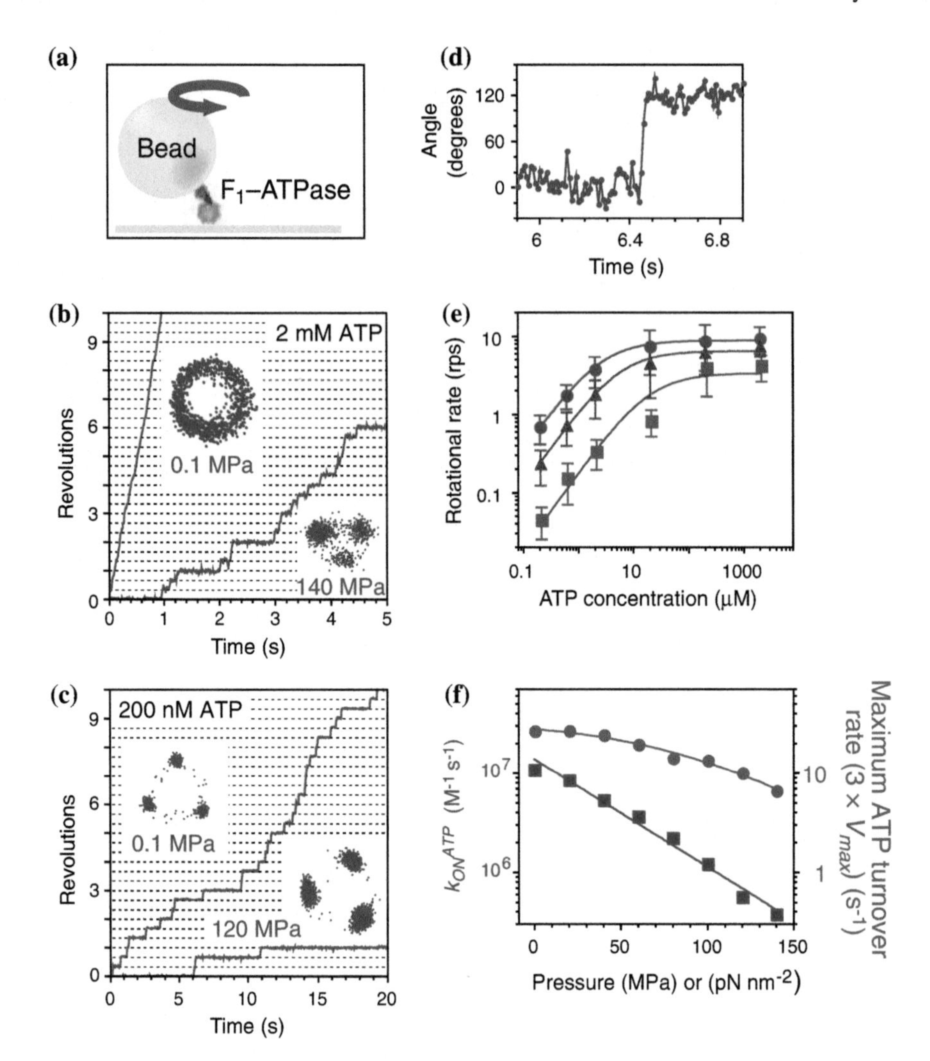

Fig. 19.4 Stepping rotation of F_1-ATPase (mean ± SD). **a** Experimental systems. **b** and **c** Time courses and their centroid traces of the rotation of F_1-ATPase at two ATP concentrations. [ATP] = 2 mM (**a**) and 200 nM (**b**). **d** Rising phase of a single 120° step on an expanded scale at 120 MPa and 200 nM ATP. **e** Rotational rates as a function of ATP concentration at 0.1 (*circles*), 60 (*triangles*), and 120 MPa (*squares*) (n = 12–102). Curves show the line of best fit with the Michaelis–Menten equation; $V_{max} = 8.9 \pm 0.3$, 6.5 ± 0.5, and $3.4 \pm 1.1\ s^{-1}$ and $K_m = 2.4 \pm 0.1$, 5.3 ± 0.6, and 18 ± 8 μM at 0.1, 60, and 120 MPa, respectively. **f** Pressure dependence of the apparent ATP-binding rate, k_{ON}^{ATP} ($= 3 \times V_{max}/K_m$) (*squares*) and maximum ATP turnover rate, $3 \times V_{max}$ (*circles*). The plots of k_{ON}^{ATP} were fitted with k_{ON}^{ATP} (0.1) = $14 \pm 2 \times 10^6\ M^{-1}s^{-1}$ and $V^{\neq} = +\,100 \pm 10$ Å^3. The plots of $3 \times V_{max}$ were fitted with $k_{p\text{-}ind} = 33 \pm 5\ s^{-1}$, $k_0 = 180 \pm 90\ s^{-1}$, and $V^{\neq} = +\,88 \pm 15$ Å^3

Next, we checked whether a high hydrostatic pressure affects the driving torque for each 120° step. In the rotation assay, we attached a single bead to the γ-subunit of a single F_1 motor for monitoring the rotational motion. The size of each bead was much larger than that of each individual F_1 motor, under which condition the torque is balanced with the hydrodynamic friction on the rotating probe, and the value is proportional to the angle velocity and drag coefficient. This idea was supported by the rotational friction coefficients of the bead attached to F_1-ATPase being in good agreement (Toyabe et al. 2011; Watanabe-Nakayama et al. 2008). Figure 19.4d displays the typical time course of a single 120° step at 120 MPa and 200 nM ATP on an expanded timescale. Most of the 120° steps took place rapidly and were independent of pressure. This means that the application of pressure did not seriously affect the angle velocity at the rising phase of the 120° step. It suggests that even under high-pressure conditions, F_1 retains a high conformational stability and exhibits its native properties. An attempt was made to find the rotation of the 80° and 40° substeps under high-pressure conditions. However, the data showed no evidence of substeps, suggesting that one or more pressure-sensitive reactions occurred at each ATP-binding angle, but not at each catalytic angle. This is strongly supported by results of a rotation assay using a mutant F_1 (ßE190D); this mutation significantly retards the hydrolysis step, causing an evident pause at the catalytic angle (Shimabukuro et al. 2003). Under ambient conditions, F_1 (ßE190D) clearly showed six pauses at each turn. They correspond to three ATP-binding angles and three catalytic angles. The application of pressure lengthened the pause time at each ATP-binding angle, but not at each catalytic angle (Okuno et al. 2013).

A systematic analysis of the rotational rate of single wild-type F_1 motors was performed under various concentrations of ATP and various hydrostatic pressures (Fig. 19.4e). The rotational rate gradually decreased as pressure increased at each concentration of ATP. The rotational rate at each pressure followed normal Michaelis–Menten kinetics. When the pressure was increased from 0.1 to 120 MPa, the V_{max} decreased from 8.5 to 3.4 s^{-1}, whereas the K_m increased from 2.4 to 18 μM. F_1 motors hydrolyze three ATP molecules at every turn. Thus, the ATP turnover rate could be calculated from the rotational rate multiplied by 3. The apparent ATP-binding rate, k_{ON}^{ATP} (= 3 × V_{max}/K_m), and maximum ATP turnover rate (= 3 × V_{max}) decreased with increased pressure (Fig. 19.4f). The activation volumes determined from the pressure dependence of the rate constants were +100 $Å^3$ for ATP binding and +88 $Å^3$ for the other pressure-sensitive reaction.

Recently, it has become possible to perform large-scale molecular dynamics simulations. In silico, high-pressure conditions could be simply achieved by reducing the volume of the box that contains the research target and water molecules (Wakai et al. 2014). Notably, the simulation would demonstrate the difference in the properties of water molecules between ambient pressure and high-pressure conditions. Comparison of the experimental results of high-pressure microscopy with molecular dynamics calculations could reveal the mechanism by which molecular motors collaborate with water molecules to generate directional motion.

The F_1 motor could become an ideal research target, since this molecular motor has been studied using experimental (Noji et al. 2017) and computational techniques (Ito and Ikeguchi 2014).

19.5 Future Direction

We have demonstrated novel motility assays that change the structure and function of ATP-driven molecular motors. The pressure-induced changes could be caused by modifications of the intermolecular interactions between proteins and water molecules. The results described here suggest that pressure-induced effects on cell morphology and activity are directly caused by the enhancement of the hydration of proteins by water molecules (Hayashi et al. 2016). In other words, the behavior of molecular machines in cells or tissues can be controlled by applying different pressures. The final goal might be the development and differentiation of high-pressure microscopy for research on higher-order living beings. High-pressure microscopy could become a powerful tool for examining the mechanisms underlying the functions of molecular machines in complex systems.

Acknowledgements We would like to thank Yoshifumi Kimura for developing a prototype of the high-pressure chamber for optical microscopy, Daichi Okuno and Hiroyuki Noji for the measurement of F_1-ATPase, and Eiro Muneyuki, Shoichi Toyabe, Shou Furuike, Masahide Terazima, Yoshie Harada, and Akitoshi Seiyama for discussion. This work was supported by Grants-in-Aid for Scientific Research on the Innovative Area "Water plays the main role in ATP energy transfer, Group Leader; Prof. Makoto Suzuki" (Nos. 21118511 and 23118710), Grants-in-Aid for Scientific Research from MEXT (Nos. 16 K04908 and 17H05880), and the Takeda Science Foundation, Research Foundation for Opto-Science and Technology, Shimadzu Science Foundation, and Nakatani Foundation for Advancement of Measuring Technologies in Biomedical Engineering.

References

Abe F (2015) Effects of high hydrostatic pressure on microbial cell membranes: structural and functional perspectives. Subcell Biochem 72:371–381. https://doi.org/10.1007/978-94-017-9918-8_18

Adachi K, Oiwa K, Nishizaka T, Furuike S, Noji H, Itoh H, Yoshida M, Kinosita K Jr (2007) Coupling of rotation and catalysis in F1-ATPase revealed by single-molecule imaging and manipulation. Cell 130:309–321. https://doi.org/10.1016/j.cell.2007.05.020

Adachi K, Oiwa K, Yoshida M, Nishizaka T, Kinosita K Jr (2012) Controlled rotation of the F_1-ATPase reveals differential and continuous binding changes for ATP synthesis. Nat Commun 3:1022. https://doi.org/10.1038/ncomms2026

Akasaka K (2006) Probing conformational fluctuation of proteins by pressure perturbation. Chem Rev 106:1814–1835

Bartlett DH (2002) Pressure effects on in vivo microbial processes Bba-Protein Struct M 1595:367–381

Block SM (2007) Kinesin motor mechanics: binding, stepping, tracking, gating, and limping. Biophys J 92:2986–2995. https://doi.org/10.1529/biophysj.106.100677

Boonyaratanakornkit BB, Park CB, Clark DS (2002) Pressure effects on intra- and intermolecular interactions within proteins. Biochem Biophys Acta 1595:235–249
Chaplin M (2006) Do we underestimate the importance of water in cell biology? Nat Rev Mol Cell Biol 7:861–866. https://doi.org/10.1038/nrm2021
Cross RA (2016) Review: Mechanochemistry of the kinesin-1 ATPase. Biopolymers 105:476–482. https://doi.org/10.1002/bip.22862
Dzwolak W, Kato M, Taniguchi Y (2002) Fourier transform infrared spectroscopy in high-pressure studies on proteins. Biochem Biophys Acta 1595:131–144
Fourme R, Girard E, Akasaka K (2012) High-pressure macromolecular crystallography and NMR: status, achievements and prospects. Curr Opin Struct Biol 22:636–642. https://doi.org/10.1016/j.sbi.2012.07.007
Fujii S, Masanari-Fujii M, Kobayashi S, Kato C, Nishiyama M, Harada Y, Wakai S, Sambongi Y (2018) Commonly stabilized cytochromes c from deep-sea *Shewanella* and *Pseudomonas*. Biosci Biotechnol Biochem. https://doi.org/10.1080/09168451.2018.1448255
Fujisawa T (2015) High pressure small-angle X-ray scattering. Subcell Biochem 72:663–675. https://doi.org/10.1007/978-94-017-9918-8_30
Furuike S, Adachi K, Sakaki N, Shimo-Kon R, Itoh H, Muneyuki E, Yoshida M, Kinosita K Jr (2008) Temperature dependence of the rotation and hydrolysis activities of F1-ATPase. Biophys J 95:761–770. https://doi.org/10.1529/biophysj.107.123307
Hayashi M, Nishiyama M, Kazayama Y, Toyota T, Harada Y, Takiguchi K (2016) Reversible morphological control of tubulin-encapsulating giant liposomes by hydrostatic pressure. Langmuir 32:3794–3802. https://doi.org/10.1021/acs.langmuir.6b00799
Ishii Y, Nishiyama M, Yanagida T (2004) Mechano-chemical coupling of molecular motors revealed by single molecule measurements. Curr Protein Pept Sci 5:81–87. https://doi.org/10.2174/1389203043486838
Ito Y, Ikeguchi M (2014) Molecular dynamics simulations of F1-ATPase. Adv Exp Med Biol 805:411–440. https://doi.org/10.1007/978-3-319-02970-2_17
Kawaguchi K, Ishiwata S (2000) Temperature dependence of force, velocity, and processivity of single kinesin molecules. Biochem Biophys Res Commun 272:895–899. https://doi.org/10.1006/bbrc.2000.2856
Kitahara R (2015) High-Pressure NMR spectroscopy reveals functional sub-states of ubiquitin and ubiquitin-like proteins. Subcell Biochem 72:199–214. https://doi.org/10.1007/978-94-017-9918-8_10
Kojima H, Muto E, Higuchi H, Yanagida T (1997) Mechanics of single kinesin molecules measured by optical trapping nanometry. Biophys J 73:2012–2022. https://doi.org/10.1016/S0006-3495(97)78231-6
Kuffel A, Zielkiewicz J (2013) Properties of water in the region between a tubulin dimer and a single motor head of kinesin. Phys Chem Chem Phys 15:4527–4537. https://doi.org/10.1039/c3cp43828g
Luong TQ, Kapoor S, Winter R (2015) Pressure-A gateway to fundamental insights into protein solvation. Dyn Func Chemphyschem 16:3555–3571
Maeno A, Akasaka K (2015) High-pressure fluorescence spectroscopy. Subcell Biochem 72:687–705. https://doi.org/10.1007/978-94-017-9918-8_32
Mazumdar M, Cross RA (1998) Engineering a lever into the kinesin neck. J Biol Chem 273:29352–29359. https://doi.org/10.1074/jbc.273.45.29352
Miyamoto Y, Muto E, Mashimo T, Iwane AH, Yoshiya I, Yanagida T (2000) Direct inhibition of microtubule-based kinesin motility by local anesthetics. Biophys J 78:940–949. https://doi.org/10.1016/S0006-3495(00)76651-3
Mozhaev VV, Heremans K, Frank J, Masson P, Balny C (1996) High pressure effects on protein structure and function. Proteins-Struct Func Genet 24:81–91
Nishiyama M (2017) High-pressure microscopy for tracking dynamic properties of molecular machines. Biophys Chem. https://doi.org/10.1016/j.bpc.2017.03.010

Nishiyama M, Higuchi H, Ishii Y, Taniguchi Y, Yanagida T (2003) Single molecule processes on the stepwise movement of ATP-driven molecular motors. Biosystems 71:145–156. https://doi.org/10.1016/s0303-2647(03)00122-9
Nishiyama M, Higuchi H, Yanagida T (2002) Chemomechanical coupling of the forward and backward steps of single kinesin molecules. Nat Cell Biol 4:790–797. https://doi.org/10.1038/ncb857
Nishiyama M, Kimura Y, Nishiyama Y, Terazima M (2009) Pressure-induced changes in the structure and function of the kinesin-microtubule complex. Biophys J 96:1142–1150. https://doi.org/10.1016/j.bpj.2008.10.023
Nishiyama M, Kojima S (2012) Bacterial motility measured by a miniature chamber for high-pressure microscopy. Int J Mol Sci 13:9225–9239. https://doi.org/10.3390/ijms13079225
Nishiyama M, Muto E, Inoue Y, Yanagida T, Higuchi H (2001) Substeps within the 8-nm step of the ATPase cycle of single kinesin molecules. Nat Cell Biol 3:425–428. https://doi.org/10.1038/35070116
Nishiyama M, Sowa Y (2012) Microscopic analysis of bacterial motility at high pressure. Biophys J 102:1872–1880. https://doi.org/10.1016/j.bpj.2012.03.033
Nishiyama M, Sowa Y (2013) Manipulation of cell motility with water molecules in living cells Kagaku. Jpn J 68:33–38
Nishiyama M, Sowa Y, Kimura Y, Homma M, Ishijima A, Terazima M (2013) High hydrostatic pressure induces counterclockwise to clockwise reversals of the *Escherichia coli* flagellar motor. J Bacteriol 195:1809–1814. https://doi.org/10.1128/jb.02139-12
Noji H, Ueno H, McMillan DGG (2017) Catalytic robustness and torque generation of the F1-ATPase. Biophys Rev 9:103–118 https://doi.org/10.1007/s12551-017-0262-x
Noji H, Yasuda R, Yoshida M, Kinosita K Jr (1997) Direct observation of the rotation of F1-ATPase. Nature 386:299–302. https://doi.org/10.1038/386299a0
Okuno D, Nishiyama M, Noji H (2013) Single-molecule analysis of the rotation of F1-ATPase under high hydrostatic pressure. Biophys J 105:1635–1642. https://doi.org/10.1016/j.bpj.2013.08.036
Payne VA, Matubayasi N, Murphy LR, Levy RM (1997) Monte Carlo study of the effect of pressure on hydrophobic association. J Phys Chem B 101:2054–2060. https://doi.org/10.1021/jp962977p
Roche J, Dellarole M, Royer CA, Roumestand C (2015) Exploring the protein folding pathway with high-pressure NMR: steady-state and kinetics studies. Subcell Biochem 72:261–278. https://doi.org/10.1007/978-94-017-9918-8_13
Schnitzer MJ, Block SM (1997) Kinesin hydrolyses one ATP per 8-nm step. Nature 388:386–390. https://doi.org/10.1038/41111
Shimabukuro K, Yasuda R, Muneyuki E, Hara KY, Kinosita K Jr, Yoshida M (2003) Catalysis and rotation of F1 motor: cleavage of ATP at the catalytic site occurs in 1 ms before 40° substep rotation. Proc Natl Acad Sci U S A 100:14731–14736. https://doi.org/10.1073/pnas.2434983100
Svoboda K, Block SM (1994) Force and velocity measured for single kinesin molecules. Cell 77:773–784. https://doi.org/10.1016/0092-8674(94)90060-4
Svoboda K, Schmidt CF, Schnapp BJ, Block SM (1993) Direct observation of kinesin stepping by optical trapping interferometry. Nature 365:721–727. https://doi.org/10.1038/365721a0
Taniguchi Y, Nishiyama M, Ishii Y, Yanagida T (2005) Entropy rectifies the Brownian steps of kinesin. Nat Chem Biol 1:342–347. https://doi.org/10.1038/nchembio741
Toyabe S, Watanabe-Nakayama T, Okamoto T, Kudo S, Muneyuki E (2011) Thermodynamic efficiency and mechanochemical coupling of F1-ATPase. Proc Natl Acad Sci U S A 108:17951–17956. https://doi.org/10.1073/pnas.1106787108
Vale RD, Milligan RA (2000) The way things move: looking under the hood of molecular motor proteins. Science 288:88–95
Vass H, Black SL, Herzig EM, Ward FB, Clegg PS, Allen RJ (2010) A multipurpose modular system for high-resolution microscopy at high hydrostatic pressure. Rev Sci Instrum 81 https://doi.org/10.1063/1.3427224

Wakai N, Takemura K, Morita T, Kitao A (2014) Mechanism of deep-sea fish alpha-actin pressure tolerance investigated by molecular dynamics simulations. PLoS ONE 9:e85852. https://doi.org/10.1371/journal.pone.0085852
Watanabe N (2015) High pressure macromolecular crystallography. Subcell Biochem 72:677–686 https://doi.org/10.1007/978-94-017-9918-8_31
Watanabe TM et al (2013) Glycine insertion makes yellow fluorescent protein sensitive to hydrostatic pressure. PLoS ONE 8 https://doi.org/10.1371/journal.pone.0073212
Watanabe-Nakayama T, Toyabe S, Kudo S, Sugiyama S, Yoshida M, Muneyuki E (2008) Effect of external torque on the ATP-driven rotation of F1-ATPase. Biochem Biophys Res Commun 366:951–957. https://doi.org/10.1016/j.bbrc.2007.12.049
Webb JN, Webb SD, Cleland JL, Carpenter JF, Randolph TW (2001) Partial molar volume, surface area, and hydration changes for equilibrium unfolding and formation of aggregation transition state: high-pressure and cosolute studies on recombinant human IFN-gamma. Proc Natl Acad Sci U S A 98:7259–7264. https://doi.org/10.1073/pnas.131194798
Winter R (2002) Synchrotron X-ray and neutron small-angle scattering of lyotropic lipid mesophases, model biomembranes and proteins in solution at high pressure. Biochem Biophys Acta 1595:160–184
Winter R (2015) Pressure effects on the intermolecular interaction potential of condensed protein solutions. Subcell Biochem 72:151–176. https://doi.org/10.1007/978-94-017-9918-8_8
Yasuda R, Noji H, Yoshida M, Kinosita K Jr, Itoh H (2001) Resolution of distinct rotational substeps by submillisecond kinetic analysis of F1-ATPase. Nature 410:898–904. https://doi.org/10.1038/35073513

Chapter 20
Modulation of the Sliding Movement of Myosin-Driven Actin Filaments Associated with Their Distortion: The Effect of ATP, ADP, and Inorganic Phosphate

Kuniyuki Hatori and Satoru Kikuchi

Abstract The motility of actin filaments interacting with myosin motors during ATP hydrolysis can be evaluated using an in vitro motility assay. Motility assays with fluorescence imaging techniques allow us to measure the sliding velocity as an index of motility and fluctuations of actin filaments at nanometer accuracy. Because actin filaments are flexible, distortions such as deformation of their filamentous structure are also observed during the sliding movement. This chapter discusses an imaging analysis of velocity fluctuation and distortion of actin filaments in the case of myosin II motors derived from skeletal fast muscle. The relationship between the velocity and distortion is discussed. In addition, the effect of ADP and inorganic phosphate (Pi), which are products of ATP hydrolysis, on this relationship is explained through the kinetics of myosin–actin binding. Considering the fluctuation studies conducted to date, we review our concept with respect to the coordination of motion along single actin filaments, wherein distortions alter the geometric features surrounding actin–myosin in terms of water behavior.

Keywords Actomyosin · In vitro motility assay · Fluorescence microscopy Fluctuation · Flexibility

20.1 Introduction

Protein stability depends on enthalpic and entropic effects in relation to water behavior. Because, at physiological temperature, the enthalpy change for protein unfolding is amply compensated with an entropy change, a modest free-energy

K. Hatori (✉) · S. Kikuchi
Department of Bio-Systems Engineering, Graduate School of Science and Engineering, Yamagata University, Yonezawa, Japan
e-mail: khatori@yz.yamagata-u.ac.jp

© Springer Nature Singapore Pte Ltd. 2018

M. Suzuki (ed.), *The Role of Water in ATP Hydrolysis Energy Transduction by Protein Machinery*, https://doi.org/10.1007/978-981-10-8459-1_20

change provides for the marginal stability of protein structures (Makhatadze and Privalov 1995). The preferential hydration of proteins also tends to stabilize their structures. Meanwhile, rigid stability may hamper the transition between the conformational states of protein machinery, such as ATPases. Apart from tertiary structure, both stability and flexibility are necessary for the functions of complex protein assemblies, such as the cytoskeletal proteins actin and tubulin. These proteins reversibly assemble into microfilaments or microtubules depending on the situation, and their morphological changes are implicated in cell motility.

The flexibility of actin filaments has been widely accepted since visualization of single filaments with fluorescent dyes under a fluorescence microscope (Yanagida et al. 1984). Flexibility is often evaluated by an index such as persistence length, which is expressed as flexural rigidity over thermal energy. In solutions, long actin filaments considerably bend because of thermal fluctuation and have several micrometers of persistence length (Isambert et al. 1995). Some actin-binding proteins, including the myosin motor, are known to modulate flexibility and polymerization kinetics, and the implication has been considered with regard to physiological roles (Crevenna et al. 2015). In addition to flexibility measurements, the motility of actin–myosin at the molecular level is usually examined using an in vitro motility assay, in which fluorescently labeled actin filaments move on a myosin-coated glass slide in the presence of ATP (Kron and Spudich 1986). The motility is referred to as sliding or gliding movement. Various types of myosin can be tested in this assay. In particular, myosin II derived from skeletal muscle and myosin V from non-muscle have been extensively examined. Myosin V is a processive motor that is capable of moving a single molecule along an actin filament by tightly coupling between one step of 36 nm and one ATP hydrolysis (Walker et al. 2000; Yildiz et al. 2003). In contrast, skeletal myosin II is a non-processive motor, which is faster than myosin V because of its short time of attachment to actin (Uyeda et al. 1990). The binding strength of actin–myosin alters in sequential processes during ATP hydrolysis as follows: strong binding of actin–myosin without ATP (rigor binding), dissociation from actin in the ATP–myosin state, weak binding in the ADP-Pi-myosin state, and strong binding in the ADP–myosin state (Taylor 1991). Sliding force is produced in the Pi release process following the ADP-Pi state. Despite long-term studies on myosin II, its sliding mechanism is still highly debated (Cooke 2004; Oosawa 2008; Batters et al. 2014).

In the in vitro motility system, in the absence of ATP, actin filaments are tightly bound to a myosin-coated surface and exhibit tiny thermal fluctuation (14 nm) without sliding. During sliding, in the presence of sufficient ATP, actin filaments exhibit meandering rather than straight shapes, and the rear end (formally barbed end) of actin filaments almost follows the same pathway as that of the leading end (pointed end). However, the pathway somewhat varies throughout the overall filament pass, and transversal fluctuation against the direction of movement appears to be at 65 nm (Hatori et al. 1996a). A method based on spot labeling along actin filaments has revealed that local fluctuation propagates from pointed end to barbed end along the filament at a faster rate than the average velocity (Hatori et al. 1998). Based on this active fluctuation, actin filaments suffer distortions such as compression, buckling, and bending by forces imposed by myosin motors as well as thermal energy. Additionally, the distortions may affect the configuration of the interface between actin and myosin, wherein the excluded volume is altered. This change in excluded volume has the potential to modulate binding affinity and drive unidirectional movement with an entropic effect (Amano et al. 2010). Therefore, a question arises regarding the connection between filamentous distortion and the coordinating sliding movement. Because the ATPase cycle and velocity can be modulated by the presence of ADP or Pi, the effect on these additives may provide insight into the role of distortion and fluctuation.

20.2 Analyses of Fluctuations of Fluorescence and Motility

20.2.1 Fluorescence Labeling for Actin Filaments

To observe actin filaments under a fluorescence microscope, tetramethyl-rhodamine-phalloidin is usually used because it specifically binds to the filaments up to a molar ratio of 1:1 and stabilizes filament formation without impairing motility (Yanagida et al. 1984). Another option is direct labeling of the dye to actin monomers by covalently binding the dye with maleimide and succinimide groups to Cys374 and any lysine residue in actin, respectively (Honda et al. 1986). The motility of actin is retained at up to twofold labeling of the dye per actin in a molar ratio (Matsushita and Hatori 2012). Various lengths of filaments (up to several ten micrometers long) are observed. At a labeling ratio of 1:1, each filament is clearly visible over the entire length. Therefore, it is difficult to determine a certain position within a long actin filament with a length of above several micrometers with high accuracy. However, for short filaments of <1 μm, the position can be readily determined by calculating the center of mass or Gaussian distribution of intensity (Sect. 20.2.2). When block copolymer composed of unlabeled and labeled short filaments is prepared via annealing (speckled actin filament), labeled portions serve as markers for the determination of positions (Hatori et al. 1998).

20.2.2 Precise Determination of Displacements in the Motility Assay

Fluorescently labeled actin filaments perform the sliding movement on a myosin-coated glass slide, which can be observed under a fluorescence microscope. Heavy meromyosin (HMM), which possesses two motor heads but not the tail portion responsible for assembly into thick filaments, is usually used instead of full-length myosin molecules.

In order to evaluate small displacements, the positions of the actin filament can be determined by Gaussian function interpolation among discrete image data (i.e., pixels), resulting in sub-pixel accuracy of <8 nm (in the case of actin filament fixed on a glass slide). Currently, the accuracy achieved after using single fluorescence dyes is valid for measuring the step size of myosin (Yildiz et al. 2003) and for high-resolution microscopy, such as stochastic optical reconstruction microscopy (Rust et al. 2006). In a previous study, we detected lateral thermal fluctuation of 14 nm of single actin filaments bound to a myosin-coated glass in the absence of ATP when the intensity profile of the filament image fitted a Gaussian function perpendicular to the long axis of the filament (Hatori et al. 1996a). For short filaments, the displacement in an *x-y* plane can also be determined by 2*D* Gaussian fitting as depicted in Fig. 20.1a and b. The acquisition rate of images is limited by camera sensitivity and the exposure time necessary to obtain sufficiently bright images. We can readily take adequate images at intervals of 1/30 s using a fluorescence microscope with a 100 × objective (NA 1.3), a 100 W mercury lamp, and a 1/2-inch EM-CCD camera. Because actin filaments move at 4 μm/s on average under standard conditions (Fig. 20.1), the average displacement at 1/30 s is roughly 130 nm, which corresponds to a 15-fold larger magnitude compared to the level of minimum detection. Therefore, velocity fluctuation can be analyzed with sufficient resolution.

20.2.3 Distribution Profile of Fluorescence Images

In addition to the determination of the position, we have further developed image processing for evaluating the distortion of actin filaments (Hatori et al. 2009). An actin filament image has a 0.5 μm width due to light interference, which does not reflect the real width of the filament (ca. 8 nm). At the same time, changes in the distortion of the filament would appear in the intensity profile. Figure 20.1a shows the fluctuation in the fluorescence intensity of a short actin filament during movement. We define two parameters as the peak intensity and spreading width of the intensity distribution. The former is the intensity (F_o) at the center, and the latter is the standard deviations (S_L or S_T), respectively (Fig. 20.1b), when the distribution fits a 2*D* Gaussian function in longitudinal (*L*) and transversal (*T*) directions toward the sliding movement. Therefore, the spreading values in the *L-T* coordination, peak intensity, and displacement (or velocity) are used to calculate the cross-correlation between distortion and movement.

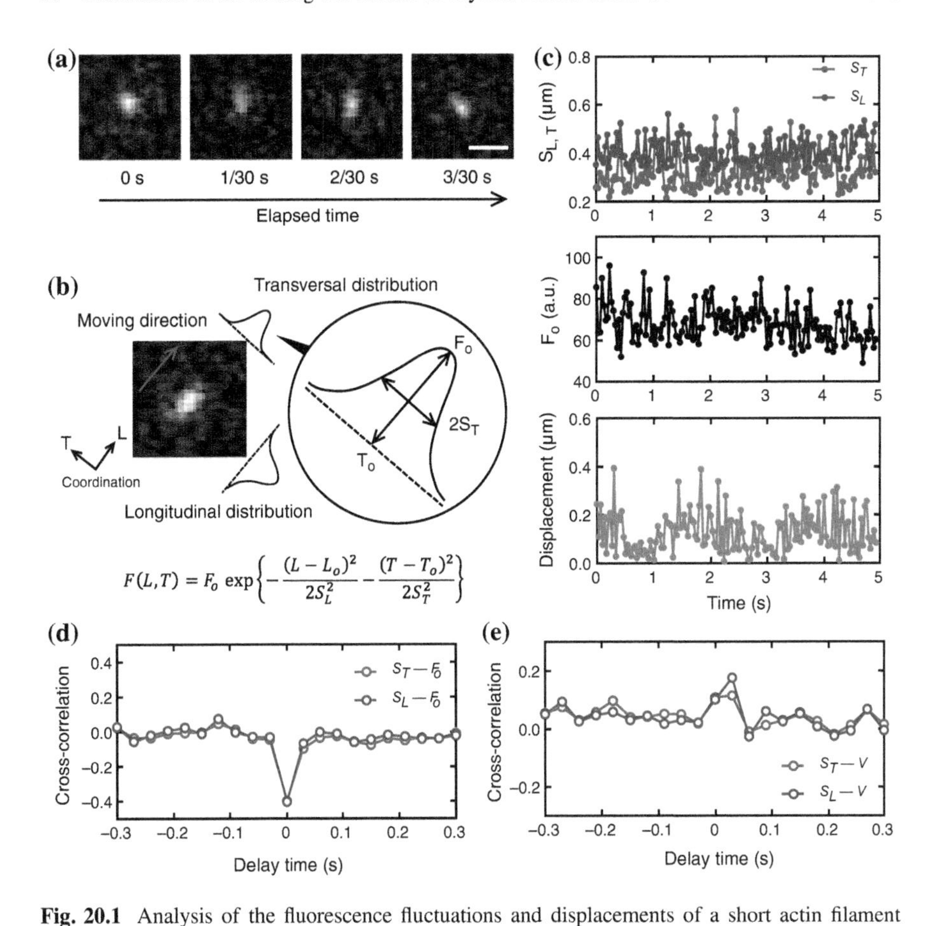

Fig. 20.1 Analysis of the fluorescence fluctuations and displacements of a short actin filament driven by HMM molecules. Conditions: 25 mM KCl, 25 mM imidazole-HCl (pH 7.4), 4 mM $MgCl_2$, 2 mM ATP, 0.5% 2-mercaptoethanol, 3 mg/mL glucose, 0.02 mg/mL catalase, and 0.1 mg/mL glucose oxidase. HMM at 0.05 mg/mL was perfused on a collodion-treated glass slide (Kron and Spudich 1986). The assay was performed at 23 °C. **a** Sequential images of rhodamine-phalloidin-bound actin filament during movement at intervals of 1/30 s. The average length of evaluated filaments was 0.4 μm. Long filaments were fragmented using an ultra-sonication. The scale bar indicates 1 μm. **b** A sketch of fluorescence intensity distributions of a single actin filament, which were fitted to a 2*D* Gaussian function with respect to longitudinal direction (moving direction) and its transversal direction. The center position, peak intensity, and standard deviation (spreading width) of the fluorescence profile were determined. F_o, S_T, and S_L denote peak intensity, standard deviations in the transversal and longitudinal directions, respectively. Displacement was defined as the difference between center positions at intervals of 1/30 s. **c** Time development of S_L, S_T, F_o, and displacement of an actin filament movement. **d** Cross-correlation functions between S_T (or S_L) and F_o, which was calculated using the formula $R_{Cross}(\tau) = \overline{F_o(t)S_T(t+\tau)}$; here τ is delay time. $R_{Cross}(\tau)$ was normalized by dividing $\sqrt{\overline{F_o^2(t)S_T^2(t)}}$. A negative peak appeared at 0 s. **e** Cross-correlation functions between each S_L, S_T, and velocity (displacement at 1/30 s was multiplied by 30). Positive peak in S_L-V appeared at 1/30 s of delay time, indicating that the increase in velocity occurs after the spreading width of fluorescence becomes broad

20.3 Correlation Between Fluctuation of the Fluorescent Image and Displacement

Here, we present our results with respect to fluctuations in spreading width, peak, and displacement over time, based on data shown in Fig. 20.1c. When normalized cross-correlation functions between the spreading width (S_L or S_T) and peak intensity (F_o) were calculated with delay time (τ), the maximum negative value (−0.4) appeared at zero delay time, implying that the peak intensity simultaneously increases with a decrease in the spreading width (Fig. 20.1d). This is likely due to a density change along the filament, since the actin filament was distorted by myosin forces and thermal agitation. Because the correlation intensity was similar between F_o-S_L and F_o-S_T, the distortions may take place not only in the *x*-*y* plane but also in the *z*-direction. Unfortunately, the absolute magnitude of distortions such as compression and extension cannot be determined because of the lack of an adequate model.

Next, the cross-correlation functions between each spreading width (S_L or S_T) and velocity (V) were calculated (Fig. 20.1e). The correlation had a peak at delay time of 1/30 s in the case of S_L-V. This indicates that the change in velocity follows the change in distortion with a subtle delay. The correlation was larger for S_L-V than for S_T-V. The correlation intensity is independent of the absolute velocity because of normalized correlation functions. Schematic illustrations for this interpretation are shown in Fig. 20.2.

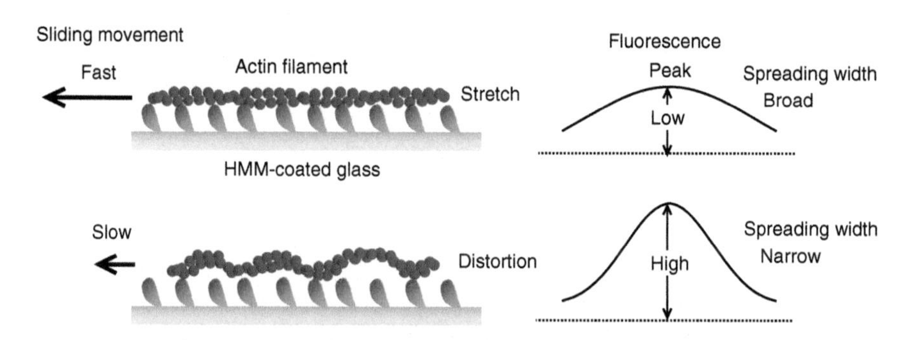

Fig. 20.2 Schematic representation of the relationship between sliding movement and distortion of an actin filament, which is estimated from changes in the distribution of fluorescence in the actin image. Cross-correlation between peak and spreading width in fluorescence distribution tends to be negative, indicating a shape change such as straight or distortion. After an actin filament stretches, filament velocity tends to be fast. However, a distorted filament tends to be slow

20.4 Effect of ADP and Pi on Motility with ATP Hydrolysis

20.4.1 Effect on Sliding Velocity and Its Variance

The motility of actin filaments depends on the kinetics of the interaction with myosin motors during ATP hydrolysis. This can be modulated by ADP and Pi, which are the products of ATP hydrolysis. Figure 20.3 shows the average velocity and its fluctuation when ADP and Pi are intentionally added to the motility assay system. The addition of ADP suppressed the velocity due to the prolonged duration of ADP–myosin binding to actin. However, the addition of Pi slightly enhanced the average velocity under our experimental conditions, although the effect of Pi on velocity is pronounced at low ATP concentrations or at low pH, as reported elsewhere (Hooft et al. 2007; Debold et al. 2011). In addition, variance of the velocity (velocity fluctuation) considerably increased in the presence of Pi, whereas ADP decreased the velocity fluctuation (Fig. 20.3c). It is likely that instability occurs in kinetics because Pi may decrease the force and the attachment time of ADP–myosin to actin (Debold et al. 2013). A similar increase is observed in the presence of AMP-PNP, which is a non-hydrolyzable ATP analogue that induces weak actin–myosin binding (Sakamaki et al. 2003). The decrease of fluctuation in the presence of ADP may be explained by prolongation of the attachment time in the post-power stroke. Interestingly, the coexistence of ADP and Pi produced a similar effect in the case of ADP, which was the decrease of average velocity and fluctuation. The predominance of the ADP–myosin state may not induce the rebinding of Pi (Debold et al. 2012).

In the absence of ATP, i.e., no sliding condition, the variance was 30-fold smaller than with sliding movement, which is similar to the level of thermal fluctuation when anchored to myosin (Hatori et al. 1999). In the presence of either ADP or Pi, there was a twofold increase in fluctuation compared to the control, presumably resulting in weaker binding strength compared to the rigorous binding of actin–myosin (Steffen et al. 2003). In view of the free-energy landscapes of actin–myosin interactions, a computer simulation revealed that the actin–myosin complex with ADP and Pi exhibits weak binding and wide fluctuations of myosin heads among multiple-basin energy landscapes (Nie et al. 2014), supporting our results. The landscapes are altered by ligand binding and myosin head configuration, inducing biased Brownian motion of myosin heads along actin filaments. Similarly, a charged particle can move unidirectionally on a head-to-tail linear array of dipoles by altering the asymmetric electric field during thermal fluctuations (Astumian and Bier 1996). The velocity fluctuation described herein is macroscopic compared with that of the above-mentioned models. However, the contribution of fluctuating landscape and asymmetric field to motility involves the implications of fluctuations in filament structure.

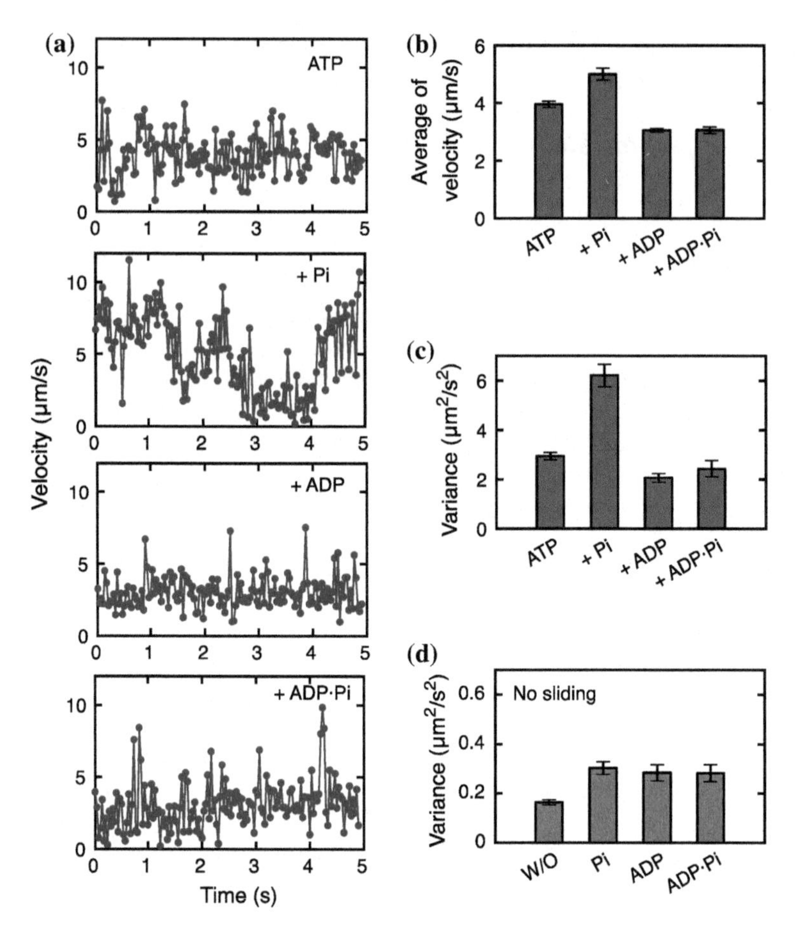

Fig. 20.3 Fluctuation of filament velocity in the presence of 4 mM ADP, 10 mM Pi, and both 4 mM ADP and 10 mM Pi combined with 2 mM ATP. Other conditions were the same as described in Fig. 20.1. **a** Time development of the velocity at intervals of 1/30 s. The present reagent is indicated in each panel. **b** The average velocity was calculated from 20 to 30 samples. Error bars indicate the standard error. **c** Variance of velocity in the presence of ATP with ADP, Pi. **d** Variance of velocity as a thermal fluctuation in the absence of ATP (no sliding movement). The scale in ordinate is magnified by a factor of 10 compared to that in (**c**). W/O denotes the case without ATP, ADP, and Pi. In each case, contaminated ATP was exhausted by addition of hexokinase and glucose

20.4.2 Effect on the Correlation Between Velocity and Distortion

Figure 20.4 summarizes the cross-correlations between each S_L, S_T, F_o, and velocity (V) as described in Sect. 20.3. During sliding movement, the positive correlation of S_L-V became marked rather than that of S_T-V (Fig. 20.4a). In addition,

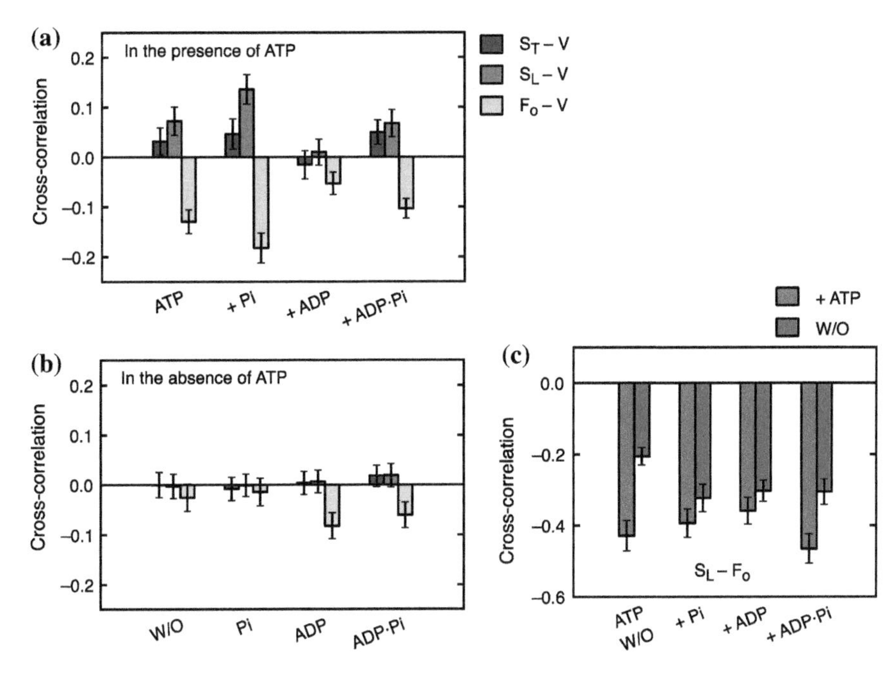

Fig. 20.4 Normalized cross-correlations between filament velocity (V) and S_L, S_T, and F_o, which are explained in Fig. 20.1. **a** In the presence of 2 mM ATP with 4 mM ADP, 10 mM Pi, and both 4 mM ADP and 10 mM Pi. The cross-correlation of S_T-V, S_L-V, and F_o-V was taken at 1/30 s of delay time because the peak was obtained at the delay time. Data from 20 to 30 samples were averaged, and the error bars indicate the standard error. **b** In the absence of ATP with ADP, Pi, and both ADP and Pi (no sliding conditions). The error bars indicate the standard error. **c** Cross-correlation of S_L-F_o at delay time zero at each additive. Red and blue bars indicate the case in the presence and absence of ATP, respectively

the correlation coefficient peaked at a time delay of 1/30 s, indicating that the increase in velocity occurs after the increase in the longitudinal spreading width of the filament. The appearance of a negative correlation of F_o-V ensured that the distortion decreased the velocity. The presence of Pi enhanced the phenomena of stretched filaments tending to increase the velocity, whereas distortion depressing the velocity. Because Pi decreases the attachment time of actin–myosin (Debold et al. 2011) and stabilizes the filamentous structure (Murakami et al. 2010), it is likely that mitigation of the distortion occurs and increases the velocity. In contrast, the presence of ADP eliminated each correlation. Prolonging the attachment time may invalidate the relationship between distortion and velocity, even when the actin filament stretches. However, the combination of ADP and Pi maintains this effect similar to the level of ATP only. Because velocity and its variance, in the case of ADP, were similar to that of ADP and Pi (Fig. 20.3b, c), the difference in correlation may result in a distinct effect on kinetics rather than the magnitude of velocity and variance.

To verify whether the phenomena regarding the sliding movement are distinct, the same analysis was applied to samples in the absence of ATP (Fig. 20.4b). In all conditions, the correlations of S_L-V and S_T-V were almost zero. Therefore, thermal agitation provides filaments with non-directional motions independent of distortion. Similarly, the correlation of F_o-V diminished in the presence of Pi. The presence of ADP still left a negative correlation of F_o-V. While the fluctuation increased due to instability of the ADP–myosin state, F_o might increase during a contingent reduction of motions corresponding to the transition to the rigor state.

In the context of distortion, Fig. 20.4c shows a comparison between S_L-F_o at each condition. The negative correlation was smaller in the absence of ATP than in the presence of only ATP. It is likely that rigor binding of actin–myosin impedes the fluctuation and distortion of filament, whereas the flexibility of the filament can be restored during ATP hydrolysis. In the absence of ATP, the presence of ADP and Pi increased the negative correlation, resulting in loose binding of actin–myosin for the rebinding ability of ADP and Pi to myosin. However, in the presence of ATP, the influence of Pi and ADP on the negative correlation was obscure. The distortion during sliding may reach an upper limit because an excess of deformation above the binding energy between constituted actin monomers results in filament breakage (Stewart et al. 2013).

20.5 The Concept of Coordination with Distortions

The transition from random motion to unidirectional movement is an important issue concerning the thermodynamics of motor proteins as mentioned in Sect. 20.4.1. Similarly, the transition from the individual motion of each element to the coordinated movement of many elements is an interesting topic. Observation of the sliding movement leads to the fundamental question of how the filament can trace the same path from pointed end to barbed end without conflict among local motions, particularly coordination of movement. Unlike myosin V, myosin II derived from skeletal muscles requires multiple motors for sliding. Kinetics such as ADP dissociation from myosin is modulated in response to the direction of imposed force, while ATP hydrolysis supplies the energy for production of the force (Oguchi et al. 2008). Furthermore, the flexibility and conformation of actin filaments are found to be altered by actin-binding proteins including myosin (McCullough et al. 2008; Vikhorev et al. 2008). Conversely, conformational changes in actin filaments can affect the binding (Ngo et al. 2016). This highlights the mutual dependence between the actin filament and myosin motor, which may bring about cooperativity and adaptation of kinetics to motility. While individual myosin motors are connected along an actin filament, in which fluctuation and distortion occur, each myosin motor should be subject to the filament movement it produces.

A coordinating effect is unveiled at low ATP concentrations in which many myosin motors bind to an actin filament long term, yet there is a small amount of force generation. The possible mode of transition of sliding movement is inferred

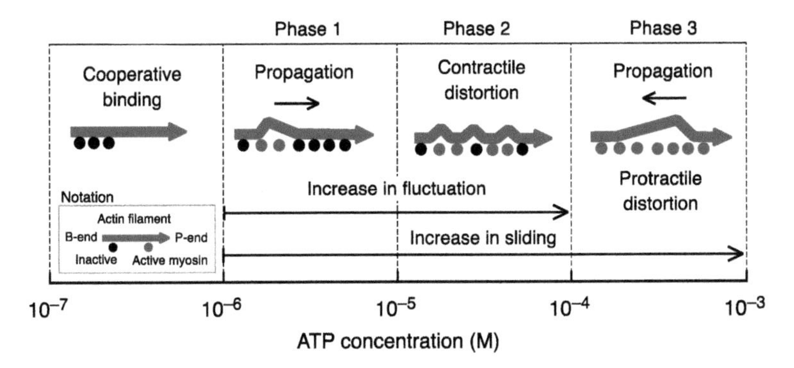

Fig. 20.5 A schematic representation of mode change in an actin filament interacting with myosin motors dependent on ATP concentration. The coordination of actions along each filament may occur via propagation of distortion. The thick red arrow, filled black circle, and filled red circle represent the actin filament, inactive myosin, and active myosin, respectively

from findings obtained so far (Fig. 20.5). At sub-micromolar amounts of ATP, cooperative binding of HMM to actin filaments takes place through a conformational change of the filament (Tokuraku et al. 2009). At micromolar amounts of ATP (phase 1), local fluctuation within an actin filament on a myosin-coated surface propagates toward its pointed end (Hatori et al. 1996b). Further increase in ATP concentration allows the filament to gradually increase fluctuation, reaching a maximum at 100 μM (phase 2) while the velocity develops (Hatori et al. 2004). In addition, studies using locally labeled filaments (speckled actin filaments) revealed that filaments tend to shrink (Honda et al. 1999), and the correlation length for velocity fluctuation along the filament is shorter than the persistence length (Shimo and Mihashi 2001). At sufficient ATP concentrations (phase 3), actin filaments tend to stretch, and local velocity fluctuation becomes propagated toward the barbed end with long correlation length (Hatori et al. 1998).

One possible reason for these distortions is to mitigate the excess load imposed on local portions. If a local distortion within an actin filament induced by a large load suppresses the local sliding motion, the portion serves as a buffer zone for adjusting the motion. At low ATP concentrations, portions close to the distortion may influence neighbor binding myosin motors, inducing the modulation of myosin action with cooperativity. At high ATP concentrations, a follow-up in changes from the pointed end to the barbed end is fulfilled by propagation of buffering along the filament. Although the binding kinetics during ATP hydrolysis governs the development of motility, the relationship between the driving force and distortion may underlie the coordination for fast movement.

In the context of distortion, the importance of the structural flexibility of actin and myosin for motility is demonstrated in experiments using co-solvents. Urea induces protein unfolding via the selective solvation of dehydrated proteins, whereas trimethylamine N-oxide (TMAO) stabilizes the tertiary structures of proteins through their preferential hydration (Cho et al. 2011). Both motility and

ATPase activity of actin–myosin are suppressed by TMAO as well as urea (Kumemoto et al. 2012). These findings indicate that adequate actin–myosin stability in terms of hydration is necessary for their machinery. Furthermore, motility may be more sensitive to changes in hydration than ATPase activity (Wazawa et al. 2013). In the case of actin dynamics, TMAO enhances the polymerization of actin monomers, although not end-to-end annealing between actin filaments (Hatori et al. 2014). This may be due to a difference in water behavior between monomers and filaments (Kabir et al. 2003).

The implication of distortion in cytoskeletons is also recognized in a more complicated contractile system. Severing via buckling of actin filaments is necessary for coordination in contraction of actin–myosin networks (Murrell and Gardel 2012). Buckling of single actin filaments as well as its bundles is often observed in cell motilities (Leijnse et al. 2015). The stretching of actin filaments can increase the affinity of myosin motors for the filaments in vivo (Uyeda et al. 2011), and actin filaments can act as tension sensors (Galkin et al. 2012). Furthermore, in more rigid microtubules, bending distortions by force difference between microtubules driven by dynein motors appear to be responsible for regulated beating of eukaryotic flagella (Lindemann and Lesich 2010; Yoke and Shingyoji 2017).

In the fundamental sense, the shape of proteins is crucial for binding. For instance, an excluded volume effect suggests that the depletion force depends on the shape of the contact area, which associates with the excluded volume (Asakura and Oosawa 1954; Munakata and Hatori 2013). When distortions are created within an actin filament, the filament configuration is altered, inducing changes in excluded volume. Additionally, the local distribution of electric charge along the filament may be altered as expected in a myosin head (Ohnuki et al. 2016), suggesting the modulation of asymmetric potential fields for a Brownian model (Astumian and Bier 1996). Our results show that distortions within an actin filament downregulate interactions between actin components and myosin motors, resulting in the suppression of motility. Meanwhile, stretching filament facilitates the motility. Coordination of the sliding movement may be a result of the interplay between stretch and distortion fluctuating along an actin filament.

20.6 Conclusion and Perspective

Here, we discussed the behavior of fluctuating actin filaments during sliding movement induced by myosin ATPase. The morphology of actin filaments may impact the motility, such that local distortions modulate the interaction between actin and myosin. Although the method presented here has the potential to detect changes in the filamentous structure, it should be considered along with studies on the conformational changes in constituents within actin filaments in order to fully elucidate the mechanism of the coordinating sliding movement. Further studies are required to thoroughly investigate the modulation of forces through distortions in the filaments with water behavior. This knowledge will provide insight into the organization of cell motilities containing multiple filaments and motors, in which distortions and heterogeneous interactions occur.

References

Amano KI, Yoshidome T, Iwaki M, Suzuki M, Kinoshita M (2010) Entropic potential field formed for a linear-motor protein near a filament: statistical-mechanical analyses using simple models. J Chem Phys 133:045103

Asakura S, Oosawa F (1954) On interaction between two bodies immersed in a solution of macromolecules. J Chem Phys 22:1255–1256

Astumian RD, Bier M (1996) Mechanochemical coupling of the motion of molecular motors to ATP hydrolysis. Biophys J 70:637–653

Batters C, Veigel C, Homsher E, Sellers JR (2014) To understand muscle you must take it apart. Front Physiol 5:90

Cho SS, Reddy G, Straub JE, Thirumalai D (2011) Entropic stabilization of proteins by TMAO. J Phys Chem B 115:13401–13407

Cooke R (2004) The Sliding Filament Model. J Gen Physiol 123:643–656

Crevenna AH, Arciniega M, Dupont A, Mizuno N, Kowalska K, Lange OF, Wedlich-Söldner R, Lamb DC (2015) Side-binding proteins modulate actin filament dynamics. Elife 4:1–18

Debold EP, Longyear TJ, Turner MA (2012) The effects of phosphate and acidosis on regulated thin-filament velocity in an in vitro motility assay. J Appl Physiol 113:1413–1422

Debold EP, Turner MA, Stout JC, Walcott S (2011) Phosphate enhances myosin-powered actin filament velocity under acidic conditions in a motility assay. Am J Physiol—Regul Integr Comp Physiol 300:R1401–R1408

Debold EP, Walcott S, Woodward M, Turner MA (2013) Direct observation of phosphate inhibiting the Force-generating capacity of a miniensemble of myosin molecules. Biophys J 105:2374–2384

Galkin VE, Orlova A, Egelman EH (2012) Actin filaments as tension sensors. Curr Biol 22:R96–R101

Hatori K, Iwasaki T, Wada R (2014) Effect of urea and trimethylamine N-oxide on the binding between actin molecules. Biophys Chem 193–194:20–26

Hatori K, Honda H, Matsuno K (1996a) ATP-dependent fluctuations of single actin filaments in vitro. Biophys Chem 58:267–272

Hatori K, Honda H, Matsuno K (1996b) Communicative interaction of myosins along an actin filament in the presence of ATP. Biophys Chem 60:149–152

Hatori K, Honda H, Shimada K, Matsuno K (1998) Propagation of a signal coordinating force generation along an actin filament in actomyosin complexes. Biophys Chem 75:81–85

Hatori K, Honda H, Shimada K, Matsuno K (1999) Onset of the sliding movement of an actin filament on myosin molecules: From isotropic to anisotropic fluctuations. Biophys Chem 82:29–33

Hatori K, Matsui M, Omote Y (2009) Slowly modulating fluctuations as mesoscopic distortions occurring on an actin filament. BioSystems 96:14–18

Hatori K, Sakamaki J, Honda H, Shimada K, Matsuno K (2004) Transition from contractile to protractile distortions occurring along an actin filament sliding on myosin molecules. Biophys Chem 107:283–288

Honda H, Hatori K, Igarashi Y, Shimada K, Matsuno K (1999) Contractile and protractile coordination within an actin filament sliding on myosin molecules. Biophys Chem 80:137–141

Honda H, Nagashima H, Asakura S (1986) Directional movement of F-actin in vitro. J Mol Biol 191:131–133
Hooft AM, Maki EJ, Cox KK, Baker JE (2007) An accelerated state of myosin-based actin motility. Biochemistry 46:3513–3520
Isambert H, Venier P, Maggs AC, Fattoum A, Kassab R, Pantaloni D, Carlier MF (1995) Flexibility of actin filaments derived from thermal fluctuations. Effect of bound nucleotide, phalloidin, and muscle regulatory proteins. J Biol Chem 270:11437–11444
Kabir SR, Yokoyama K, Mihashi K, Kodama T, Suzuki M (2003) Hyper-mobile water is induced around actin filaments. Biophys J 85:3154–3161
Kron SJ, Spudich JA (1986) Fluorescent actin filaments move on myosin fixed to a glass surface. Proc Natl Acad Sci U S A 83:6272–6276
Kumemoto R, Yusa K, Shibayama T, Hatori K (2012) Trimethylamine N-oxide suppresses the activity of the actomyosin motor. Biochim Biophys Acta 1820:1597–1604
Leijnse N, Oddershede LB, Bendix PM (2015) Helical buckling of actin inside filopodia generates traction. Proc Natl Acad Sci U S A 112:136–141
Lindemann CB, Lesich KA (2010) Flagellar and ciliary beating: the proven and the possible. J Cell Sci 123:519–528
Makhatadze GI, Privalov PL (1995) Energetics of protein structure. Adv Protein Chem 47: 307–425
Matsushita S, Hatori K (2012) Insight into force transmission along actin filaments: Sliding movement of actin filaments containing inactive components on myosin molecules. In: Consuelas VA, Minas DJ (eds) Actin: structure, functions and disease. Nova Science Publishers, Hauppauge NY, pp 257–269
McCullough BR, Blanchoin L, Martiel JL, De La Cruz EM (2008) Cofilin increases the bending flexibility of actin filaments: Implications for severing and cell mechanics. J Mol Biol 381: 550–558
Nie QM, Togashi A, Sasaki TN, Takano M, Sasai M, Terada TP (2014) Coupling of lever arm swing and biased Brownian motion in actomyosin. PLoS Comput Biol 10:1–13
Munakata S, Hatori K (2013) The excluded volume effect induced by poly(ethylene glycol) modulates the motility of actin filaments interacting with myosin. FEBS J 280:5875–5883
Murakami K, Yasunaga T, Noguchi TQP, Gomibuchi Y, Ngo KX, Uyeda TQP, Wakabayashi T (2010) Structural basis for actin assembly, activation of ATP hydrolysis, and delayed phosphate release. Cell 143:275–287
Murrell MP, Gardel ML (2012) F-actin buckling coordinates contractility and severing in a biomimetic actomyosin cortex. Proc Natl Acad Sci U S A 109:20820–20825
Ngo KX, Umeki N, Kijima ST, Kodera N, Ueno H, Furutani-Umezu N, Nakajima J, Noguchi TQP, Nagasaki A, Tokuraku K, Uyeda TQP (2016) Allosteric regulation by cooperative conformational changes of actin filaments drives mutually exclusive binding with cofilin and myosin. Sci Rep 6:35449
Oguchi Y, Mikhailenko SV, Ohki T, Olivares AO, De La Cruz EM, Ishiwata S (2008) Load-dependent ADP binding to myosins V and VI: Implications for subunit coordination and function. Proc Natl Acad Sci U S A 105:7714–7719
Ohnuki J, Sato T, Takano M (2016) Piezoelectric allostery of protein. Phys Rev E 94:12406
Oosawa F (2008) The unit event of sliding of the chemo-mechanical enzyme composed of myosin and actin with regulatory proteins. Biochem Biophys Res Commun 369:144–148
Rust MJ, Bates M, Zhuang X (2006) Stochastic optical reconstruction miscroscopy (STORM) provides sub-diffraction-limit image resolution. Nat Methods 3:793–795
Sakamaki J, Honda H, Imai E, Hatori K, Shimada K, Matsuno K (2003) Enhancement of the sliding velocity of actin filaments in the presence of ATP analogue: AMP-PNP. Biophys Chem 105:59–66
Shimo R, Mihashi K (2001) Fluctuation of local points of F-actin sliding on the surface-fixed H-meromyosin molecules in the presence of ATP. Biophys Chem 93:23–35
Steffen W, Smith D, Sleep J (2003) The working stroke upon myosin-nucleotide complexes binding to actin. Proc Natl Acad Sci U S A 100:6434–6439

Stewart TJ, Jackson DR, Smith RD, Shannon SF, Cremo CR, Baker JE (2013) Actin sliding velocities are influenced by the driving forces of actin-myosin binding. Cell Mol Bioeng 6: 26–37

Taylor EW (1991) Kinetic studies on the association and dissociation of myosin subfragment 1 and actin. J Biol Chem 266:294–302

Tokuraku K, Kurogi R, Toya R, Uyeda TQP (2009) Novel mode of cooperative binding between myosin and Mg^{2+}-actin filaments in the presence of low concentrations of ATP. J Mol Biol 386:149–162

Uyeda TQP, Iwadate Y, Umeki N, Nagasaki A, Yumura S (2011) Stretching actin filaments within cells enhances their affinity for the myosin II motor domain. PLoS ONE 6:e26200

Uyeda TQP, Kron SJ, Spudich JA (1990) Myosin step size. estimation from slow sliding movement of actin over low densities of heavy meromyosin. J Mol Biol 214:699–710

Vikhorev PG, Vikhoreva NN, Månsson A (2008) Bending flexibility of actin filaments during motor-induced sliding. Biophys J 95:5809–5819

Walker ML, Burgess SA, Sellers JR, Wang F, Hammer JA, Trinick J, Knight PJ (2000) Two-headed binding of a processive myosin to F-actin. Nature 405:804–807

Wazawa T, Yasui S, Morimoto N, Suzuki M (2013) 1,3-Diethylurea-enhanced Mg-ATPase activity of skeletal muscle myosin with a converse effect on the sliding motility. Biochim Biophys Acta 1834:2620–2629

Yanagida T, Nakase M, Nishiyama K, Oosawa F (1984) Direct observation of motion of single F-actin filaments in the presence of myosin. Nature 307:58–60

Yildiz A, Forkey JN, McKinney SA, Ha T, Goldman YE, Selvin PR (2003) Myosin V walks hand-over-hand: single fluorophore imaging with 1.5-nm localization. Science 300:2061–2065

Yoke H, Shingyoji C (2017) Effects of external strain on the regulation of microtubule sliding induced by outer arm dynein of sea urchin sperm flagella. J Exp Biol 220:1122–1134

Printed by Printforce, the Netherlands